AF356445

Geological Hazards
and Risk Management

Geological Hazards
and Risk Management

Editors

Jian Chen
Chong Xu

Basel • Beijing • Wuhan • Barcelona • Belgrade • Novi Sad • Cluj • Manchester

Editors
Jian Chen
School of Engineering and
Technology
China University of
Geosciences
Beijing
China

Chong Xu
National Institute of Natural
Hazards
Ministry of Emergency
Management of China
Beijing
China

Editorial Office
MDPI
St. Alban-Anlage 66
4052 Basel, Switzerland

This is a reprint of articles from the Special Issue published online in the open access journal *Sustainability* (ISSN 2071-1050) (available at: www.mdpi.com/journal/sustainability/special_issues/ geological_hazards_and_risk_management).

For citation purposes, cite each article independently as indicated on the article page online and as indicated below:

Lastname, A.A.; Lastname, B.B. Article Title. *Journal Name* **Year**, *Volume Number*, Page Range.

ISBN 978-3-7258-1012-3 (Hbk)
ISBN 978-3-7258-1011-6 (PDF)
doi.org/10.3390/books978-3-7258-1011-6

Contents

About the Editors

Jian Chen

Dr. Jian Chen serves as a professor at the Department of Geoengineering, China University of Geosciences (Beijing). He graduated from the Institute of Geology and Geophysics, China Academy of Sciences and received his PhD in geological engineering in 2005. At CUGB, he teaches and conducts research on geohazards, engineering geology and paleo-landslides. He also serves as the Vice Chairman of the Ecological Restoration Committee of the China Society of Forestry and Environmental Promotion, Deputy Secretary-General of the Mining Damage and Ecological Restoration Committee of the Chinese Society for Rock Mechanics and Engineering, a member of the Earthquake Disaster Chain Committee of the Chinese Seismological Society, a member of the Landslide and Debris Flow Prevention Committee of the Chinese Soil and Water Conservation Society, and an Associate Editor for international journals such as *Natural Hazards Review* and the *International Journal of Sediment Research*.

Chong Xu

Chong Xu is a Research Professor at the National Institute of Natural Hazards, Ministry of Emergency Management of China (formerly Institute of Crustal Dynamics, China Earthquake Administration) and has worked at the Institute of Geology, China Earthquake Administration. He attained his Ph.D. at the Institute of Geology and Geophysics, Chinese Academy of Sciences in 2010. His research interests include earthquake- and rainfall-induced landslide mechanisms and hazard analysis using GIS, remote sensing and statistical analysis techniques. He chaired four NSFC projects (Youth, General, and International Joint Research) and four sub-projects of the National Key R&D Program of China. His research achievements, mainly relating to landslides triggered by recent large earthquakes and strong rainfall events, have extended to about 100 research institutes and universities worldwide and are adopted directly by more than 200 research teams, significantly elevating the international status of China in the international community of landslide research. He has published more than 360 papers, with about 13000 citations. He won dozens of awards, such as the "Outstanding Young Scientist Award" in 2019 from the ICGDR and the "AOGS SE Distinguished Lecture" award in 2023.

Preface

In a world shaped by both natural processes and human activities, geological hazards undoubtedly pose a major threat to society. The management of geological hazards encompasses a sophisticated framework and methodology aimed at deriving viable and efficacious strategies for risk prevention and mitigation. This endeavor necessitates a harmonious balance between human endeavors and geological hazards, establishing geo-hazard risk management as a fundamental pillar of sustainable societal development. It serves as a crucial safeguard, mitigating the loss of life, injury, socioeconomic damage and environmental impacts attributable to geological disasters and their consequent secondary catastrophes.

Globally, the exacerbation of geological hazards can be attributed to climate change, earthquakes and anthropogenic influences, manifesting as increased occurrences of ice avalanches, mass movements, glacial lake outbursts and alpine debris flows, particularly in high mountainous regions. This increase in geological hazards enhances the exposure and vulnerability of mountainous communities. Moreover, catastrophic geological events often trigger a domino effect, where one disaster precipitates another, compounding environmental and societal harm. Addressing the proliferation of secondary hazards presents a formidable challenge for the scientific community. Despite the development and application of diverse methodologies and strategies in geo-hazard field investigations, risk assessments, early warnings, prolonged monitoring and mitigation efforts, there is a continuous demand for advancements in sustainable risk management practices.

This reprint is a collection of research papers that shed light on recent advancements in the theoretical foundations, methodologies, and practical applications of geo-hazard risk management, focused on a singular nation or spanning various global regions. This collection covers issues such as risk assessment, early warning, monitoring and mitigation strategies for geohazards. This special issue aspires to serve as a conduit for disseminating the latest insights and developments to both academic and practical stakeholders engaged in geo-hazard risk management at the international level.

Jian Chen and Chong Xu
Editors

sustainability

MDPI

Editorial

Geological Hazards and Risk Management

Jian Chen [1] and Chong Xu [2,*]

1   School of Engineering and Technology, China University of Geosciences, Beijing 100083, China; jianchen@cugb.edu.cn
2   National Institute of Natural Hazards, Ministry of Emergency Management of China, Beijing 100085, China
*   Correspondence: xc11111111@126.com

1. Introduction

The occurrence of geological hazards is widespread, particularly in mountainous regions. Globally, the escalation of geological hazards can be attributed to factors such as climate variations, river dynamics, seismic activities, and human interventions. The increase in the human and economic consequences of geological hazards comes as a result of population expansion and infrastructural developments encroaching upon high-risk zones, thereby amplifying vulnerability to potential catastrophic geological phenomena. Recent years have seen notable development in various technologies, including artificial intelligence, remote sensing, geophysical surveys, and big data analytics, contributing significantly to the detection, monitoring, and warning of geological hazards (alongside evaluations of the risks they pose). These advancements offer invaluable insights into decision-making processes for the purpose of preventing and mitigating geological hazards in vulnerable regions.

To showcase the latest progress in this domain, a Special Issue titled "Geological Hazards and Risk Management" has been compiled. This issue contains 13 papers covering various topics such as predicting susceptibility to landslides, risk assessment, inventory and distribution of rainfall-induced landslides, slope stability analyses, the mechanisms of oil and gas pipeline disasters, the spatial distribution of urban flooding, and volcanic eruption preparedness.

2. Predicting Susceptibility to and Assessing the Risk of Regional Landslides

Landslide susceptibility prediction (LSP) serves as a fundamental pillar of risk management and plays a pivotal role in ensuring social resilience. However, the modeling of LSP is hindered by various factors. Ma et al. delved into LSP, focusing on landslides within the Yinghu Lake basin in Shaanxi (Contribution 1). They compiled an initial inventory of landslides (totaling 46 incidents) and updated it (totaling 46 + 176 incidents) through data compilation, remote sensing interpretation, and field surveys. Employing slope units as mapping units, they selected twelve conditioning factors, including elevation, slope gradient, aspect, topographic relief, elevation fluctuation coefficient, slope structure, lithology, normalized difference vegetation index (NDVI), normalized difference built-up index (NDBI), proximity to roads, proximity to rivers, and rainfall. The modeling utilized random forest (RF) and artificial neural network (ANN) machine learning techniques. The findings revealed RF's superior predictive performance, offering a foundational reference for uncertain LSP analysis and managing the risk of regional landslides.

Numerous machine learning techniques have been employed in landslide susceptibility mapping (LSM). However, utilizing models for susceptibility prediction that lack interpretability could bring significant challenges in practical scenarios. Fang et al. conducted a thorough assessment of LSM in Nayong, Guizhou, China, employing explainable artificial intelligence techniques (Contribution 7). This study integrated remote sensing data, field surveys, geographic information system methodologies, and interpretable machine learning approaches to analyze landslide sensitivity and compare it with conventional

Citation: Chen, J.; Xu, C. Geological Hazards and Risk Management. *Sustainability* **2024**, *16*, 3286. https://doi.org/10.3390/su16083286

Received: 2 April 2024
Accepted: 10 April 2024
Published: 15 April 2024

models. The study highlighted the effectiveness of the GAMI-net-based model, demonstrating strong predictive capability and enhanced interpretability, both of which are pivotal in facilitating landslide management strategies and decision-making processes.

The conventional infinite slope model, pioneered by Newmark, remains a prevalent tool for evaluating coseismic landslide hazards. However, it is limited in that it overlooks the impact of rock mass structure on slope stability. Introducing a novel approach, Li et al. considered the roughness of potential slide surfaces within the slope, offering a fresh perspective on coseismic landslide hazard mapping. The authors refined their methodology using datasets from the 2013 Lushan earthquake (Contribution 3), incorporating geological data, peak ground acceleration (PGA) measurements, and high-resolution digital elevation models (DEMs). These components were integrated into the infinite slope model, incorporating Newmark's permanent deformation analysis. The outcome is a hazard map delineating zones prone to coseismic landslides, thereby offering valuable insights for both infrastructure development and post-earthquake reconstruction efforts.

The accurate evaluation of landslide hazards, vulnerabilities, and risks is of huge importance when devising effective mitigation strategies, planning changes in land use, and executing developmental projects. However, such evaluations are often challenging in complex and data-scarce regions. Shah et al. present an integrated approach to assessing the risk of landslides by leveraging freely available geospatial data and semi-quantitative techniques in landslide-prone regions within the Hindukush mountain ranges of northern Pakistan (Contribution 8). The resulting landslide risk index map aids in identifying areas prone to landslide risks, thereby facilitating subsequent efforts to mitigate and reduce risk.

3. Rainfall-Induced Landslide Inventory and Distribution

As climate change continues to escalate, heavy rainfall events have become increasingly frequent, leading to a surge in rainfall-triggered landslides, which rank among the most prevalent geological disasters globally. Employing high-resolution remote sensing imagery both before and after these events, Xie et al. conducted visual interpretations, mapping in detail the distribution of rainfall-induced landslides in Jiexi County, Guangdong Province, China (Contribution 2). Their findings revealed 1844 instances of rainfall-induced landslides within Jiexi County during one specific rainfall event; they were primarily concentrated in the northeastern, central, and southwestern regions, thereby aligning with the distribution pattern of rainfall intensity. To delve deeper into the impact of the regional environment on the occurrence of landslides, the study assessed eight influencing factors, including elevation, slope aspect, slope angle, topographic wetness index (TWI), topographic relief, lithology, distance to the river, and accumulated rainfall. Through statistical analysis conducted on a data analysis platform, the study unveiled the relationship between landslide distribution and the triggering factors associated with this event. These findings enhance our collective understanding of trends in the regional growth in rainfall-induced landslides and contribute to disaster prevention and mitigation efforts.

NASA's Global Landslide Catalog (GLC) provides a comprehensive compilation of rainfall-triggered landslide events sourced from media reports, academic publications, and existing databases worldwide. Dandridge et al. conducted an assessment of global landslide reporting patterns using data from the GLC (Contribution 6). This assessment comprises an examination of the spatial and temporal distribution of landslide events globally, including associated casualties, and comparisons with other landslide inventories. The GLC serves as a valuable tool for identifying landslide hotspots, analyzing geographical patterns in landslides, and training and validating landslide models on both a local and global scale.

4. Characteristics and Formation Mechanisms of Geological Hazards

In periods of drought, the occurrence of cracks in loess due to shrinkage is common, resulting in the precipitation and accumulation of salt on the surface. Understanding the influence of soluble salt content on the cracking characteristics and mechanisms of loess is of great importance in engineering projects and for preventing geological issues and disasters in salinized loess regions. Wei et al. conducted a study involving the desiccation of loess samples with varying NaCl concentrations (Contribution 5). Utilizing scanning electron microscopy (SEM) and energy-dispersive spectrum (EDS) methods, they analyzed the microstructure and elemental distribution of their samples. Their findings revealed the impact of NaCl concentrations on the cracking characteristics and mechanisms of loess, indicating that higher NaCl concentrations led to reduced evaporation rates and increased residual water contents, thereby restraining crack expansion.

Massif rupture does not always occur under saturated conditions, meaning analyses of unsaturated phenomena in specific scenarios are required. Costa et al. employed a probabilistic approach to exploring unsaturated and transient conditions, aiming to elucidate the roles of physical and hydraulic parameters within slope stability (Contribution 4). Their model, founded on the infinite slope method and a novel unsaturated constitutive shear strength model introduced by Cavalcante and Mascarenhas in 2021, demonstrated promising outcomes. It emphasized a shift from deterministic to probabilistic analyses, accounting for numerous stochastic variables. This model aids in comprehending the impact of the moisture content on slope stability, offering potential utility in managing the risk of natural disasters.

Landslides may be particularly damaging to oil and gas pipelines, with varying cross-cutting relationships playing a pivotal role in the occurrence of pipeline landslides. Thus, it is crucial that we study the stress and deformation characteristics of pipelines with various cross-cutting relationships. A et al. developed an optimal pipe–soil interaction model for oblique crossing landslides near oil and gas pipelines; they did so using FLAC3D (Contribution 9). Through analyzing a typical pipeline obliquely crossing landslide, they investigated how pipeline burial depth, sliding body displacement, and different intersection angles between the landslide and pipeline influence deformation and stress patterns. Their findings highlighted the intricate stress distribution in pipeline oblique crossing landslides, with notable concentrations of stress at the shear outlets and trailing edges of landslides. Additionally, they found that pipeline burial depth significantly affects displacement and stress, emphasizing depths of 3–3.5 m as particularly risky. Moreover, they found that the intersection angle between the pipeline and landslide notably impacts pipeline stress, offering crucial insights into how we might improve oil and gas pipeline landslide prevention and control measures.

Lemenkova and Debeir examined 2000 earthquake events from the IRIS seismic database over a 25-year period (1997–2021) in order to map seismic activity (Contribution 10). Their approach, combining GMT scripts with advanced GIS techniques, proved effective in spatial dataset mapping and swift data processing through iterative methods. This study lays the groundwork for predictive seismic analysis in geologically vulnerable regions of Venezuela.

5. Spatial Distribution, Influencing Factors, and the Evaluation of Geological Hazards

Rapid urbanization has led to a multitude of environmental challenges, with urban water security emerging as a prominent concern. Liu et al. conducted simulations of urban flooding in various land use and drainage system scenarios, focusing on Handan City (Contribution 11). Their study elucidated the impact of historical ground and underground constructions on waterlogging patterns. They found that changes in land use, particularly the expansion of sealed surfaces, amplified flood distribution and volume in Handan City; the drainage system, on the other hand, acted to alleviate flooding. Over time, shifts in flooding patterns indicated reduced risk in many areas due to enhanced drainage infrastructure. However, certain regions experienced exacerbations due to an increase in the number of impermeable surfaces, rapid pipe drainage, and inadequate outlets. This study highlights how alterations in both land use and drainage networks can shape the distribution of urban flooding, offering crucial guidance for bolstering urban water security and the sustainable management of water resources.

As volcanic eruption activity escalates worldwide, its disruptive impact on communities is becoming increasingly pronounced. In spite of this, there remains a gap in the literature concerning preparedness for volcanic eruptions. German et al. employed a machine learning ensemble to evaluate communities' preparedness for volcanic eruptions (Contribution 12). Their ensemble, comprising random forest classifiers and artificial neural networks, demonstrated high prediction accuracy (93% and 98.86%, respectively). Notably, they identified the media as the most significant factor influencing preparedness; its role in disseminating information significantly shapes the readiness of communities. Understanding the paramount influence of the media alongside other contributing factors such as personal experiences, social networks, and preparation through infrastructure reinforces the importance of the media in fostering communities that are well prepared for volcanic eruptions. This methodology is advantageous in assessing human behavior and predicting the factors that affect peoples' preparedness in the face of various natural disasters, offering valuable insights that are globally applicable.

Equations for predicting ground motion are pivotal in the assessment of seismic hazards, yet Pakistan lacks a dedicated equation derived from local seismic data. To address this research gap, Waseem et al. conducted an evaluation of equations for predicting global ground motion in shallow active regions using a Pakistani seismic ground motion database (Contribution 13). Their study assessed the applicability of thirteen equations using a dataset comprising peak ground accelerations from 27 shallow earthquakes in Pakistan. Through residual analysis and goodness-of-fit procedures, the study concluded that global ground motion prediction equations can be utilized to great effect in Pakistan's shallow active regions for studies of seismic hazards. These findings provide a foundation for implementing measures to mitigate the risk of disasters in Pakistan.

6. Conclusions and Future Perspectives

The articles featured in this Special Issue have made substantial contributions to advancing our understanding of geological hazards, offering innovative insights and paving the way for future developments and collaborations. We extend our appreciation to the *Sustainability* team for their invaluable support in organizing this Special Issue and overseeing the review process. Additionally, we express gratitude to all authors for their insightful contributions and to reviewers for their constructive feedback, which has enhanced the quality of the manuscripts. We look forward to continued progress and collaboration in addressing geological hazards for a sustainable future.

Funding: This study was supported by the National Key Research and Development Program of China (2023YFC3007201).

Conflicts of Interest: The authors declare no conflicts of interest.

List of Contributions:

1. Ma, S.; Chen, J.; Wu, S.; Li, Y. Landslide susceptibility prediction using machine learning methods: A case study of landslides in the Yinghu lake basin in Shaanxi. *Sustainability* **2023**, *15*, 15836. https://doi.org/10.3390/su152215836.
2. Xie, C.; Huang, Y.; Li, L.; Li, T.; Xu, C. Detailed inventory and spatial distribution analysis of rainfall-induced landslides in Jiexi county, Guangdong province, China in August 2018. *Sustainability* **2023**, *15*, 13930. https://doi.org/10.3390/su151813930.
3. Li, G.; Zang, M.; Qi, S.; Bo, J.; Yang, G.; Liu, T. An infinite slope model considering unloading joints for spatial evaluation of coseismic landslide hazards triggered by a reverse seismogenic fault: A case study of the 2013 Lushan earthquake. *Sustainability* **2023**, *16*, 138. https://doi.org/10.3390/su16010138.
4. Costa, K.R.C.B.; Dantas, A.P.N.; Cavalcante, A.L.B.; Assis, A.P. Probabilistic approach to transient unsaturated slope stability associated with precipitation event. *Sustainability* **2023**, *15*, 15260. https://doi.org/10.3390/su152115260.
5. Wei, X.; Dong, L.; Chen, X.; Zhou, Y. Influence of soluble salt NaCl on cracking characteristics and mechanism of loess. *Sustainability* **2023**, *15*, 5268. https://doi.org/10.3390/su15065268.

6. Dandridge, C.; Stanley, T.A.; Kirschbaum, D.B.; Lakshmi, V. Spatial and temporal analysis of global landslide reporting using a decade of the global landslide catalog. *Sustainability* **2023**, *15*, 3323. https://doi.org/10.3390/su15043323.

7. Fang, H.; Shao, Y.; Xie, C.; Tian, B.; Shen, C.; Zhu, Y.; Cuo, Y.; Yang, Y.; Chen, G.; Zhang, M. A new approach to spatial landslide susceptibility prediction in Karst mining areas based on explainable artificial intelligence. *Sustainability* **2023**, *15*, 3094. https://doi.org/10.3390/su150 43094.

8. Shah, N.A.; Shafigue, M.; Ishfaq, M.; Faisal, K.; Meijde, M.V. Integrated approach for landslide risk assessment using Geoinformation tools and field data in Hindukush mountain ranges, Northern Pakistan. *Sustainability* **2023**, *15*, 3102. https://doi.org/10.3390/su15043102.

9. Fa-You, A.; Chen, T.H.; Yang, C.; Wu, Y.F.; Yan, S.Q. Study on disaster mechanism of oil and gas pipeline oblique crossing landslide. *Sustainability* **2023**, *15*, 3102. https://doi.org/10.3390/su1 5043102.

10. Lemenkova, P.; Debeir, O. Seismotectonics of shallow-focus earthguakes in Venezuela with links to gravity anomalies and geologic heterogeneity mapped by a GMT scripting language. *Sustainability* **2022**, *14*, 15966. https://doi.org/10.3390/su142315966.

11. Liu, B.; Xu, C.; Yang, J.; Lin, S.; Wang, X. Effect of land use and drainage system changes on urban flood spatial distribution in Handan City: A case study. *Sustainability* **2022**, *14*, 14610. https://doi.org/10.3390/su142114610.

12. German, J.D.; Redi, A.A.N.P.; Ong, A.K.S.; Prasetyo, Y.T.; Sumera, B.L.M. Predicting factors affecting preparedness of volcanic eruption for a sustainable community: A case study in the Philippines. *Sustainability* **2022**, *14*, 11329. https://doi.org/10.3390/su141811329.

13. Waseem, M.; Rehman, Z.U.; Sabetta, F.; Arshad, I.; Ahmad, M.; Sabri, M.M.S. Evaluation of the predictive performance of regional and global ground motion predictive equations for shallow active regions in Pakistan. *Sustainability* **2022**, *14*, 8152. https://doi.org/10.3390/su14138152.

sustainability

MDPI

Article

Landslide Susceptibility Prediction Using Machine Learning Methods: A Case Study of Landslides in the Yinghu Lake Basin in Shaanxi

Sheng Ma, Jian Chen *, Saier Wu and Yurou Li

School of Engineering and Technology, China University of Geosciences, Beijing 100083, China; masheng263@163.com (S.M.); cielo_wu@cugb.edu.cn (S.W.); liyurou2000@163.com (Y.L.)
* Correspondence: jianchen@cugb.edu.cn

Abstract: Landslide susceptibility prediction (LSP) is the basis for risk management and plays an important role in social sustainability. However, the modeling process of LSP is constrained by various factors. This paper approaches the effect of landslide data integrity, machine-learning (ML) models, and non-landslide sample-selection methods on the accuracy of LSP, taking the Yinghu Lake Basin in Ankang City, Shaanxi Province, as an example. First, previous landslide inventory (totaling 46) and updated landslide inventory (totaling 46 + 176) were established through data collection, remote-sensing interpretation, and field investigation. With the slope unit as the mapping unit, twelve conditioning factors, including elevation, slope, aspect, topographic relief, elevation variation coefficient, slope structure, lithology, normalized difference vegetation index (NDVI), normalized difference built-up index (NDBI), distance to road, distance to river, and rainfall were selected. Next, the initial landslide susceptibility mapping (LSM) was obtained using the K-means algorithm, and non-landslide samples were determined using two methods: random selection and semi-supervised machine learning (SSML). Finally, the random forest (RF) and artificial neural network (ANN) machine-learning methods were used for modeling. The research results showed the following: (1) The performance of supervised machine learning (SML) (RF, ANN) is generally superior to unsupervised machine learning (USML) (K-means). Specifically, RF in the SML model has the best prediction performance, followed by ANN. (2) The selection method of non-landslide samples has a significant impact on LSP, and the accuracy of the SSML-based non-landslide selection method is controlled by the ratio of the number of landslide samples to the number of mapping units. (3) The quantity of landslides has an impact on how reliably the results of LSM are obtained because fewer landslides result in a smaller sample size for LSM, which deviates from reality. Although the results in this dataset are satisfactory, the zoning results cannot reliably anticipate the recently added landslide data discovered by the interpretation of remote-sensing data and field research. We propose that the landslide inventory can be increased by remote sensing in order to achieve accurate and impartial LSM since the LSM of adequate landslide samples is more reasonable. The research results of this paper will provide a reference basis for uncertain analysis of LSP and regional landslide risk management.

Keywords: landslide susceptibility prediction; machine learning; landslide inventory; non-landslide

Citation: Ma, S.; Chen, J.; Wu, S.; Li, Y. Landslide Susceptibility Prediction Using Machine Learning Methods: A Case Study of Landslides in the Yinghu Lake Basin in Shaanxi. *Sustainability* **2023**, *15*, 15836. https://doi.org/10.3390/su152215836

Academic Editor: Maurizio Lazzari

Received: 10 October 2023
Revised: 8 November 2023
Accepted: 9 November 2023
Published: 10 November 2023

1. Introduction

Landslides are one of the most widely developed geological disasters in the world, posing a great threat to human life and property safety [1,2]. Property losses can be effectively reduced by accurately predicting the potential location of landslides [3]. Landslide susceptibility prediction (LSP) is an effective tool often used for regional disaster prevention and reduction. LSP may be defined as a measure of regional landslide susceptibility, and it is the foundation of landslide risk assessment [3,4]. However, LSP modeling is influenced by various factors [5–7]; thus, obtaining accurate and reliable landslide susceptibility mapping (LSM) has become a challenge.

The acquisition of landslide data and the selection of evaluation models are important factors that constrain the accuracy of LSP, bringing to it significant uncertainty [8]. The landslide inventory is the foundation of LSP analysis [9], and a more reliable LSM can be obtained through a complete and reliable landslide inventory. However, the cost of obtaining a complete landslide inventory is relatively high, especially for earthquake-induced landslides [10]. For this reason, researchers sometimes directly use landslide data from government surveys [11]. In addition, the number of landslides is limited by the focus of the survey and the geological environment, thus making the landslide data incomplete or even spatially aggregated, in turn causing difficulties in identifying the development of regional landslides and conducting landslide susceptibility assessment [12,13]. In recent years, some scholars have carried out relevant research on the integrity of landslide data. For example, Xu [14] conducted modeling using the incomplete landslide data of the 2008 Wenchuan earthquake and found that newly recorded landslides could be accurately predicted, with a prediction rate of up to 86.93%. Lin [15] reached different conclusions, and through comparative experiments found that incomplete landslide data may cause errors in LSP. Therefore, exploring the impact of landslide inventory integrity on LSP and processing incomplete landslide data is of great significance.

The selection of evaluation models is one of the important steps in LSP modeling, and, currently, heuristic models, statistical models, and machine-learning models (ML) have been widely used in LSP [16,17]. The accuracy of heuristic-model (e.g., AHP) evaluation is greatly influenced by human subjective factors, and statistical models (e.g., FR) require data to follow a normal distribution [18–20]. In comparison, ML overcomes these difficulties and does not need to consider the statistical laws of the data; for this reason, the application of ML in LSP has become a research hot spot. ML can be divided into supervised machine learning (SML) and unsupervised machine learning (USML), depending on whether labels are required for data [3]. SML data are required to have labels, and common SML includes support vector machine (SVM) [21], decision tree (DT) [22], random forest (RF) [23,24], and artificial neural networks (ANN) [25]. USML data do not require labels, and common algorithms include K-means [26], self-organizing mapping neural network (SOM) [16], and the Kohonen model [3]. Overall, the prediction accuracy of USML is inferior to SML, because USML does not consider prior knowledge of the data. However, USML has strong scalability and fast running speed [3]. Therefore, in recent years some scholars have even applied semi-supervised machine learning (SSML) concepts to LSP [27].

At present, there is no universally recognized conclusion regarding the selection of ML models, and the applicability of the models is likely related to the study area. In addition, even though these ML models have little difference in prediction accuracy, there may be significant differences in the distribution characteristics of landslide susceptibility indexes [28]. In summary, taking the Yinghu Lake Basin in Ankang City, Shaanxi Province, as an example and based on two sets of landslide inventories with different quantities, K-means, RF, and ANN machine-learning algorithms are used to explore the uncertain impact on LSP of landslide data integrity, machine-learning algorithms, and negative landslide sampling strategies.

The results of the present study serve as the theoretical basis for selecting regional models, thereby improving the accuracy of landslide susceptibility prediction and providing important scientific guidance for risk management in the region.

2. Study Area

The Yinghu Lake Basin is located in the southern part of Hanbin District, Ankang City, Shaanxi Province, and has a total area of approximately 195.13 km^2, with coordinates ranging from 108°46′21.89″ E–108°57′8.33″ E, 32°30′13.01″ N–32°40′16.44″ N (Figure 1). The Yinghu Lake Basin is a section of the Han River Basin, with low- and medium-relief terrain in the north and south and valley terraces and bench terraces in the middle. The study area is in the northern subtropical region and has a humid and semi-humid monsoon climate, characterized by four distinct seasons, adequate sunshine, and abundant rainfall. The average annual temperature here is 14.52 °C, and the average annual rainfall is 798.2 mm. The Quaternary, Middle-Upper Paleozoic, and Lower Paleozoic strata are exposed in the study area, mainly consisting of limestone, slate, and phyllite (Figure 2). The study area is located at the intersection of the South Qinling Indosinian Fold Belt and North Daba Mountain Caledonian Fold Belt in the Qinling Fold System and the geological structure is relatively complex, but the faults and seismic activity in the study area are quite weakly developed. The basic seismic intensity is VI and the ground motion peak acceleration is 0.05 g throughout the area, with little impact on the development of landslides.

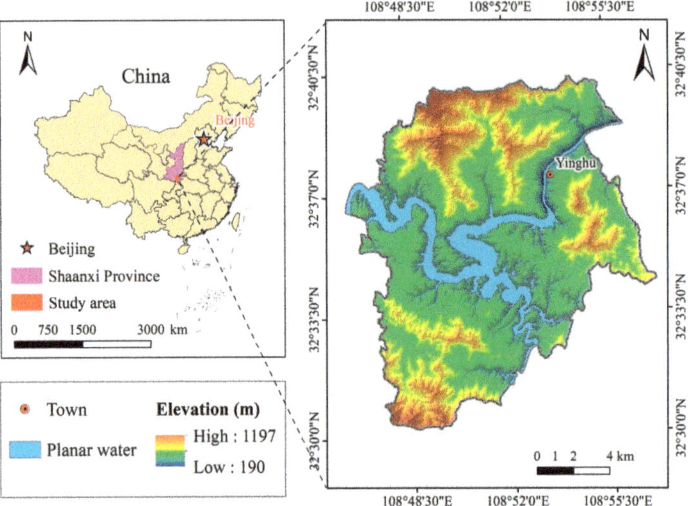

Figure 1. Map showing the geographic location of the study area.

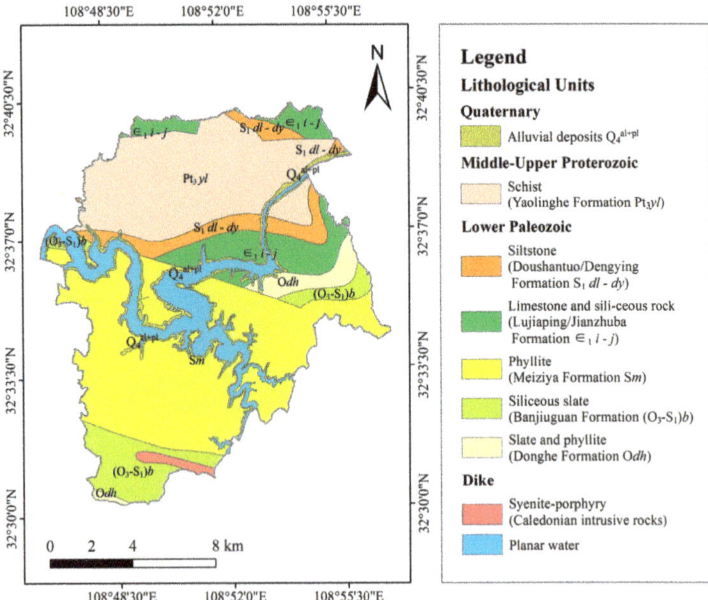

Figure 2. Geological map of the study area.

3. Data and Methods

3.1. Procedure of Landslide Susceptibility Modeling

The modeling process for landslide susceptibility using ML algorithms and considering landslide inventories of different quantities is shown in Figure 3. The major steps are as follows:

(1) Preparation of landslide inventories and conditioning factors: The landslide inventories can be divided into updated landslide inventory (46 + 176) and previous landslide inventory (46), with data mainly sourced from government data collection and field investigations. Next, based on GIS and RS, slope units are used to characterize five types of landslide conditioning factors (topography, geological structure, meteorology and hydrology, land cover, and human activities). The hydrological analysis method is adopted for the division of slope units [29].

(2) Non-landslide sampling based on the K-means semi-supervised machine-learning algorithm (SSML): The conditioning factor classification frequency ratio (FR) is calculated, factor correlation testing is performed, then the initial LSM is obtained using the K-means algorithm, and finally random selection of non-landslide samples equivalent to the number of landslides from the very low-susceptibility zones of LSM is conducted, so as to form the SML dataset.

(3) LSP based on supervised machine learning (SML): For comparison with SSML, a random negative sampling strategy is used to form the dataset. Finally, all datasets (totaling four groups) are divided into training and testing sets in a 7:3 ratio, and the final LSM is obtained using the ANN and RF algorithms.

(4) Result comparison: The LSP results are compared using ROC curves, FR accuracy, and the distribution pattern of the landslide susceptibility index (LSI), then the uncertainty of landslide susceptibility modeling is explored.

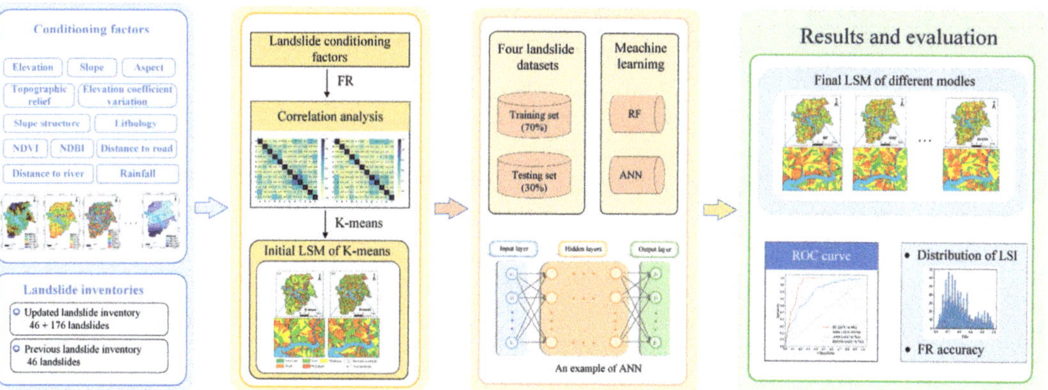

Figure 3. The flowchart of this study.

3.2. Data Sources

The main data needed for this study are landslide inventory and conditioning factors (Table 1). The landslide inventory is obtained from data collection, remote-sensing interpretation, and field survey. The digital elevation model (DEM) is ASTER GDEM version 1 (V1), which is mainly used for slope unit delineation and topographic, geomorphological, and river factor extraction. The Landsat 8 images are taken for the extraction of the normalized vegetation index (NDVI) and the normalized building index (NDBI). A detailed description of conditioning factor information is shown in Section 4.

Table 1. Data sources in landslide susceptibility prediction.

Data	Spatial Resolution	Source	Purpose
Landslide inventory	--	Data collection, remote-sensing interpretation, and field survey	Landslide dataset
DEM	30 m	https://www.gscloud.cn/sources/accessdata/310?pid=302 (accessed on 1 January 2009)	Slope unit delineation and topographic, geomorphological, and river factors extraction
Landsat 8 TM	30 m	USGS	Extraction for NDVI and NDBI
Road network	--	Manual sketching	Extraction for distance to road
Strata chronology	1:50,000	http://www.ngac.org.cn/	Lithological distribution

3.3. Landslide Inventory

The landslide inventory reflects information such as spatial distribution, geometric size, and the attributes of landslides [30]. In this study, the landslide inventory can be divided into two categories:

(1) Previous landslide inventory. This is mainly composed of existing recorded landslide data. As of 2022 year, a total of 46 landslide disasters were collected at the Natural Resources Bureau of Ankang City, and the events occurred mainly before 2010. These records also served as the landslide inventory used for the susceptibility assessment of Yinghu Town by Ma [31]. The location, scale, and inducing factors of landslides were recorded in this inventory, with Figure 4d,e showing photos of two landslides. The statistical data showed that the minimum landslide area in the inventory was approximately 2391 m^2, the maximum area was approximately 124,689 m^2, and the average area was approximately 28,189 m^2. All landslides fell into the category of soil landslides, and the main inducing factor was rainfall. However, for the purpose of risk control, the landslides in the reservoir were mainly large-scale landslides threatening the safety of human life and property, and consequently the landslide inventory was incomplete.

(2) Updated landslide inventory. The updated landslide inventory consists of 176 landslides determined through field investigation and remote-sensing interpretation (new landslide dataset), along with 46 landslides recorded in the previous inventory (totaling 46 + 176). The newly added landslides occurred from 2011 to 2022, and their scale was smaller than the landslides in the previous landslide inventory, but the data were relatively complete and reflected the basic situation of regional landslide development. The statistical data showed that the minimum area of newly added landslides was approximately 140 m^2, the maximum area was approximately 14,692 m^2, and the average area was approximately 1992 m^2. Both soil and rock landslides were developed and mainly distributed near rivers and highways, and the main inducing factor was artificial cut slope (Figure 4b,c).

In this study, landslide susceptibility assessment was conducted using ML algorithms and these two sets of landslide inventories to explore the impact of landslide data integrity on the evaluation results. Based on the evaluation results of the previous landslide inventory, the model was tested to check its ability to predict the location of new landslides.

3.4. Frequency Ratio Model

The frequency ratio (FR) can be used to characterize the nonlinear relationships between landslides and environmental factors and reflects the quantitative statistics of the linkage between the attribute intervals of factors and landslide susceptibility [32]. The FR is shown in Equation (1). The FR can better reflect the impact of environmental factors on the development of landslides in various intervals [20,33]. When $FR > 1$, this indicates that this factor interval is conducive to landslide occurrence, and when $FR < 1$ this factor interval is not conducive to landslide occurrence [3].

$$FR = \frac{N_j / N}{S_i / S},\tag{1}$$

where N is the total area/number of landslides developed in the study area, N_j is the total area/number of landslides developed under the environmental factor classification, S is the total area of evaluation units, and S_i is the area of evaluation units under the environmental factor classification (area was used in this study).

Figure 4. Distribution and illustration of landslides in the study area: (**a**) landslide distribution; (**b**,**c**) photos of new landslide; (**d**,**e**) photos of previous landslide.

3.5. Machine-Learning Model

3.5.1. RF

Random forest (RF) is an integrated classification model composed of multiple decision trees; since its introduction, RF has been widely applied for landslide susceptibility assessment and has achieved good results [34,35]. The general steps of model construction are as follows: First, randomly select n samples from the original data that have been replaced to form a sub-dataset (Bootstrap sampling); then, randomly select k attributes from the sub-dataset and select one of the optimal feature attributes as the partition node; next, repeat this step to build a sub-tree, and multiple sub-trees are integrated to form a random forest. The RF prediction result is determined by the optimal voting result of these decision trees [36].

In RF, the relative weight of landslide conditioning factors can be determined by calculating the Gini index or out-of-bag error, after which landslide susceptibility evaluation can be conducted [34,37]. The Gini index calculation is shown in Equation (2):

$$Gini(k, x_j) = \sum_{j=1}^{m} \frac{a_i}{n_s} l(k_{ui}),\tag{2}$$

where m is the number of landslides at each node k, n_s is the number of feature vectors input for training, and $l(k_{ui})$ is the distribution of class labels on nodes. The specific calculation is shown in Equation (3):

$$l(k_{ui}) = 1 - \sum_{t=0}^{c} \frac{n_{c_i}^2}{a_i^2},\tag{3}$$

where n_{ci} is the sample of c_i, with a value of u_i, and ai is the number of samples with a value of u_i at node k.

3.5.2. ANN

Artificial neural network (ANN) is a nonlinear mathematical model commonly used for regression or classification problems [38]. Its structure consists of three parts: input layer, hidden layer, and output layer (Figure 5).

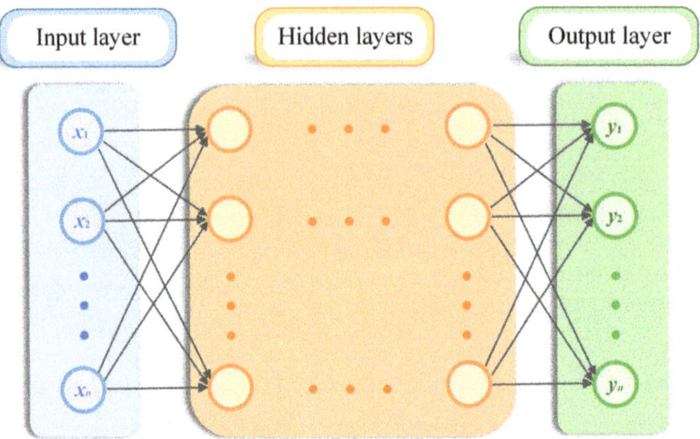

Figure 5. An example of ANN.

A back-propagation (BP) neural network is the most common and representative neural network in ANN, belonging to the supervised learning method [39]. The gradient descent algorithm is adopted to minimize the value of the error function. It aims to continuously adjust the weights and thresholds between neurons, so that the output results approximate the expected effect, as shown in Equation (4):

$$y_i = f(\sum w_{ij} x_i + b_j), j = 1, 2, 3, \ldots, k, \tag{4}$$

where w_{ij} is the weight between connector neurons i and j, b_j is the bias of neuron j, and f is the activation function in sigmoid form, as shown in Equation (5):

$$f(x) = \frac{1}{1 + e^{-x}}, \tag{5}$$

3.5.3. K-Means

The K-means clustering algorithm is an unsupervised machine-learning algorithm [26]. The major steps are as follows [40]:

(1) Determine the value of K: The number of centroids, K, can be directly determined according to actual requirements or through algorithm implementation. In this paper, the susceptibility of landslides is divided into five categories: extremely high, high, medium, low, and extremely low. Therefore, the value of K is set to 5.

(2) Divide samples: Randomly select the cluster centers of the samples, calculate the Euclidean distance (Equation (6)), and divide each sample into five clusters on the principle of nearest distance.

$$d(x, y) = \sqrt{(x_1 - y_1)^2 + (x_2 - y_2)^2 + \ldots + (x_n - y_n)^2} = \sqrt{\sum_{i=1}^{n} (x_i - y_i)^2}, \tag{6}$$

(3) Calculate new centroid: Calculate the distance based on the divided cluster samples and select the minimum or mean point as the new centroid center.

(4) End K-means classification calculation: Repeat steps 2–3 until the centroid no longer changes or reaches the given maximum number of loops, then cease the calculation.

3.6. Model Evaluation

3.6.1. ROC

ROC curve, as one of the methods for LSP accuracy testing, may effectively reduce the interference caused by testing set differences, thereby making model performance evaluation more objective [41]. The vertical axis of the ROC curve is TPR (Sensitivity) and the horizontal axis is FPR (1-Specificity) [42]. The sensitivity calculation is shown in Equation (7) and the specificity calculation is shown in Equation (8):

$$Sensitivity = \frac{TP}{TP + FN}, \tag{7}$$

$$Specificity = \frac{TN}{FR + TN}, \tag{8}$$

where TP and FN, respectively, represent the numbers of landslides correctly identified by the classifier as landslides and those incorrectly identified as non-landslides, while FP and TN, respectively, represent the numbers of non-landslides incorrectly identified by the classifier as landslides and those correctly identified as non-landslides.

The area under the ROC curve (AUC) can be used to describe the prediction performance of the model, with values ranging from 0.5 to 1.0. The larger the value, the better the prediction performance. Generally speaking, 0.5–0.7 represents a poor predictive ability of the model, 0.7–0.9 represents a high prediction performance of the model, and 0.9–1.0 represents an excellent prediction performance [43].

3.6.2. FR Accuracy Validation

FR accuracy is defined as the ratio of the total FR of very high- and high-susceptibility zones to the total FR of all zones in the results of susceptibility zoning (Equation (9)). The higher the FR accuracy, the better the prediction performance of the model [3,44].

$$FR\ accuracy = \frac{FR_1 + FR_2}{FR_1 + FR_2 + FR_3 + FR_4 + FR_5}, \tag{9}$$

where FR_1, FR_2, FR_3, FR_4, and FR_5, respectively, refer to FR in the very high-, high-, medium-, low-, and very low-susceptibility zones.

3.6.3. Distribution Patterns of LSIs

The distribution pattern of the landslide susceptibility index (LSI) is a model evaluation method that has been proposed in recent years [8], and it reflects the model's control over the overall landslide susceptibility trend in the study area. In general, a reasonable evaluation result should have the characteristics of small mean and large standard deviation in its LSI. The lower the mean of the LSI, the smaller the overall landslide susceptibility trend in the region. Meanwhile, the higher the standard deviation, the greater the difference between LSIs and the better the susceptibility zoning effect.

4. Conditioning Factors

The conditioning factors affect the development of landslides; thus, the determination of reasonable conditioning factors is an important step in landslide susceptibility prediction. Generally speaking, conditioning factors can be divided into five categories: topography, geological structure, meteorology and hydrology, land cover, and human engineering activities [4]. Starting from the above five types of factors and based on the selection of factors [31], twelve conditioning factors, including elevation, slope, aspect, topographic relief, elevation variation coefficient, slope structure, lithology, normalized difference vegetation index (NDVI), normalized difference built-up index (NDBI), distance to road, distance to river, and rainfall were selected.

In this study, the slope unit was used to characterize each type of conditioning factor, with discrete factors (aspect, slope structure, strata chronology, distance to river, and distance to road) statistically representing the mode and continuity factors statistically representing the average value [45,46]. It should be noted that, except for aspect and strata chronology factors (classified according to actual conditions) and the average annual rainfall (divided into four levels), all other factors are classified into five categories according to the natural breaks. The conditioning factors after final classification are shown in Figure 6, and the FR calculated by factor classification is shown in Table 2 (where N_i refers to the area of factor classification and S_i refers to the area of landslide development under the factor classification).

Table 2. Frequency ratios of conditioning factors.

Conditioning Factor	Level	N_i (km^2)	Updated Inventory		Previous Inventory	
			S_i (km^2)	FR	S_i (km^2)	FR
Elevation (m)	285–413	37.03	0.59	1.67	0.44	1.58
	414–510	44.26	0.59	1.39	0.52	1.57
	511–623	45.22	0.17	0.39	0.13	0.39
	624–762	33.08	0.21	0.67	0.16	0.65
	763–1073	14.98	0.11	0.76	0.05	0.48
Slope (°)	5–15	8.52	0.04	0.49	0.03	0.41
	16–19	39.27	0.63	1.69	0.58	1.98
	20–23	64.27	0.46	0.74	0.37	0.76
	24–27	49.77	0.39	0.83	0.27	0.73
	28–38	12.75	0.15	1.23	0.06	0.67
Aspect	North	14.01	0.02	0.15	0.02	0.16
	Northeast	18.83	0.26	1.44	0.25	1.75
	East	27.55	0.56	2.11	0.44	2.13
	Southeast	30.64	0.37	1.24	0.29	1.24
	South	18.40	0.15	0.86	0.12	0.85
	Southwest	22.15	0.12	0.55	0.07	0.44
	West	21.88	0.13	0.63	0.09	0.56
	Northwest	21.10	0.07	0.34	0.04	0.23
Topographic relief (m)	5–96	17.68	0.04	0.21	0.01	0.08
	97–155	41.49	0.40	1.00	0.33	1.04
	156–218	46.61	0.42	0.94	0.34	0.97
	219–297	42.19	0.59	1.45	0.47	1.47
	298–450	26.61	0.23	0.90	0.17	0.84
Elevation coefficient variation	0.004–0.049	17.47	0.03	0.17	0.01	0.10
	0.050–0.078	45.57	0.20	0.45	0.08	0.24
	0.079–0.109	51.88	0.61	1.22	0.55	1.41
	0.110–0.154	43.69	0.62	1.47	0.50	1.52
	0.155–0.248	15.85	0.22	1.46	0.17	1.40
Slope structure	Bedding slope	30.32	0.36	1.25	0.32	1.41
	Diagonal slope	58.83	0.63	1.11	0.47	1.05
	Horizon slope	56.53	0.45	0.84	0.35	0.82
	Anti-dip slope	28.89	0.23	0.84	0.18	0.81

Table 2. *Cont.*

Conditioning Factor	Level	N_i (km^2)	Updated Inventory		Previous Inventory	
			S_i (km^2)	FR	S_i (km^2)	FR
Lithology	Slate and phyllite	6.25	0.14	2.34	0.13	2.65
	Syenite-porphyry	1.36	0.01	0.55	0.00	0.00
	Siliceous slate	15.55	0.14	0.92	0.07	0.62
	Phyllite	73.51	0.59	0.84	0.42	0.75
	Limestone and siliceous rock	18.48	0.12	0.68	0.10	0.73
	Siltstone	11.53	0.21	1.93	0.19	2.20
	Schist	47.89	0.46	1.01	0.41	1.13
NDVI	0.249–0.549	1.77	0.00	0.00	0.00	0.00
	0.550–0.687	10.03	0.19	1.93	0.15	1.97
	0.688–0.751	31.87	0.65	2.14	0.52	2.15
	0.752–0.792	66.72	0.43	0.68	0.35	0.70
	0.793–0.857	64.18	0.40	0.66	0.30	0.61
NDBI	0.402–0.510	33.26	0.10	0.32	0.05	0.18
	0.511–0.531	57.85	0.31	0.56	0.22	0.51
	0.532–0.556	55.27	0.62	1.17	0.51	1.23
	0.557–0.591	22.91	0.59	2.69	0.49	2.84
	0.592–0.705	5.28	0.05	1.06	0.04	1.11
Distance to road (m)	0–200	106.61	1.33	1.31	1.05	1.31
	201–400	44.18	0.29	0.69	0.26	0.77
	401–600	16.50	0.02	0.10	0.00	0.00
	601–800	5.83	0.02	0.32	0.01	0.13
	>800	1.44	0.02	1.07	0.00	0.00
Distance to river (m)	0–200	77.44	0.99	1.34	0.82	1.40
	201–300	54.06	0.49	0.95	0.35	0.87
	301–400	26.13	0.09	0.37	0.06	0.32
	401–500	10.38	0.03	0.25	0.01	0.12
	>500	6.57	0.07	1.13	0.07	1.43
Rainfall (mm)	850–900	81.30	1.08	1.39	0.97	1.58
	901–950	59.66	0.33	0.58	0.24	0.53
	951–1000	16.79	0.05	0.32	0.01	0.09
	1001–1050	16.83	0.21	1.29	0.10	0.77

4.1. Topographic and Geomorphological Factors

Based on the GIS platform, the digital elevation model (DEM) was adopted to extract topographic and geomorphological factors, e.g., elevation, slope, aspect, topographic relief, and elevation variation coefficient (Figure 6a–e). Then, two sets of landslide inventories were used to calculate the FR of factor classification intervals, with the results shown in Table 2. The statistical trends presented by the two sets of inventories were basically consistent. However, to make the statistical rules more objective, the statistical results of the updated landslide inventories were used for explanation (as with the other factors).

The elevation affects the potential energy of the slope, the climate distribution of small regions, vegetation types, and human engineering activities [42]. The FR in the study area was greater than 1 at elevations of 285–413 m and 414–510 m, and less than 1 in the rest of the range. The main reason for this was that human activities were more intense in the low-altitude areas, resulting in the development of landslides. Slope, an important factor affecting the development of landslides, changes the stress conditions, sliding force, and sliding resistance force of slopes [47]. According to the statistical results, as the slope increased, the FR basically showed an upward trend, and FR > 1 within the slope range of 16–19° and 28–38°. The aspect controls the intensity of solar radiation, affecting vegetation cover, evaporation, slope erosion, and slope stability [41]. The statistical results showed that FR > 1 in the northeastern, eastern, and southeastern parts of the study area.

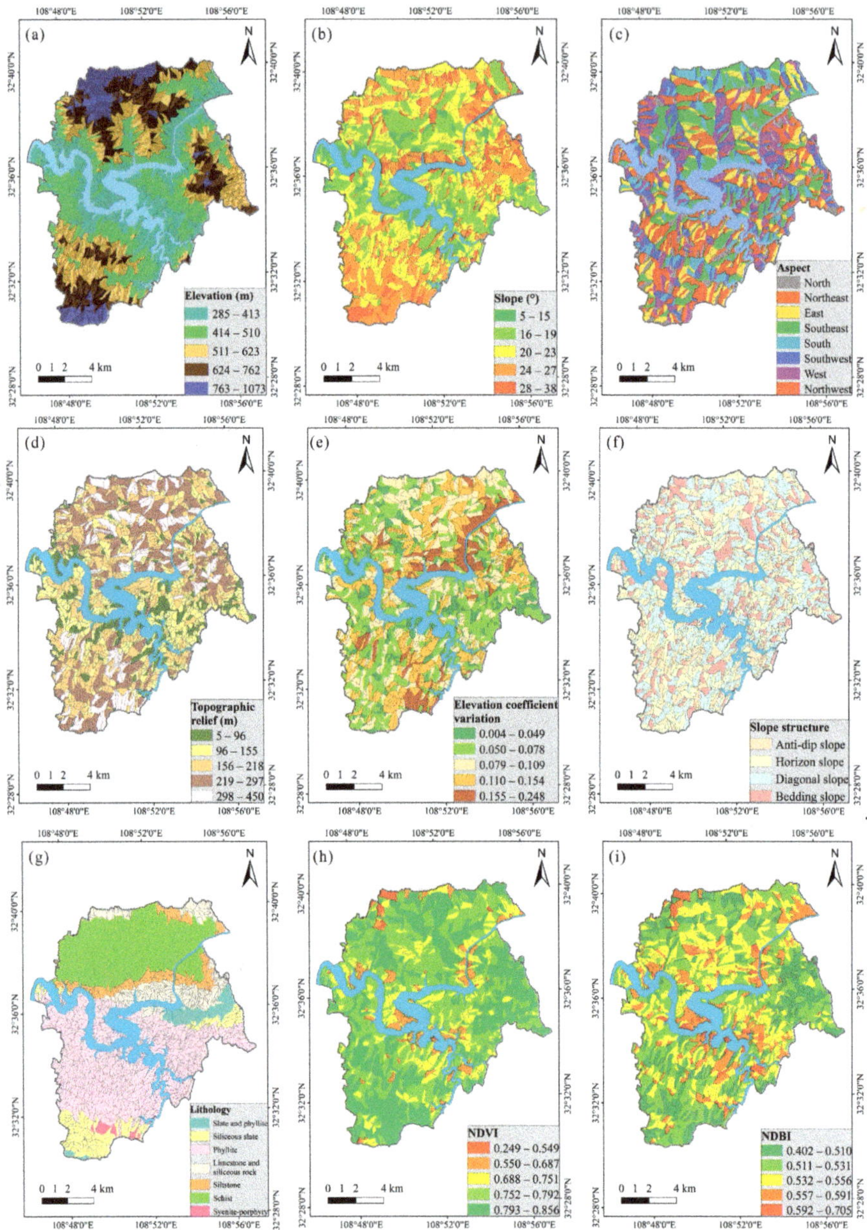

Figure 6. *Cont.*

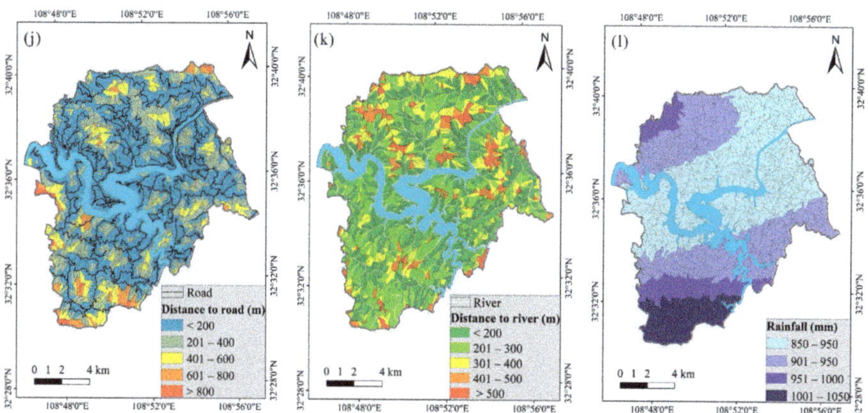

Figure 6. Conditioning factors of landslide in study area: (**a**) elevation; (**b**) slope; (**c**) aspect; (**d**) topographic relief; (**e**) elevation coefficient of variation; (**f**) slope structure; (**g**) strata chronology; (**h**) NDVI; (**i**) NDBI; (**j**) distance to river; (**k**) distance to road; (**l**) rainfall.

The topographic relief refers to the difference between the highest and lowest elevation values in a certain area, which can be used to quantitatively describe the characteristics of micro-topography in the region [27]. The statistical results showed that FR > 1 within the elevation range of 219–297 m. The elevation variation coefficient is the ratio of the standard deviation of elevation to the average elevation, and it is used to reflect the erosion cutting situation and terrain fluctuation changes in the area. The larger the coefficient, the higher the erosion intensity [48]. The statistical results showed a high consistency between landslide development and the elevation variation coefficient, with FR > 1 in the coefficient ranges of 0.079–0.109, 0.110–0.154, and 0.155–0.248.

4.2. Geological Factors

The slope structure plays an important role in the development of landslides. According to the relationship between the dip angle and aspect, the slope structure can be divided into bedding slope, diagonal slope, horizon slope, and anti-dip slope [49]. Each slope structure in the study area was classified based on the field investigation results. The statistical results showed that slopes in the study area were mainly diagonal and horizon slopes, but the FR of the dip and diagonal slopes was greater than 1.

Lithology is an important influencing factor for the development of landslides, and the formation lithology varies among different strata chronology units. Landslides are often well developed in areas prone to sliding. The distribution of lithological units in the study area is shown in Figure 6g. Table 2 shows that slate and phyllite, siltstone, and schist are more prone to sliding, with FR > 1.

4.3. Meteorological and Hydrological Factors

Rivers erode the foot of slopes, creating free surfaces and altering the stress distribution of slopes, in turn increasing the possibility of landslides [50]. In this paper, a buffer zone analysis of rivers was conducted based on GIS (Figure 6j), and the statistical results indicated that as the distance to the river increased, the development of landslides also showed a weakening trend. FR > 1 was observed in the distance ranges of <200 m and >500 m. Specifically, landslide development in the distance range of >500 m may not be controlled by the river, hence FR > 1.

Rainfall is an inducing factor of landslide development. According to the statistics collected by the Natural Resources Bureau of Ankang City, 46 landslides in the previous landslide inventory occurred during the rainstorm period. In this paper, multi-year rainfall data from various meteorological stations in Ankang City were collected and

Kriging interpolation calculations were performed in GIS to obtain rainfall contour maps (Figure 6l). The rainfall in the study area showed a trend of being high in the northern and southern parts and low in the middle part, with FR > 1 within the rainfall ranges of 851–900 and 1001–1050 mm. One possible reason for the high development of landslides in the 851–900 mm range is that this area was located on both sides of the Han River, which has a high population and strong human engineering activities.

4.4. Land Cover Factors

The normalized difference vegetation index (NDVI) and normalized difference built-up index (NDBI) are two types of land cover factors. NDVI reflects vegetation growth and cover, while NDBI is used to characterize the distribution of surface buildings [3]. The higher the NDVI, the denser the vegetation cover, and the higher the NDBI, the greater the number of buildings. Landsat 8 images taken on 2 August 2021 were used for NDVI and NDBI extraction in ENVI, and the results are shown in Figure 6h,i.

It was found that on 2 August 2021, the vegetation in the study area was dense and the NDVI was high. As the NDVI increased, the development of landslides showed a convex function trend, with FR > 1 in the ranges of 0.550–0.687 and 0.688–0.751. The NDBI showed the same trend, and the larger the NDBI, the more developed the landslide, with FR > 1 in the ranges of 0.532–0.556, 0.557–0.591, and 0.592–0.705.

4.5. Human Activity Impact Factors

In this study, road construction is used as a representation of human engineering activities; the construction of highways is always accompanied by slope cutting behavior and slope excavation will form steep slopes and create free surfaces, thus affecting slope stability. Here, a road buffer zone was generated based on the GIS platform (Figure 6k); the statistical results showed that as the distance to the road increased, the development of landslides weakened. Landslides were most developed when the distance to the road was <200 m, with an FR reaching 1.33. However, when the distance was >800 m, the FR was also greater than 1, possibly due to other factors controlling the development of landslides within this range.

4.6. Correlation of Conditioning Factors

Before conducting landslide susceptibility evaluation, it is necessary to conduct correlation tests on conditioning factors. The redundancy of conditioning factors may lead to overfitting of the model, thereby reducing modeling accuracy. Here, the factor correlation was tested using the Pearson correlation coefficient method based on Python. The factor correlation coefficient matrices of two sets of landslide inventories are shown in Figure 7. Generally speaking, if the correlation coefficient $r > 0.5$, it is considered that the correlation is too high [51]. According to Figure 6a, the results obtained based on the updated landslide inventory indicated a high correlation between NDVI and NDBI (0.67). Considering that the vegetation in the study area was dense, the influence of vegetation on landslides might not be significant; thus, removing the NDVI factor was considered. The results of the previous landslide inventory in Figure 6b showed a high correlation among the elevation and distance to the river (0.53), the NDVI and NDBI (0.7). Based on the research results of Ma [31], the elevation and NDVI factors were ultimately excluded.

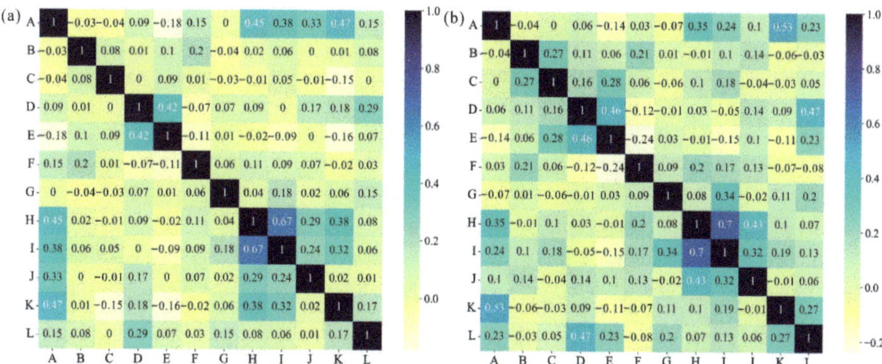

Figure 7. Output results of the Pearson correlation matrix: (**a**) updated landslide inventory; (**b**) previous landslide inventory (A, elevation; B, slope; C, aspect; D, topographic relief; E, elevation coefficient of variation; F, slope structure; G, strata chronology; H, NDVI; I, NDBI; J, distance to river; K, distance to road; L, rainfall).

5. Results

5.1. LSP Based on the K-Means Model

Before performing K-means clustering, it is necessary to calculate the FR of conditioning factors for the two types of landslide inventories separately as input values of the model. Next, the model parameters are set: 5 for the number of clusters, 100 for the maximum number of iterations, and default values for other parameters. Finally, five types of partition results for K-means were obtained, and the FR values for each partition were calculated and arranged in descending order. The final results correspond to very high, high, medium, low, and very low susceptibility [3]. The results of K-means landslide susceptibility zoning are shown in Figure 8 and the statistical results of the zoning are shown in Table 3.

5.2. LSP Based on RF Model

When training the RF model, non-landslide data were also required in addition to landslide data. In this study, two non-landslide sample-selection strategies were used to compare the differences between negative sampling methods. The first sample-selection strategy was random selection outside the landslide, that is, selecting non-landslide samples within the slope units where no landslide had occurred. The second strategy was the semi-supervised machine-learning strategy (SSML) proposed by Huang [27], in which non-landslide samples were randomly selected outside the landslide area in the very low susceptibility zone of the initial LSM of K-means. The ratio of number of non-landslides to number of landslides selected by the above two methods was 1:1 [52].

Two landslide datasets were obtained, and the data were divided into a training set and a testing set in a 7:3 ratio; the training set was used to construct the model and the testing set was used to evaluate the model. Finally, RF and semi-supervised RF (SSRF) were used to obtain the LSM. To compare the impact of different models and non-landslide selection methods on the LSM, the classification method proposed by Pradhan was adopted [21]. The landslide susceptibility index (LSI) was classified into five categories: very high (10%), high (20%), moderate (40%), low (20%), and very low (10%). The final LSM is shown in Figure 9 and the zoning statistical results are shown in Table 3.

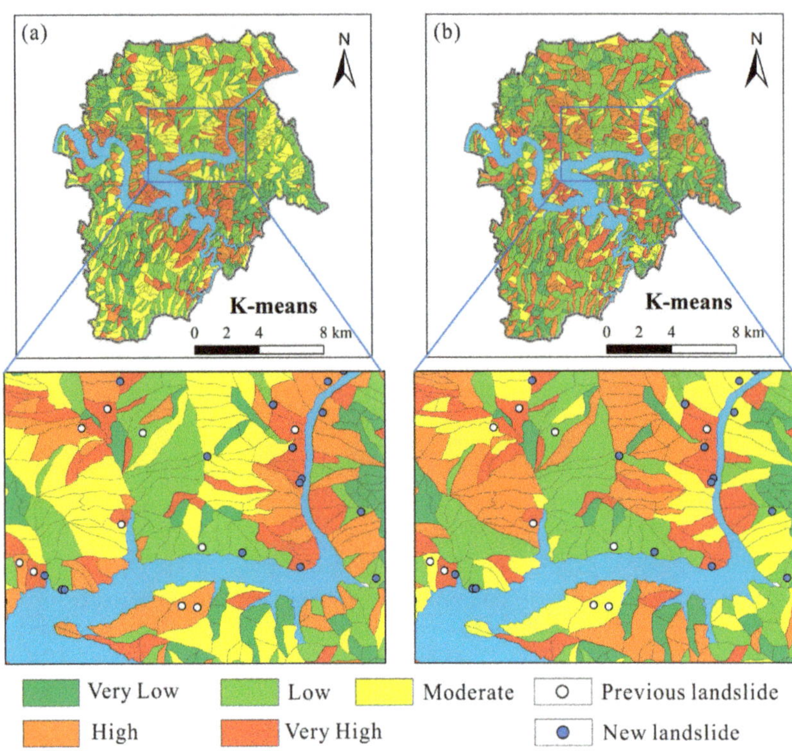

Figure 8. LSMs of K-means model with updated landslide inventory (**a**) and previous landslide inventory (**b**).

Table 3. Frequency ratios of the five susceptibility classes of different models with different inventories.

Model	Inventory	Class	Area of Slope Unit (km^2)	Proportion of Slope Unit (%)	Area of Landslide (km^2)	Proportion of Landslide (%)	FR	FR Accuracy
K-means	Updated	Very High	22.78	13.05	0.59	35.28	2.70	
		High	21.22	12.16	0.25	14.85	1.22	
		Moderate	49.04	28.09	0.55	33.02	1.18	0.68
		Low	47.38	27.14	0.23	13.99	0.52	
		Very Low	34.14	19.56	0.05	2.86	0.15	
	Previous	Very High	22.70	13.00	0.49	37.30	2.87	
		High	39.17	22.44	0.36	27.39	1.22	
		Moderate	20.85	11.95	0.16	11.97	1.00	0.70
		Low	57.65	33.03	0.29	22.03	0.67	
		Very Low	34.20	19.59	0.02	1.30	0.07	
RF	Updated	Very High	17.18	9.84	0.75	44.90	4.56	
		High	34.35	19.67	0.76	45.69	2.32	
		Moderate	68.72	39.37	0.12	7.17	0.18	0.96
		Low	36.41	20.86	0.03	1.85	0.09	
		Very Low	17.91	10.26	0.01	0.39	0.04	
	Previous	Very High	17.37	9.95	0.88	67.36	6.77	
		High	34.04	19.50	0.40	30.78	1.58	
		Moderate	70.40	40.33	0.02	1.42	0.04	0.99
		Low	34.86	19.97	0.01	0.45	0.02	
		Very Low	17.89	10.25	0.00	0.00	0.00	

Table 3. *Cont.*

Model	Inventory	Class	Area of Slope Unit (km^2)	Proportion of Slope Unit (%)	Area of Landslide (km^2)	Proportion of Landslide (%)	FR	FR Accuracy
SSRF	Updated	Very High	16.15	9.25	0.70	41.75	4.51	
		High	34.94	20.02	0.51	30.55	1.53	
		Moderate	70.96	40.65	0.43	25.49	0.63	0.89
		Low	34.66	19.86	0.04	2.20	0.11	
		Very Low	17.86	10.23	0.00	0.01	0.00	
	Previous	Very High	16.93	9.70	0.72	54.99	5.67	
		High	34.88	19.98	0.53	40.74	2.04	
		Moderate	70.10	40.16	0.04	2.85	0.07	0.98
		Low	35.54	20.36	0.02	1.42	0.07	
		Very Low	17.13	9.81	0.00	0.00	0.00	
ANN	Updated	Very High	17.34	9.93	0.64	38.49	3.87	
		High	35.18	20.15	0.47	27.80	1.38	
		Moderate	69.78	39.97	0.49	29.55	0.74	0.84
		Low	34.91	20.00	0.05	3.14	0.16	
		Very Low	17.36	9.95	0.02	1.02	0.10	
	Previous	Very High	17.41	9.98	0.39	30.07	3.01	
		High	34.84	19.96	0.37	28.20	1.41	
		Moderate	68.90	39.47	0.54	41.14	1.04	0.81
		Low	36.11	20.68	0.01	0.59	0.03	
		Very Low	17.31	9.92	0.00	0.00	0.00	
SSANN	Updated	Very High	17.74	10.16	0.44	26.58	2.62	
		High	34.65	19.85	0.72	43.25	2.18	
		Moderate	69.89	40.03	0.47	27.83	0.70	0.85
		Low	34.89	19.98	0.04	2.17	0.11	
		Very Low	17.41	9.97	0.00	0.18	0.02	
	Previous	Very High	17.34	9.93	0.55	41.73	4.20	
		High	34.92	20.00	0.49	37.17	1.86	
		Moderate	69.74	39.95	0.28	21.10	0.53	0.92
		Low	35.15	20.14	0.00	0.00	0.00	
		Very Low	17.42	9.98	0.00	0.00	0.00	

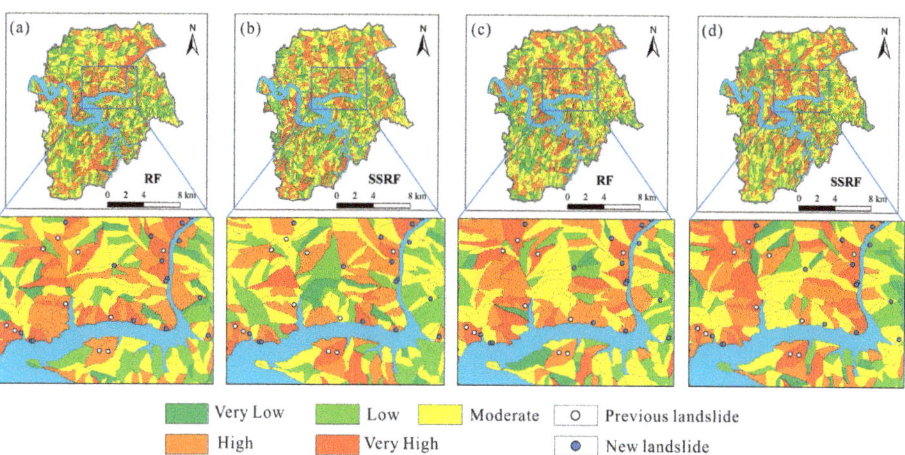

Figure 9. LSMs of RF and SSRF with updated landslide inventory (**a**,**b**) and previous landslide inventory (**c**,**d**).

5.3. LSP Based on the ANN Model

The selection of non-landslide samples, division of training and testing sets, and susceptibility zoning methods were the same as those in Section 5.2; thus, no further description is made here. It is important to note that in this study, a trial-and-error method was used to determine the number of hidden layers. It was found that this works better with two layers; the model was prone to overfitting when there were too many layers and underfitting when there were too few. Finally, the optimal number of hidden layers was determined by the minimum prediction error method [32]; the number of hidden layer neurons in the updated landslide data was (13, 11) and the previous landslide inventory was (14, 12). The LSM of the final ANN is shown in Figure 10 and the zoning statistical results in Table 3.

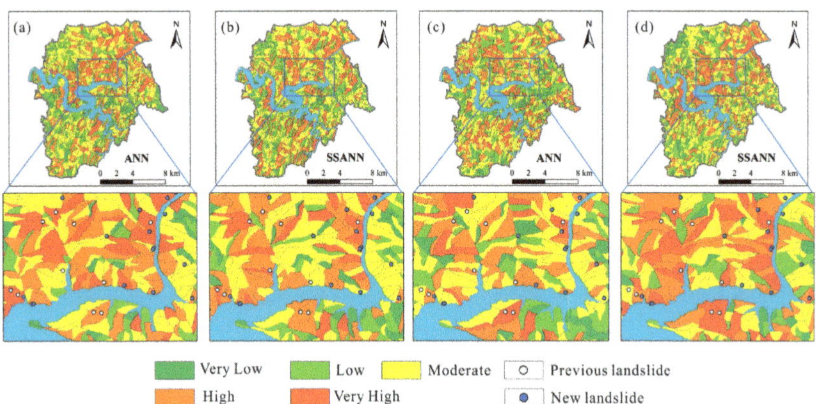

Figure 10. LSMs of ANN and SSANN with updated landslide inventory (**a,b**) and previous landslide inventory (**c,d**).

5.4. Accuracy Assessment and Validation

5.4.1. ROC Curve

Figure 11 shows the ROC curve results considering different landslide inventories and evaluation models. It is worth noting that due to the inability of ROC to evaluate the performance of unsupervised machine learning [3], the ROC curve for K-means is not plotted here. In the updated landslide inventory, the respective AUCs of RF, SSRF, ANN, and SSANN were 0.901, 0.870, 0.781, and 0.731, while in the previous landslide inventory the values were 0.830, 0.879, 0.722, and 0.793. The AUCs of all models were observed to be greater than 0.7, indicating that the evaluation results were reliable. Specifically, RF was observed to have higher prediction accuracy than ANN in both the new and old landslide databases.

5.4.2. Frequency Ratio Accuracy Validation

FR accuracy can be used to evaluate SML and USML models. Higher FR accuracy generally means better model performance. Tables 2 and 3 show the FR values and accuracy of the susceptibility zone in each model.

For the updated landslide inventory, the respective FR accuracies of K-means, RF, SSRF, ANN, and SSANN were 0.68, 0.96, 0.89, 0.84, and 0.85, while for the previous landslide inventory, the values were 0.70, 0.99, 0.98, 0.80, and 0.92. It was found in the FR accuracy test that the accuracy of SML was generally higher than that of USML, while that of RF was higher than that of ANN. Therefore, it can be confirmed that SML has better prediction performance and zoning results than USML. Meanwhile, in SML, RF performs better than ANN and its zoning is more reasonable.

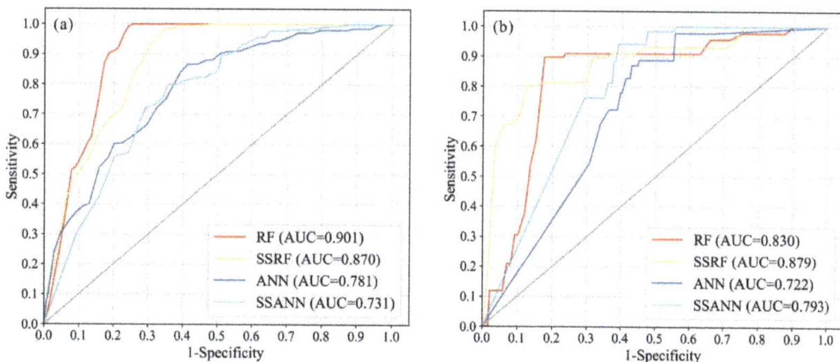

Figure 11. ROC under different inventories: (**a**) updated landslide inventory, (**b**) previous landslide inventory.

5.4.3. Distribution Features of Landslide Susceptibility Indexes

A new evaluation method proposed by Huang was adopted [27]. This method has been applied in various scenarios since its introduction. The basic idea is to evenly divide the landslide susceptibility index (LSI) interval (0–1) into 80 grade ranges, then count the number of evaluation units within each range, draw a histogram, and calculate the mean and standard deviation. In general, a smaller mean and larger standard deviation represent better model performance.

Figures 12 and 13 show the LSI distribution of different landslide inventories and the evaluation results of different machine-learning models. Significant differences were observed in the LSI distribution among the different models. The distribution of LSI in ANN was relatively concentrated on both sides of the interval and relatively uniform in the middle, while that of RF approximately followed a normal distribution, mainly concentrated in the middle. The mean of each model was calculated, with a general result of RF < ANN. For standard deviation, RF > ANN; however, for SSRF and SSANN, an opposite result was observed.

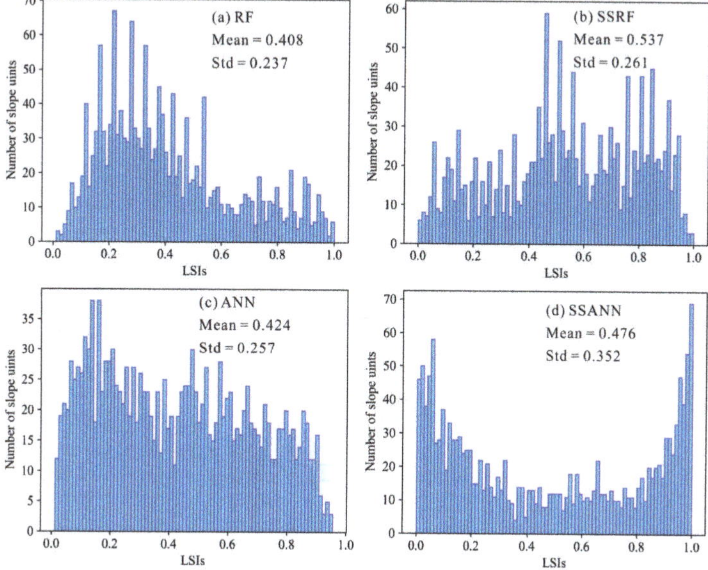

Figure 12. LSIs distribution features of RF (**a**), SSRF (**b**), ANN (**c**), and SSANN (**d**), with updated landslide inventory.

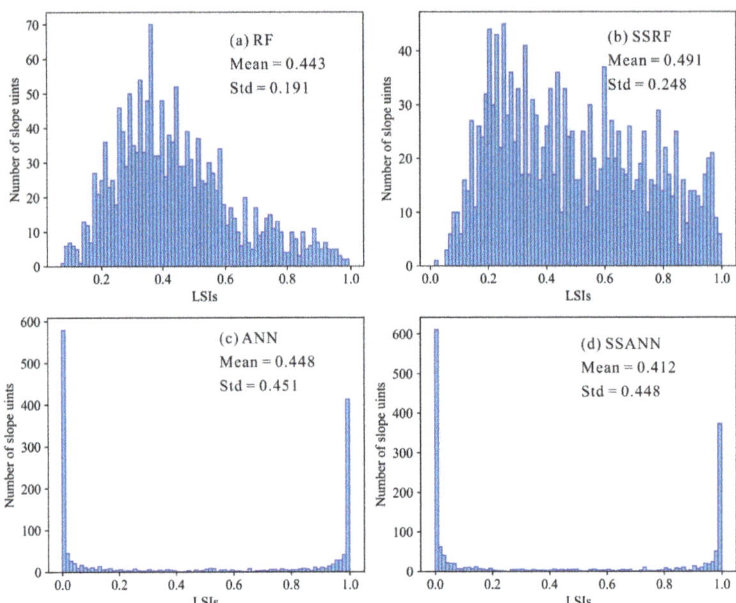

Figure 13. LSIs distribution features of RF (**a**), SSRF (**b**), ANN (**c**), and SSANN (**d**), with previous landslide inventory.

6. Discussion

6.1. Effects of Non-Landslide Sample Selection

Non-landslide samples might suppress overestimation of landslide susceptibility during modeling, and selecting reasonable non-landslide samples is a hot topic in landslide susceptibility research [52]. At present, the negative sample-selection methods for landslides include random sampling outside the landslide [53], buffer zone and target space [54,55], low slope sampling [56], information value sampling [57], and semi-supervised machine learning (SSML) [27]. The most-used random sampling and newly proposed SSML sampling method were used here to explore the effects of non-landslide sample selection on the results.

From Table 3 and Figure 11, it can be seen that the selection method of non-landslide samples in different landslide inventories had an impact on the evaluation model, and differences were observed in the pattern of landslide inventories, detailed as follows: (1) Regardless of landslide inventory type, the AUC value and FR accuracy of RF and SSRF, ANN and SSANN differed, thus confirming that the selection method of non-landslide samples had an impact on the results. (2) In the updated landslide inventory, the AUC results showed that RF (0.901) > SSRF (0.870) and ANN (0.781) > SSANN (0.731), and the use of SSML reduced the prediction accuracy of the model. (3) In the previous landslide inventory, the AUC results showed that RF (0.830) < SSRF (0.879) and ANN (0.722) < SSANN (0.793), and the use of SSML improved the prediction accuracy of the model.

Previous research results had shown that the use of the SSML non-landslide selection method in SSML might reduce the uncertainty of the model and improve its prediction accuracy [27,58,59]. However, the conclusions obtained in this study differed from previous ones. Semi-supervised learning was observed to have applicable conditions, and there were differences in different landslide inventories. The number of mapping units remained the same in this study; what differed between the updated landslide inventory and the previous landslide inventory was the number of landslides. Therefore, it was inferred that the ratio of the number of landslide samples to the number of mapping units was an important factor influencing the method. Table 4 shows a case where this method was applied to different study areas; when the ratio of the number of landslide samples to the number of mapping units was small, the model accuracy was improved after the SSML method was applied. Previously, scholars conducted evaluation mainly based on grid units; thus, this ratio was generally quite small. However, slope unit was adopted in this paper, and the ratio was quite large. Furthermore, the use of the previous landslide inventory was observed to improve the accuracy and the use of the updated landslide inventory reduced the accuracy. Therefore, we believe that the accuracy of the SSML non-landslide selection method is controlled by the ratio of the number of landslide samples to the number of mapping units.

Table 4. Frequency ratios of the five susceptibility classes of different models.

Test	Huang [27] (2020)	Zhang [58] (2022)	Yang [59] (2023)	This Paper 1	This Paper 2
Number of landslide sample (A)	3235	622	349	222	46
Number of mapping units (B)	2,565,787	2,191,350	7,422,530	1594	1594
A/B (%)	0.00126	0.00028	0.00005	13.92723	2.88582
Whether to improve accuracy	Yes	Yes	Yes	No	Yes

The main reason for the above phenomenon lies in the fact that the non-landslide samples were mostly collected from very low-susceptibility zones. When the ratio of the number of landslide samples to the number of mapping units was large, this meant that non-landslide samples were more concentrated in very low-susceptibility zones, covering a narrow combination of factors, and the model could not obtain more knowledge [14]. On the contrary, the factor combination covered was wider and the effect was better by randomly selecting samples outside the landslide. Therefore, when slope unit evaluation was adopted and there were many landslides (especially earthquake-induced ones), then non-landslide random sampling might yield better prediction results.

It can be inferred that thresholds influence SSML effectiveness, and threshold range needs further exploration in future research.

6.2. Comparison of Different ML Models

Considering different modeling principles, there are differences in the susceptibility prediction results obtained by ML models [8]. To describe the uncertainty of evaluation models, ROC, FR accuracy, and LSI distribution patterns are adopted to characterize the differences between models.

From Table 3, it can be seen that in the FR accuracy test, the accuracy of the USML model (represented by K-means) (0.68, 0.70) was lower than that of the SML model. According to the zoning results, the FR values of SML decreased in order from the very high-susceptibility zone to very low-susceptibility zone, and FR > 1 only occurred in the very high- and high-susceptibility zones, indicating that the zoning results were reasonable. However, the FR value of USML in the moderate susceptibility zone was ≥ 1, resulting in poor zoning results and high uncertainty. The main reason for this phenomenon was the difference between SML and USML: the core of SML is prediction and classification, while that of USML is cluster analysis [3]; thus, they are essentially different. For this reason, SML performs better than USML.

The performance of SML models (RF, ANN) also varies, detailed as follows:

(1) From the perspective of the accuracy evaluation of the ROC curve (Figure 10), for the updated landslide inventory, RF (0.901) > ANN (0.781) and SSRF (0.870) > SSANN (0.731); for the previous landslide inventory, RF (0.830) > ANN (0.722) and SSRF (0.879) > SSANN (0.793). It was observed that under different combination conditions, the AUC value of RF was greater than that of ANN, thus confirming that RF had lower uncertainty, higher model robustness, and higher prediction accuracy than ANN.

(2) From the perspective of the FR accuracy test (Table 3), for the updated landslide inventory, RF (0.96) > ANN (0.84) and SSRF (0.89) > SSANN (0.85); for the previous

landslide inventory, RF (0.99) > ANN (0.80) and SSRF (0.98) > SSANN (0.92). It was observed that under different combination conditions, the FR accuracy of RF was greater than that of ANN, thus confirming that the RF zoning effect was more reasonable than ANN.

(3) From the perspective of the distribution characteristics of LSI (Figures 12 and 13), if the mean and standard deviation discrimination criteria proposed by Huang were adopted, then the regularity and reliability of the results may not be particularly high, but they still provided important reference value. First, in the same landslide inventory, the LSIs of RF and ANN showed basically the same distribution trend. For non-landslide random selection methods, Mean (RF) < Mean (ANN) was observed in different inventories, and for for SSML, Mean (RF) > Mean (ANN) was observed; regardless of the combination conditions, Std (ANN) > Std (RF) was observed. However, the distribution trend of LSI indicated that ANN was more concentrated in the high- and low-susceptibility index intervals, while RF was more concentrated in the middle, thus explaining Std (ANN) > Std (RF). Viewed from the perspective of susceptibility zoning, the LSI distribution characteristics of ANN were not very reasonable due to its overestimation of the number of very high- and low-susceptibility zones, while the frequency of the intermediate areas was basically the same, posing significant difficulties for zoning. Comparatively speaking, the differences within each area of RF were observed to be significant, and the zoning was also more reasonable and reliable.

Overall, the performance of machine-learning models is ranked as follows: RF > ANN > K-means. The prediction accuracy of USML is inferior to SML, and in SML, RF performance is superior to ANN, with less uncertainty.

6.3. Effects of Landslide Inventory Completeness

The acquisition of landslide data is an important step in LSP, and in this study two sets of landslide inventories are used to explore the impact of landslide data integrity on the LSP results.

(1) Identification of landslide main control factors: Here, the model with the best prediction performance (RF for the updated landslide inventory and SSRF for the previous landslide inventory) is selected from two sets of inventories for landslide main control factor identification. In RF, the importance of conditioning factors is measured by the percentage of the average Gini index reduction to the total average Gini reduction of all environmental factors. The weights of 11 conditioning factors, namely elevation (ELE), slope (SLO), aspect (ASP), topographic relief (TR), elevation coefficient variation (ECV), slope structure (SLOS), strata chronology (SC), NDBI, distance to road (DTRO), distance to river (DTRI), and rainfall (RAIN), are shown in Figure 14. (Since ELE is not considered in the previous landslide inventory, its weight is 0.)

From Figure 14, the importance of aspect and NDBI factors ranked among the top three in both types of inventories, showing consistency. However, the slope structure factor presented different results, ranking second in the previous landslide inventory, yet only seventh in the updated landslide inventory. We believe the reason for this phenomenon is that the attitude of rock in the slope structure must be considered. However, the field investigation data is limited. Therefore, interpolation is used to expand the data, thus leading to the bias between the calculation results and actual situation. This means that, as the number of landslide samples increases in the updated inventory, additional bias caused by the interpolation of slope structure is introduced into the model. As a result, the slope structure factor presents a different importance in the two inventories.

In general, there were certain differences in the results of the two inventories, imposing a certain impact on understanding the causes of regional landslides. The main reason for this phenomenon is the differences in the data samples. The updated landslide inventory had a greater number of landslides compared to the previous landslide inventory, and more complete statistical information, which in turn affected the model input values, ultimately leading to differences in the results of the model to identify the importance of the factors.

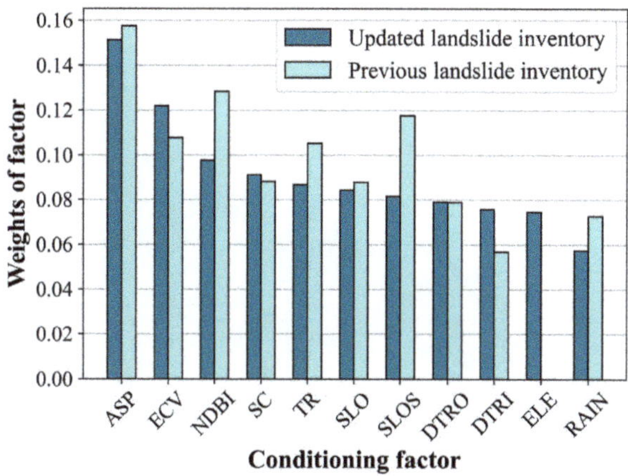

Figure 14. Conditioning factor weights under different landslide inventories.

(2)   Prediction of newly added landslides with incomplete data: Here, based on the evaluation results of the previous landslide inventory SSRF, the number of newly added landslides, FR values, and FR accuracy in each susceptibility zone were calculated (Table 5), and the prediction rate curve was plotted (Figure 15). From Table 5 and Figure 15, it can be observed that the results were not very satisfactory. First, the FR values in the very high- to very low-susceptibility zones did not show a downward trend. On the contrary, the FR value in the high-susceptibility zone was higher than that in the very high-susceptibility zone, and the FR value in the very low-susceptibility zone was higher than that in the low-susceptibility zone, which was unreasonable. Second, the FR accuracy test result was 0.65 and the AUC of the prediction rate curve was only 0.638. Based on the above results, we believe that the use of recorded landslide data alone is insufficient to characterize the landslide susceptibility trend in the study area, and more complete landslide data should be used [14].

Table 5. Frequency ratios of the five susceptibility classes of SSRF using the previous landslide inventory to predict newly added landslides.

Model	Class	Area of Slope Unit (km^2)	Proportion of Slope Unit (%)	Area of Newly Added Landslide (km^2)	Proportion of Landslide (%)	FR	FR Accuracy
	Very High	16.93	9.70	0.06	16.57	1.71	
	High	34.88	19.98	0.13	35.52	1.78	
SSRF	Moderate	70.10	40.16	0.12	34.57	0.86	0.65
	Low	35.54	20.36	0.02	6.64	0.33	
	Very Low	17.13	9.81	0.02	6.70	0.68	

It is held in this study that a relatively complete landslide inventory must be considered when conducting LSP. The sample size of landslide data recorded by data collection are small; thus, it is not possible to have a good understanding of the development of regional landslides and effectively conduct risk assessment work. At present, some scholars have conducted research on the issue of missing landslide data and have found that the use of nonlinear mixed effect models will reduce the impact of incomplete landslide data on the final results [15,60]. Moreover, the landslide inventory can be expanded based on multi-source remote sensing to improve the model's accuracy.

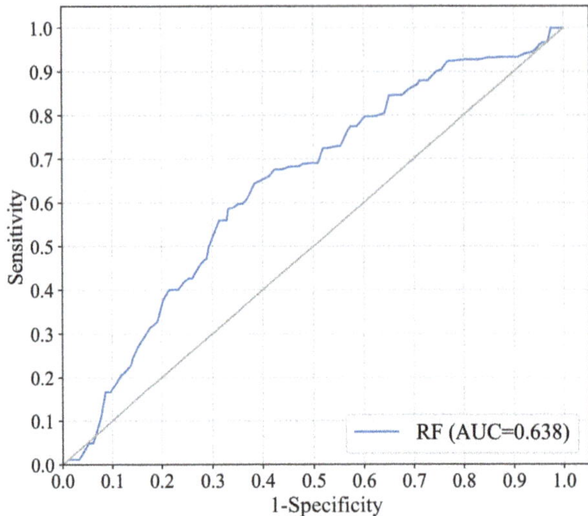

Figure 15. Prediction rate curve using previous landslides to predict newly added landslides.

The comparison of landslide inventories can be further refined in future studies. This paper compares the experimental results of the updated landslide inventory with the previous one. The newly added landslides can also be used as the third landslide inventory, based on which exploring the differences between the three landslide inventories would make the conclusions more complete.

7. Conclusions

LSP is the basis for risk management; thus, it is important to obtain accurate and reliable LSM. This study explored the LSP modeling uncertainty from three perspectives: machine-learning models, non-landslide sample selection, and landslide data integrity. The results can provide important theoretical references for regional risk prevention and control. The following understandings are achieved:

(1) The ML model can achieve good results when applied to LSM, and its overall performance is RF > ANN > K-means. The USML model performs poorly when compared to the other models, and the zoning outcomes are less than ideal, while SML performs better. Especially in the SML model, RF has superior prediction performance and stronger robustness compared to ANN.

(2) The selection method of non-landslide samples has a significant impact on LSP. The ratio of the number of landslide samples to the number of mapping units is crucial in the SSML non-landslide selection method, which combines USML and SML. When the ratio is low, the non-landslide selection results of SSML are better than random selection, and when this ratio is high, the non-landslide selection results of SSML are inferior to random selection.

(3) The quantity of landslides influences how reliably the results of LSM are obtained because fewer landslides result in a smaller sample size for LSM, which deviates from reality. Although the results are satisfactory on this dataset, the recently added landslide data discovered by field research and remote-sensing interpretation cannot be reliably predicted by the zoning results. On the contrary, the LSM of sufficient landslide samples is more reasonable.

(4) Among the results of the two landslide inventories, aspect and NDBI are all significant conditioning factors for landslide development. However, due to the differences in the base data, the importance of other factors is not consistent between two inventories.

(5)   The slope structure information is obtained by interpolation, which leads to the deviation of the calculation results from the on-site measured values. Therefore, as the number of landslides increases, the modeling calculation error increases, resulting in a difference in the slope structural importance presented in the two inventories.

Author Contributions: S.M. and J.C. collected and processed the data; S.M. and Y.L. proposed the model and analyzed the results; S.M. and J.C. drafted the manuscript; and J.C. and S.W. revised the manuscript. All authors have read and agreed to the published version of the manuscript.

Funding: This research was funded by the National Science and Technology Infrastructure Center of China (2019FY101605) and the National Natural Science Foundation of China (41571012).

Institutional Review Board Statement: Not applicable.

Informed Consent Statement: Not applicable.

Data Availability Statement: Data are contained within the article.

Acknowledgments: The authors are grateful to the editors and reviewers for their enthusiastic help and valuable comments, and for the support of the National Science and Technology Infrastructure Center of China (2019FY101605) and the data support given by Shaanxi Nuclear Industry Engineering Survey Institute Co., Ltd.

Conflicts of Interest: The authors declare no conflict of interest.

References

1.   Turner, D.; Lucieer, A.; De Jong, S. Time series analysis of landslide dynamics using an unmanned aerial vehicle (UAV). *Remote Sens.* **2015**, *7*, 1736–1757. [CrossRef]
2.   Shao, X.; Ma, S.; Xu, C.; Zhang, P.; Wen, B.; Tian, Y.; Zhou, Q.; Cui, Y. Planet image-based inventorying and machine learning-based susceptibility mapping for the landslides triggered by the 2018 Mw6. 6 Tomakomai, Japan Earthquake. *Remote Sens.* **2019**, *11*, 978. [CrossRef]
3.   Chang, Z.; Du, Z.; Zhang, F.; Huang, F.; Chen, J.; Li, W.; Guo, Z. Landslide susceptibility prediction based on remote sensing images and GIS: Comparisons of supervised and unsupervised machine learning models. *Remote Sens.* **2020**, *12*, 502. [CrossRef]
4.   Reichenbach, P.; Rossi, M.; Malamud, B.D.; Mihir, M.; Guzzetti, F. A review of statistically-based landslide susceptibility models. *Earth-Sci. Rev.* **2018**, *180*, 60–91. [CrossRef]
5.   Ayalew, L.; Yamagishi, H. The application of GIS-based logistic regression for landslide susceptibility mapping in the Kakuda-Yahiko Mountains, Central Japan. *Geomorphology* **2005**, *65*, 15–31. [CrossRef]
6.   Lan, H.; Zhou, C.; Wang, L.; Zhang, H.; Li, R. Landslide hazard spatial analysis and prediction using GIS in the Xiaojiang watershed, Yunnan, China. *Eng. Geol.* **2004**, *76*, 109–128. [CrossRef]
7.   Pradhan, A.M.S.; Lee, S.-R.; Kim, Y.-T. A shallow slide prediction model combining rainfall threshold warnings and shallow slide susceptibility in Busan, Korea. *Landslides* **2019**, *16*, 647–659. [CrossRef]
8.   Huang, F.; Cao, Z.; Guo, J.; Jiang, S.-H.; Li, S.; Guo, Z. Comparisons of heuristic, general statistical and machine learning models for landslide susceptibility prediction and mapping. *Catena* **2020**, *191*, 104580. [CrossRef]
9.   Hong, H.; Pradhan, B.; Sameen, M.I.; Kalantar, B.; Zhu, A.; Chen, W. Improving the accuracy of landslide susceptibility model using a novel region-partitioning approach. *Landslides* **2018**, *15*, 753–772. [CrossRef]
10. Xu, C.; Xu, X.; Dai, F.; Wu, Z.; He, H.; Shi, F.; Wu, X.; Xu, S. Application of an incomplete landslide inventory, logistic regression model and its validation for landslide susceptibility mapping related to the May 12, 2008 Wenchuan earthquake of China. *Nat. Hazards* **2013**, *68*, 883–900. [CrossRef]
11. Chang, Z.; Huang, F.; Huang, J.; Jiang, S.-H.; Liu, Y.; Meena, S.R.; Catani, F. An updating of landslide susceptibility prediction from the perspective of space and time. *Geosci. Front.* **2023**, *14*, 101619. [CrossRef]
12. Lazzari, M.; Gioia, D.; Anzidei, B. Landslide inventory of the Basilicata region (Southern Italy). *J. Maps* **2018**, *14*, 348–356. [CrossRef]
13. Lazzari, M.; Gioia, D. Regional-scale landslide inventory, central-western sector of the Basilicata region (Southern Apennines, Italy). *J. Maps* **2016**, *12*, 852–859. [CrossRef]
14. Xu, C.; Dai, F.; Xu, S.; Xu, X.; He, H.; Wu, X.; Shi, F. Application of logistic regression model on the Wenchuan earthquake triggered landslide hazard mapping and its validation. *Hydrogeol. Eng. Geol.* **2013**, *40*, 98–104.
15. Lin, Q.; Lima, P.; Steger, S.; Glade, T.; Jiang, T.; Zhang, J.; Liu, T.; Wang, Y. National-scale data-driven rainfall induced landslide susceptibility mapping for China by accounting for incomplete landslide data. *Geosci. Front.* **2021**, *12*, 101248. [CrossRef]
16. Huang, F.; Yin, K.; Huang, J.; Gui, L.; Wang, P. Landslide susceptibility mapping based on self-organizing-map network and extreme learning machine. *Eng. Geol.* **2017**, *223*, 11–22. [CrossRef]

17. Maurizio, L.; Maria, D. A multi temporal kernel density estimation approach for new triggered landslides forecasting and susceptibility assessment. *Disaster Adv.* **2012**, *5*, 100–108.

18. He, S.; Pan, P.; Dai, L.; Wang, H.; Liu, J. Application of kernel-based Fisher discriminant analysis to map landslide susceptibility in the Qinggan River delta, Three Gorges, China. *Geomorphology* **2012**, *171*, 30–41. [CrossRef]

19. Ruff, M.; Czurda, K. Landslide susceptibility analysis with a heuristic approach in the Eastern Alps (Vorarlberg, Austria). *Geomorphology* **2008**, *94*, 314–324. [CrossRef]

20. Li, L.; Lan, H.; Guo, C.; Zhang, Y.; Li, Q.; Wu, Y. A modified frequency ratio method for landslide susceptibility assessment. *Landslides* **2017**, *14*, 727–741. [CrossRef]

21. Pradhan, B. A comparative study on the predictive ability of the decision tree, support vector machine and neuro-fuzzy models in landslide susceptibility mapping using GIS. *Comput. Geosci.* **2013**, *51*, 350–365. [CrossRef]

22. Thai Pham, B.; Tien Bui, D.; Prakash, I. Landslide susceptibility modelling using different advanced decision trees methods. *Civ. Eng. Environ. Syst.* **2018**, *35*, 139–157. [CrossRef]

23. Sun, D.; Wen, H.; Wang, D.; Xu, J. A random forest model of landslide susceptibility mapping based on hyperparameter optimization using Bayes algorithm. *Geomorphology* **2020**, *362*, 107201. [CrossRef]

24. Bui, D.T.; Tsangaratos, P.; Nguyen, V.-T.; Van Liem, N.; Trinh, P.T. Comparing the prediction performance of a Deep Learning Neural Network model with conventional machine learning models in landslide susceptibility assessment. *Catena* **2020**, *188*, 104426. [CrossRef]

25. Yilmaz, I. Landslide susceptibility mapping using frequency ratio, logistic regression, artificial neural networks and their comparison: A case study from Kat landslides (Tokat—Turkey). *Comput. Geosci.* **2009**, *35*, 1125–1138. [CrossRef]

26. Wang, Q.; Wang, Y.; Niu, R.; Peng, L. Integration of information theory, K-means cluster analysis and the logistic regression model for landslide susceptibility mapping in the Three Gorges Area, China. *Remote Sens.* **2017**, *9*, 938. [CrossRef]

27. Huang, F.; Cao, Z.; Jiang, S.-H.; Zhou, C.; Huang, J.; Guo, Z. Landslide susceptibility prediction based on a semi-supervised multiple-layer perceptron model. *Landslides* **2020**, *17*, 2919–2930. [CrossRef]

28. Huang, F.; Hu, S.; Yan, X.; Li, M.; Wang, J.; Li, W.; Guo, Z.; Fan, W. Landslide susceptibility prediction and identification of its main environmental factors based on machine learning models. *Bull. Geol. Sci. Technol.* **2022**, *41*, 79–90.

29. Xie, M.; Esaki, T.; Zhou, G. GIS-based probabilistic mapping of landslide hazard using a three-dimensional deterministic model. *Nat. Hazards* **2004**, *33*, 265–282. [CrossRef]

30. Zhuo, L.; Huang, Y.; Zheng, J.; Cao, J.; Guo, D. Landslide Susceptibility Mapping in Guangdong Province, China, Using Random Forest Model and Considering Sample Type and Balance. *Sustainability* **2023**, *15*, 9024. [CrossRef]

31. Ma, S.; Chen, J.; Wu, S. Distribution characteristics and susceptibility assessment of landslide hazard in Yinghu Town, Ankang City, Shaanxi Province. *Geoscience* **2023**, 1–17. [CrossRef]

32. Li, D.; Huang, F.; Yan, L.; Cao, Z.; Chen, J.; Ye, Z. Landslide susceptibility prediction using particle-swarm-optimized multilayer perceptron: Comparisons with multilayer-perceptron-only, bp neural network, and information value models. *Appl. Sci.* **2019**, *9*, 3664. [CrossRef]

33. Cao, Y.-M.; Guo, W.; Wu, Y.-M.; Li, L.-P.; Zhang, Y.-X.; Lan, H.-X. An hourly shallow landslide warning model developed by combining automatic landslide spatial susceptibility and temporal rainfall threshold predictions. *J. Mt. Sci.* **2022**, *19*, 3370–3387. [CrossRef]

34. Chen, W.; Peng, J.; Hong, H.; Shahabi, H.; Pradhan, B.; Liu, J.; Zhu, A.-X.; Pei, X.; Duan, Z. Landslide susceptibility modelling using GIS-based machine learning techniques for Chongren County, Jiangxi Province, China. *Sci. Total Environ.* **2018**, *626*, 1121–1135. [CrossRef]

35. Chen, W.; Zhang, S.; Li, R.; Shahabi, H. Performance evaluation of the GIS-based data mining techniques of best-first decision tree, random forest, and naïve Bayes tree for landslide susceptibility modeling. *Sci. Total Environ.* **2018**, *644*, 1006–1018. [CrossRef]

36. Bravo-López, E.; Del Castillo, T.F.; Sellers, C.; Delgado-García, J. Analysis of Conditioning Factors in Cuenca, Ecuador, for Landslide Susceptibility Maps Generation Employing Machine Learning Methods. *Land* **2023**, *12*, 1135. [CrossRef]

37. Breiman, L. Random forests. *Mach. Learn.* **2001**, *45*, 5–32. [CrossRef]

38. Saha, S.; Majumdar, P.; Bera, B. Deep learning and benchmark machine learning based landslide susceptibility investigation, Garhwal Himalaya (India). *Quat. Sci. Adv.* **2023**, *10*, 100075. [CrossRef]

39. Liu, Y.; Meng, Z.; Zhu, L.; Hu, D.; He, H. Optimizing the sample selection of machine learning models for landslide susceptibility prediction using information value models in the Dabie mountain area of Anhui, China. *Sustainability* **2023**, *15*, 1971. [CrossRef]

40. Mwakapesa, D.S.; Mao, Y.; Lan, X.; Nanehkaran, Y.A. Landslide Susceptibility Mapping Using DIvisive ANAlysis (DIANA) and RObust Clustering Using linKs (ROCK) Algorithms, and Comparison of Their Performance. *Sustainability* **2023**, *15*, 4218. [CrossRef]

41. Cantarino, I.; Carrion, M.A.; Goerlich, F.; Martinez Ibañez, V. A ROC analysis-based classification method for landslide susceptibility maps. *Landslides* **2019**, *16*, 265–282. [CrossRef]

42. Huang, F.; Tao, S.; Li, D.; Lian, Z.; Catani, F.; Huang, J.; Li, K.; Zhang, C. Landslide susceptibility prediction considering neighborhood characteristics of landslide spatial datasets and hydrological slope units using remote sensing and GIS technologies. *Remote Sens.* **2022**, *14*, 4436. [CrossRef]

43. Qin, H.; Tan, S.; Shi, Y.; Li, H.; WAng, B. Geological hazard susceptibility assessment based on CF&LR combined model:case of Ning'er Hani and Yi Autonomous County, Yunnan Province. *Yangtze River* **2022**, *53*, 119–127.

44. Tien Bui, D.; Shahabi, H.; Shirzadi, A.; Chapi, K.; Hoang, N.-D.; Pham, B.T.; Bui, Q.-T.; Tran, C.-T.; Panahi, M.; Bin Ahmad, B. A novel integrated approach of relevance vector machine optimized by imperialist competitive algorithm for spatial modeling of shallow landslides. *Remote Sens.* **2018**, *10*, 1538. [CrossRef]
45. Jacobs, L.; Kervyn, M.; Reichenbach, P.; Rossi, M.; Marchesini, I.; Alvioli, M.; Dewitte, O. Regional susceptibility assessments with heterogeneous landslide information: Slope unit-vs. pixel-based approach. *Geomorphology* **2020**, *356*, 107084. [CrossRef]
46. Chang, Z.; Catani, F.; Huang, F.; Liu, G.; Meena, S.R.; Huang, J.; Zhou, C. Landslide susceptibility prediction using slope unit-based machine learning models considering the heterogeneity of conditioning factors. *J. Rock Mech. Geotech. Eng.* **2023**, *15*, 1127–1143. [CrossRef]
47. Tao, Z.; Shu, Y.; Yang, X.; Peng, Y.; Chen, Q.; Zhang, H. Physical model test study on shear strength characteristics of slope sliding surface in Nanfen open-pit mine. *Int. J. Min. Sci. Technol.* **2020**, *30*, 421–429. [CrossRef]
48. Yang, X.; Wang, P.; Li, X.; Xie, C.; Zhou, B.; Huang, X. Application of topographic slope and elevation variation coefficient in identifying the motuo active fault zone. *Seismol. Egology* **2019**, *41*, 419–435.
49. Qin, Y.; Yang, G.; Lu, K.; Sun, Q.; Xie, J.; Wu, Y. Performance evaluation of five GIS-based models for landslide susceptibility prediction and mapping: A case study of Kaiyang County, China. *Sustainability* **2021**, *13*, 6441. [CrossRef]
50. Chen, C.-Y.; Huang, W.-L. Land use change and landslide characteristics analysis for community-based disaster mitigation. *Environ. Monit. Assess.* **2013**, *185*, 4125–4139. [CrossRef]
51. Gao, H.-X.; Yin, K.-L. Discuss on the correlations between landslides and rainfall and threshold for landslide early-warning and prediction. *Yantu Lixue Rock Soil Mech.* **2007**, *28*, 1055–1060.
52. Liu, Q.; Tang, A.; Huang, D. Exploring the uncertainty of landslide susceptibility assessment caused by the number of non–landslides. *Catena* **2023**, *227*, 107109. [CrossRef]
53. Chen, W.; Zhang, S. GIS-based comparative study of Bayes network, Hoeffding tree and logistic model tree for landslide susceptibility modeling. *Catena* **2021**, *203*, 105344. [CrossRef]
54. Lucchese, L.V.; de Oliveira, G.G.; Pedrollo, O.C. Investigation of the influence of nonoccurrence sampling on landslide susceptibility assessment using Artificial Neural Networks. *Catena* **2021**, *198*, 105067. [CrossRef]
55. Hong, H.; Miao, Y.; Liu, J.; Zhu, A.-X. Exploring the effects of the design and quantity of absence data on the performance of random forest-based landslide susceptibility mapping. *Catena* **2019**, *176*, 45–64. [CrossRef]
56. Kavzoglu, T.; Sahin, E.K.; Colkesen, I. Landslide susceptibility mapping using GIS-based multi-criteria decision analysis, support vector machines, and logistic regression. *Landslides* **2014**, *11*, 425–439. [CrossRef]
57. Xiaoting, Z.; Faming, H.; Weicheng, W.; Chuangbing, Z.; Shiyi, Z.; Lihan, P. Regional Landslide Susceptibility Prediction Based on Negative Sample Selected by Coupling Information Value Method. *Adv. Eng. Sci. Gongcheng Kexue Yu Jishu* **2022**, *54*, 25–35.
58. Zhang, Y.; Yan, Q. Landslide susceptibility prediction based on high-trust non-landslide point selection. *ISPRS Int. J. Geo-Inf.* **2022**, *11*, 398. [CrossRef]
59. Yang, N.; Wang, R.; Liu, Z.; Yao, Z. Landslide susceptibility prediction improvements based on a semi-integrated supervised machine learning model. *Environ. Sci. Pollut. Res.* **2023**, *30*, 50280–50294. [CrossRef] [PubMed]
60. Steger, S.; Brenning, A.; Bell, R.; Glade, T. The influence of systematically incomplete shallow landslide inventories on statistical susceptibility models and suggestions for improvements. *Landslides* **2017**, *14*, 1767–1781. [CrossRef]

Article

Detailed Inventory and Spatial Distribution Analysis of Rainfall-Induced Landslides in Jiexi County, Guangdong Province, China in August 2018

Chenchen Xie [1,2,3], Yuandong Huang [2,3,4], Lei Li [2,3], Tao Li [2,3] and Chong Xu [2,3,*]

[1] Institute of Disaster Prevention, Sanhe 065201, China
[2] National Institute of Natural Hazards, Ministry of Emergency Management of China, Beijing 100085, China
[3] Key Laboratory of Compound and Chained Natural Hazards Dynamics, Ministry of Emergency Management of China, Beijing 100085, China
[4] School of Emergency Management Science and Engineering, University of Chinese Academy of Sciences, Beijing 100049, China
* Correspondence: chongxu@ninhm.ac.cn or xc11111111@126.com

Abstract: In recent years, with the intensification of climate change, the occurrence of heavy rain events has become more frequent. Landslides triggered by heavy rainfall have become one of the common geological disasters around the world. This study selects an extreme rainfall event in August 2018 in Jiexi County, Guangdong province, as the research object. Based on high-resolution remote sensing images before and after the event, visual interpretation is conducted to obtain a detailed distribution map of rainfall-induced landslides. The results show that a total of 1844 rainfall-induced landslides were triggered within Jiexi County during this rainfall event. In terms of triggered scale, the total area of the landslides is 3.3884 million m^2, with the largest individual landslide covering an area of 22,300 m^2 and the smallest one covering an area of 417.78 m^2. The landslides are concentrated in the northeastern, central, and southwestern parts of the study area, consistent with the distribution trend of rainfall intensity. To investigate further the influence of the regional environment on landslide distribution, this study selects eight influencing factors, including elevation, slope aspect, slope angle, topographic wetness index (TWI), topographic relief, lithology, distance to river, and accumulated rainfall. The landslide number density (LND) and landslide area percentage (LAP) are used as evaluation indicators. Based on statistical analysis using a data analysis platform, the relationship between landslide distribution and influencing factors triggered by this event is revealed. The results of this study will contribute to understanding the development law of regional rainfall-induced landslides and provide assistance for disaster prevention and mitigation in the area. The research results show that the elevation range of 100–150 m is the high-risk zone for landslides. In addition, this study has verified previous findings that slopes in the southeast direction are more prone to landslides. The steeper the slope, the more significant its influence on landslide development. When the topographic wetness index (TWI) is less than 4, landslides tend to have a high-density distribution. Greater variation in terrain relief is more likely to trigger landslides. The instability of lithology in Mesozoic strata is the main cause of landslides. The farther away from the water system, the fewer landslides occur. An increase in cumulative rainfall leads to an increase in both the number and area of landslides.

Keywords: rainfall-induced landslides; development law; accumulated rainfall; landslide number density; landslide area percentage

Citation: Xie, C.; Huang, Y.; Li, L.; Li, T.; Xu, C. Detailed Inventory and Spatial Distribution Analysis of Rainfall-Induced Landslides in Jiexi County, Guangdong Province, China in August 2018. *Sustainability* **2023**, *15*, 13930. https://doi.org/10.3390/su151813930

Academic Editor: Antonio Miguel Martínez-Graña

Received: 1 September 2023
Revised: 14 September 2023
Accepted: 17 September 2023
Published: 19 September 2023

1. Introduction

China is one of the countries most severely affected by landslide disasters [1,2]. In particular, rainfall-induced landslides are widely present and highly destructive in mountainous and hilly areas [3–5]. Due to the large variations in topography in the southeastern coastal regions of China, which provide advantageous conditions for disaster occurrence, coupled with intense rainfall, it is widely recognized as a high-frequency area for rainfall-induced landslides [6–8]. Existing studies have shown that the association between rainfall and landslide triggers mainly lies in the fact that water infiltrates into the soil pores during rainfall, increasing the pore water pressure and weakening the shear strength of slope materials [9–12]. This provides extremely favorable conditions for landslide triggers, particularly during periods of persistent heavy rainfall. In addition, the vulnerability of the exposed population is also a factor to consider in assessing the disaster risk. The southeastern coastal regions of China are densely populated and economically developed. Once heavy rainfall occurs and triggers mountainous landslides, the resulting loss of life and economic damage can be enormous [13–15].

Researchers have conducted extensive studies on rainfall-induced landslides in the southeastern coastal regions of China, and have achieved significant results. Relevant studies indicate that the majority of rainfall-induced landslides occur during and after rainfall, with their development intensifying with increasing precipitation [16,17]. Some studies indicate a strong correlation between the development of rainfall-induced landslides and rainfall intensity, while the correlation with precipitation amount is relatively weak [18,19]. Other studies indicate that the development of rainfall-induced landslides is strictly influenced by the rainfall location, timing, and geological conditions [20–22]. The development of landslides is a complex nonlinear coupled process that is influenced by various factors to different extents [23–28]. The development law of rainfall-induced landslides can vary in different regions due to variations in internal factors such as topography and geology, as well as external factors like rainfall duration, intensity, and amount [23,24,29]. Therefore, analyzing the development law of rainfall-induced landslides in different regions holds significance in filling the gaps in research on such landslides in different areas.

The Guangdong region is frequently affected by severe rainstorm disasters, triggering a large number of rainfall-induced landslides with devastating consequences. For instance, in 2005, an extremely heavy rainstorm caused a severe collapse and deformation of the slope near the Guanghui Oil Depot in Shenzhen, directly threatening the safety of the depot and the transmission lines of the Daya Bay Nuclear Power Plant [30]. In 2010, multiple collapses, landslides, and debris flows occurred in Magui town, Maoming city, Guangdong province, due to heavy rainfall brought by a typhoon. These events resulted in at least 73 deaths, damage to 8300 buildings, and an economic loss of up to 2.2 billion yuan [31,32]. In 2019, sudden continuous heavy rainfall occurred in Longchuan County, Guangdong province, triggering numerous landslides. This event affected 1571 people, damaged 220 buildings, and resulted in a direct economic loss of up to CN¥ 1.545 billion [4,33]. These examples clearly demonstrate the significant threat posed by landslides in the Guangdong region. Therefore, analyzing the development law of landslides is of great importance for studying their occurrence and development, as well as implementing disaster prevention and mitigation measures [34–36].

In the Guangdong region, there is a scarcity of databases on landslides triggered by individual rainfall events, and limited research has been conducted on the development law of rainfall-induced landslides in this area. Therefore, this study focuses on Jiexi County, Guangdong province, selecting a major rainstorm event in this region. High-resolution remote sensing images before and after the rainfall were visually interpreted to obtain a comprehensive database of landslides triggered by this specific rainfall event. Based on this database, this study conducted an analysis of the development law of rainfall-induced landslides and their influencing factors from multiple perspectives, including topography, geology, and rainfall amounts. Through an in-depth study of the development law of rainfall-induced landslides, this study aims to gain a better understanding of their formation mechanisms and further optimize disaster prevention and mitigation measures, thereby reducing the loss caused by such disasters.

2. Study Area

Jiexi County is located in the eastern part of the Guangdong province (Figure 1a), situated in the middle-upper reaches of the Rongjiangnan river and the southern foothills of the Dabeishan mountain, a branch of the Lianhua mountain. The geographical coordinates range from 115°36′22″ E to 116°11′15″ E and from 23°18′53″ N to 23°41′13″ N. The total area of the county is 1347 square kilometers. The elevation within Jiexi County ranges from −24 to 1208 m, and the landforms mainly consist of mountains, hills, and plains, with significant topographical variations (Figure 1c). Jiexi County has a subtropical monsoon climate with a long summer and a short autumn. The spring season is often characterized by cool and rainy weather. The summers are hot and rainy, with the peak period of annual precipitation occurring from late May to early August. This period also experiences a high frequency of rainfall-induced geological disasters. The average monthly rainfall is 119.5 mm [37].

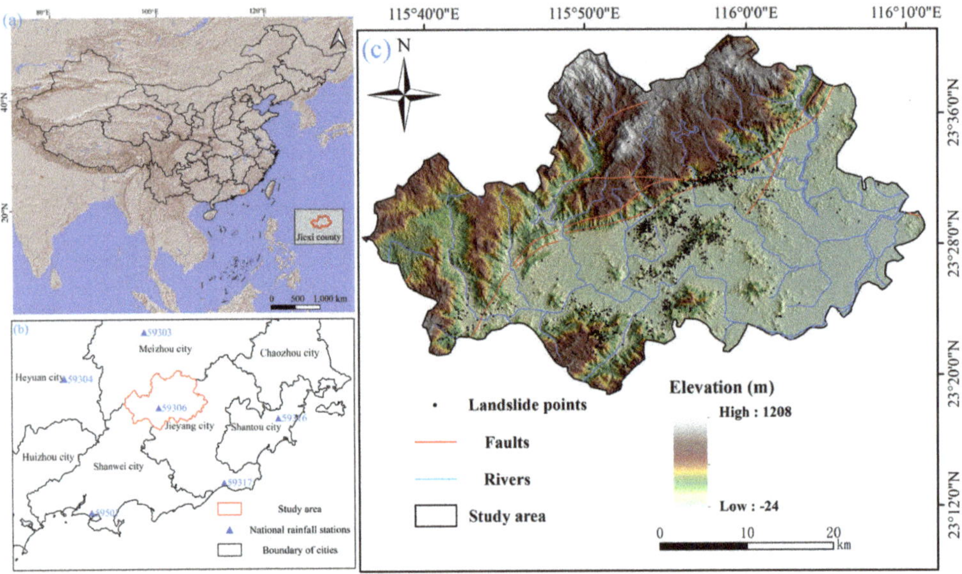

Figure 1. Study area location and other relevant information; (**a**) Administrative division of the study area; (**b**) Distribution of the study area and rainfall stations; (**c**) Overview map of the study area.

3. Data and Methods

3.1. Remote Sensing Data

High-resolution remote sensing images that closely match the rainfall events are crucial for landslide interpretation [38]. This study used Planet remote sensing image data with a resolution of 3 m (https://www.planet.com/, accessed on 30 June 2023). These images are monthly composite images, covering the entire study area. Due to high cloud cover in the study area's remote sensing images in September, this study selected July and October remote sensing images as the pre-rainfall and post-rainfall remote sensing image data (Figure 2).

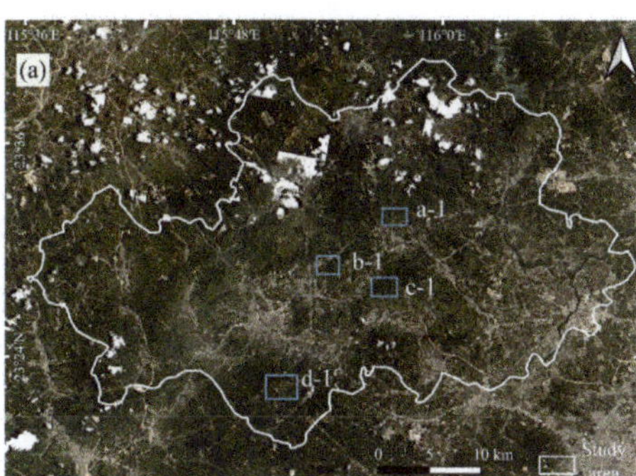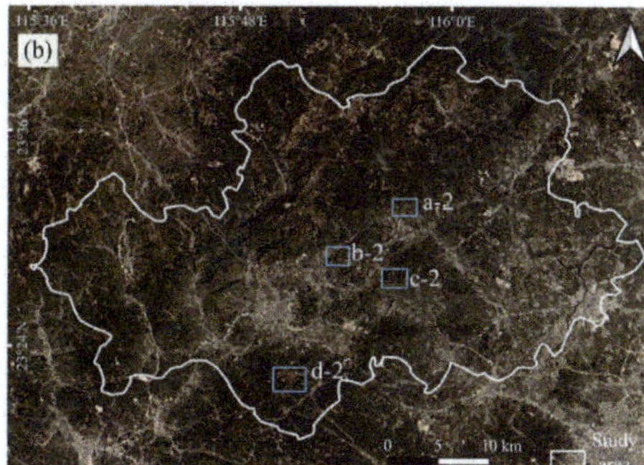

Figure 2. The figure displays the Planet remote sensing image data before and after the rainfall events; (**a**) Remote sensing image prior to rainfall in July 2018; (**b**) Remote sensing image after the rainfall in October 2018.

3.2. Topographic and Geological Data

The DEM elevation data is sourced from the ALOS PALSAR 12.5 m resolution grid (https://search.asf.alaska.edu, accessed on 30 June 2023). Based on this, the slope aspect, slope angle, topographic wetness index (TWI), and topographic relief were generated using a data processing platform. The water system data in the study area are derived from the "1:1,000,000 Public Version Basic Geographical Information Data (2021)" of the National Geographical Information Resources Catalog Service System. The water systems within the study area were extracted, and distance analysis was conducted based on the horizontal maps to obtain raster data of distances from the water systems (https://www.webmap.cn, accessed on 30 June 2023). The geological lithology data in the study area is sourced from the "National 1:200,000 Digital Geological Map Spatial Database". It was extracted and converted into the required format to obtain geological lithology data (http://www.ngac.org.cn, accessed on 30 June 2023) [39].

The age range of the exposed strata in Jiexi County is Triassic to Quaternary. The lithology mainly includes sandstone, shale, conglomerate, mudstone, and granite, which are distributed in different geological periods and regions. Among the Cenozoic strata are the Quaternary, mainly river sediments, lacustrine sediments, and marsh sediments. Mesozoic strata have a relatively wide distribution, including the Cretaceous, Jurassic, and Triassic, of which the main lithologies of the Cretaceous are conglomerate, sandstone, and mudstone; the main lithologies of the Jurassic are muddy sandstone, sandy shale, and granite; The main lithologies of the Triassic are sandstone, conglomerate, and shale. The lithologic description of the strata is shown in Table 1.

Table 1. Lithologic classification of the study area.

No.	Stratigraphic Chronology	Lithologic Classification
1	Quaternary (Q)	Mud, Siltstone, Fine sandstone, Clay
2	Cretaceous (K)	Conglomerate, Mudstone, Granite, Granodiorite, Quartz porphyry, Granite porphyry
3	Jurassic (J)	Sandstone, Siltstone, Rhyolite porphyry, Rhyolite dacite porphyry, Dacite porphyry, Carbonaceous shale, Granite, Biotite granite, Monzonitic granite, Granodiorite
4	Triassic (T)	Sandstone, Siltstone, Silty clay, Conglomerate, Feldspar quartz sandstone, Fine sandstone, Oil shale, Shale

3.3. Rainfall Data

Under the influence of monsoon low pressure, a severe rainstorm occurred in the eastern part of the Guangdong Province, China (22° N–24° N, 114° E–117° E), from 27 to 31 August 2018. This rainstorm was characterized by concentrated heavy rainfall, long duration, and large precipitation amounts [40]. According to statistics from the Department of Civil Affairs of Guangdong Province, a total of 29 districts and counties in Guangdong were affected by floods and other disasters. The number of affected people reached 1.8819 million, including two deaths and two missing persons. The direct and indirect economic losses amounted to a total of CN¥ 3.715 billion. During this rainfall event, numerous geological hazards such as landslides and debris flows were triggered [41]. Based on data from the 59306 national rainfall stations within the study area, this study selected the 24 h rainfall data from 1 July to 30 September to analyze this rainfall event (Figure 3). The results showed that during 27 to 31 August, the lowest recorded rainfall was 20.4 mm on the 28th, while the highest was 184.3 mm. Relative to the overall accumulated rainfall, the accumulated rainfall before and after the rainfall event showed a steady upward trend; only 27 to 31 August showed a steep upward trend. Therefore, this study conducted interpolation on the data from six neighboring national rainfall stations in the study area (Figure 1b) and used the inverse distance weighting method to obtain a spatial distribution map of the five-day rainfall in the study area (Figure 4) [42].

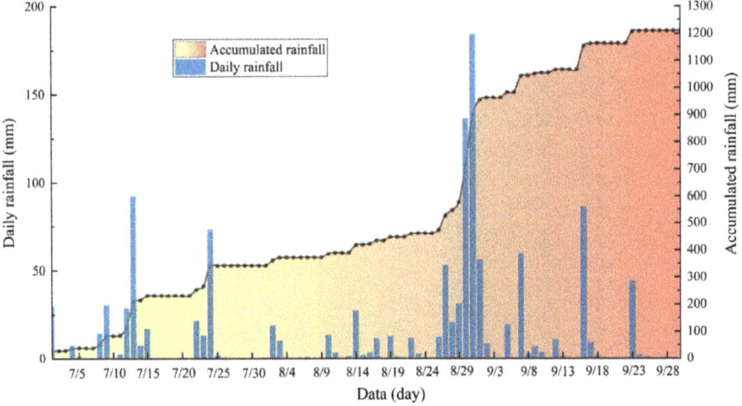

Figure 3. Daily and cumulative rainfall data from the 59306 rainfall stations within the study area from July to September.

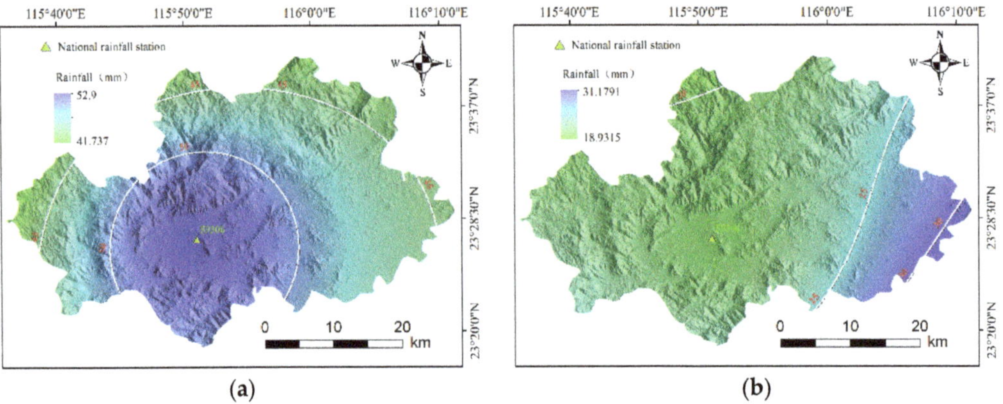

(**a**)   (**b**)

Figure 4. *Cont.*

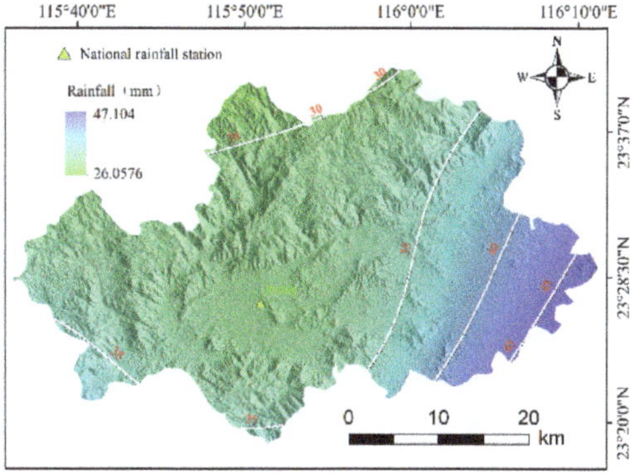

Figure 4. Daily rainfall distribution in the study area from 27 to 31 August; (**a**) 27 August 2018; (**b**) 28 August 2018; (**c**) 29 August 2018; (**d**) 30 August 2018; (**e**) 31 August 2018.

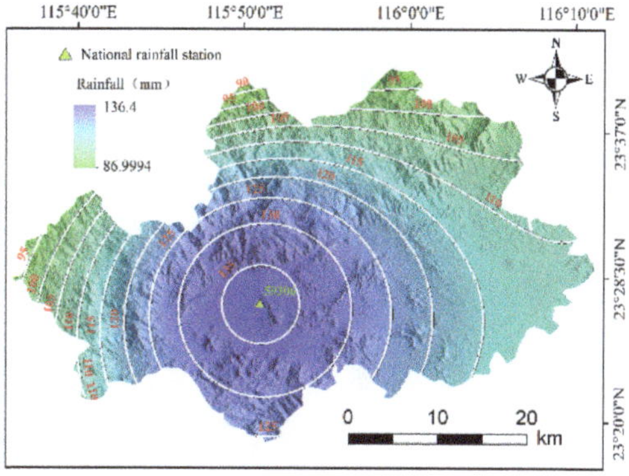

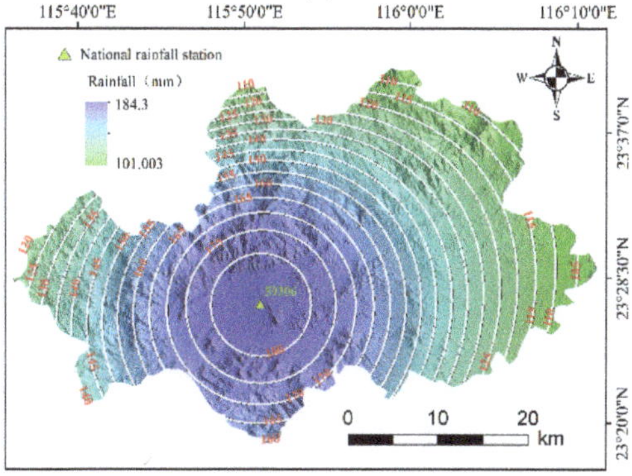

3.4. Methods

By comparing remote sensing images before and after the rainfall event, this study identified areas where there were no apparent phenomena in the remote sensing images before the rainfall, but showed no vegetation cover, changes in color tones, and flowing phenomena after the rainfall as rainfall-induced landslides [43]. This study delineated these areas as vector data along their boundaries. The four typical landslide-dense display areas in Figure 2b correspond to Figure 5.

Figure 5. *Cont.*

Figure 5. The landslide interpretation maps of the four highlighted boxes in Figure 2b, representing the four landslide-dense areas; (**a-1–d-1**) The remote sensing image before the rainfall in July 2018. (**a-2–d-2**) The remote sensing image after the rainfall in October 2018.

In this study, two commonly used indicators for analyzing landslide development law, namely landslide number density (LND) and landslide area percentage (LAP), were selected [44,45]. Using LND and LAP as measurement metrics, this study explored the relationship between rainfall-induced landslide distribution and influencing factors based on data analysis platforms. LND represents the number of landslides per square kilometer (Equation (1)), while LAP represents the percentage of landslide area within each factor category (Equation (2)). In this paper's analysis, the study primarily relied on these two indicators and the landslide number to comprehensively assess the degree of landslide development within the delineated intervals [46–49]. In the analysis of results, if LND is large and LAP is small, it indicates that within this factor category, the landslide scale is relatively large, but they are all small area landslides. If LND is small and LAP is large, it indicates that within this factor category, the landslide scale is relatively small, but they are all large area landslides. If both LND and LAP are large, then within this factor category, the landslide development is severe. Of course, the magnitude of LND and LAP values is related to the size of the categorized interval areas, so we can conduct a comprehensive analysis by combining the number of landslides.

$$\text{LND} = \frac{\text{Landslide Number}}{\text{Influence Factor Class Area}} \tag{1}$$

$$\text{LAP} = \frac{\text{Landslide Area}}{\text{Influence Factor Class Area}} \tag{2}$$

4. Results and Analysis

4.1. Landslide Inventory

A total of 1844 landslides were interpreted, with a cumulative landslide area of 3.3884 million m^2. The maximum individual rainfall-induced landslide area was 22,300 m^2, while the minimum was 417.78 m^2. Among them, 63.39% of the landslide areas ranged from 1000 to 3000 m^2. Using a search radius of 2 km, this study analyzed the landslide number density (LND) in the study area. The results showed that the highest LND value was 27.9 km^2. Geographically, landslides were primarily concentrated in the northeast, central, and southwest regions of the study area (Figure 6).

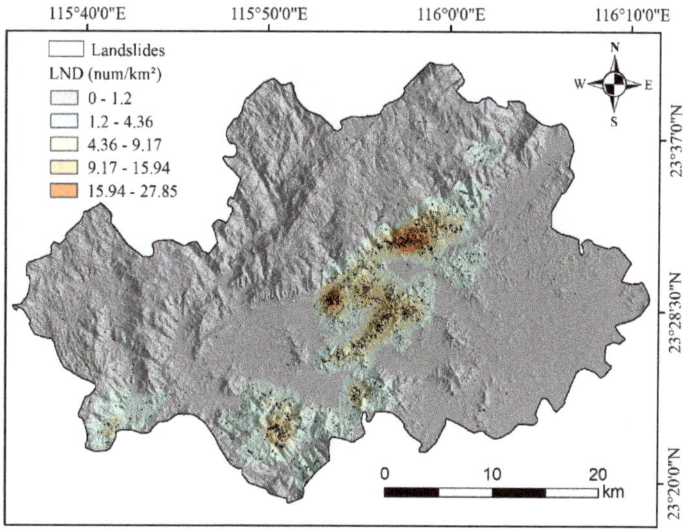

Figure 6. Landslide area distribution map and point density map.

4.2. Landslide Development Law

The elevation in Jiexi County ranges from −24 m to 1208 m. There were 38 landslides occurring below 50 m and 98 landslides occurring above 500 m, accounting for only 7% of the total. Therefore, this study divided these two ranges into separate intervals, with an interval spacing of 50 m, resulting in 11 intervals. According to the analysis in Figure 7b, it can be concluded that both LND and LAP exhibit a trend of first increasing and then decreasing, with slight fluctuations occurring in the range greater than 200 m. They reach their maximum values at 100–150 m, which are 4.67 km^{-2} and 0.88%, respectively. Within this elevation range, the total number and area of landslides peak at 408 and 0.77 km^2, respectively.

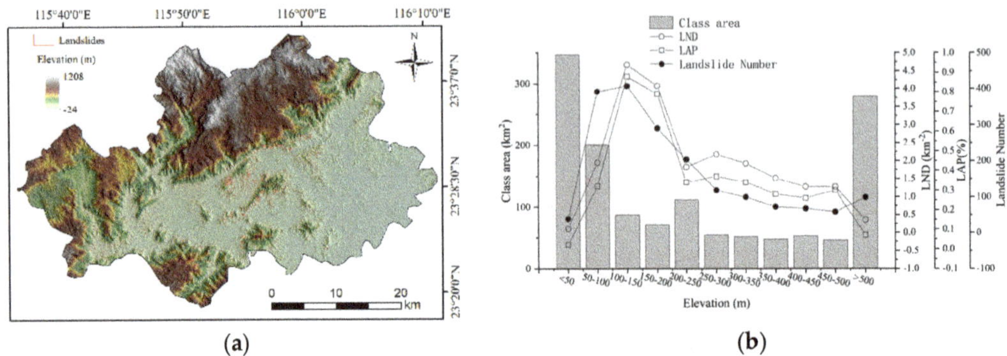

(a)   **(b)**

Figure 7. Relationship between elevation and landslide distribution; (**a**) Elevation and landslide locations in the study area; (**b**) Statistical analysis results.

This study divided the slope aspect direction into nine categories, including the flat category, based on the commonly used eight slope direction intervals. According to the analysis in Figure 8b, LND, LAP, and landslide quantities show similar trends in each interval. Among them, the LND and LAP values in the SE direction are significantly higher than those in the other eight intervals, reaching 2.23 km^{-2} and 0.4%, respectively. The number of landslides also reaches its peak at 436. The S and E directions follow closely, with LND values of 1.85 km^{-2} and 1.73 km^{-2}, respectively, LAP values are 0.32% for both, and landslide quantities are 320 and 310, respectively. Therefore, the E to S direction is the most prone direction for landslides in this area.

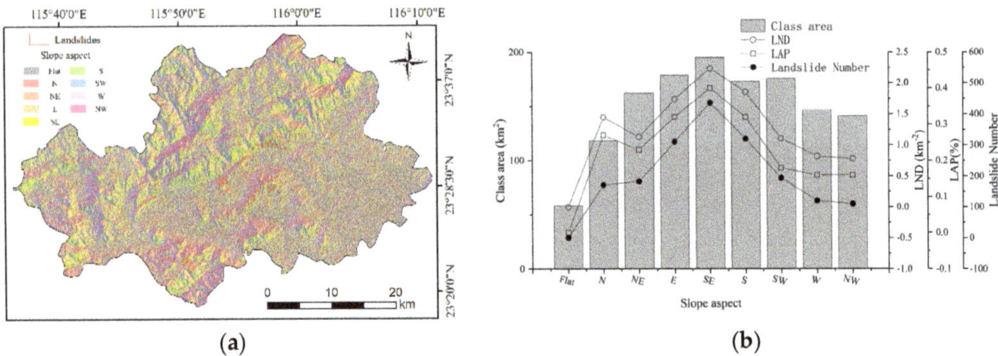

(a)   **(b)**

Figure 8. Relationship between slope aspect and landslide distribution; (**a**) Slope aspect and landslide locations in the study area; (**b**) Statistical analysis results.

The maximum slope angle in the study area is 76.09°, and there are 178 landslides with slopes greater than 30, accounting for 9% of the total landslide count. Therefore, this study defined slopes greater than 30 as one interval and divided the slope range into seven intervals with a 5° interval spacing. According to the analysis in Figure 9b, both LND and LAP increase with increasing slope angle, while the landslide quantity initially increases and then decreases. The LND reaches its peak at 2.77 km^{-2} in the 25°–30° interval, and the LAP reaches its maximum value at 0.58% in the interval greater than 30°. The landslide quantity is highest in both the 15°–20° and 20°–25° intervals, with 407 landslides each. As the area of the interval greater than 30° accounts for 5% of the study area and exhibits high values in both LND and LAP, it indicates that higher slope angle is more favorable for landslide occurrence.

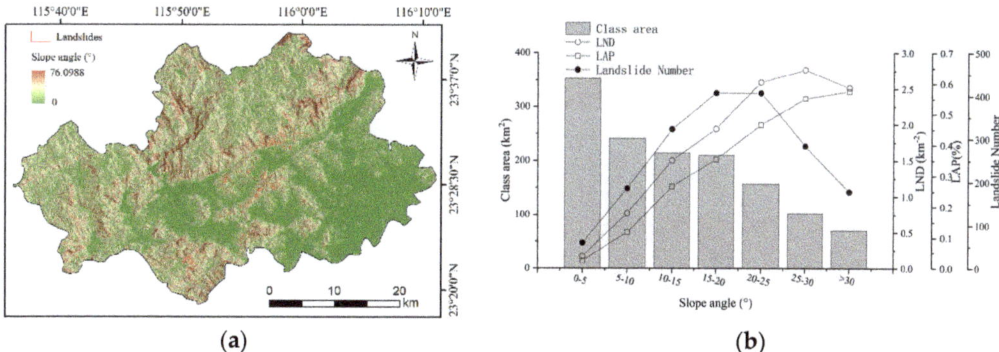

Figure 9. Relationship between slope angle and landslide distribution; (**a**) Slope angle and landslide locations in the study area; (**b**) Statistical analysis results.

The TWI within the study area ranges from 2.78 to 24.0, but most landslides are concentrated in the range of 4 to 10. Therefore, this study divided the TWI range into eight intervals with a unit interval spacing. According to the analysis in Figure 10b, although both LND and LAP reach their maximum values in the interval less than four, at 1.99 km^{-2} and 0.39%, respectively, this interval has the smallest area of 35.67 km^2, with a landslide count of 71. In contrast, the interval of five to six has the highest landslide count with 572 landslides. This indicates that the probability of landslide occurrence is higher in the TWI range of four to seven compared to the other intervals. Additionally, there is a concentration of landslides in the interval less than four.

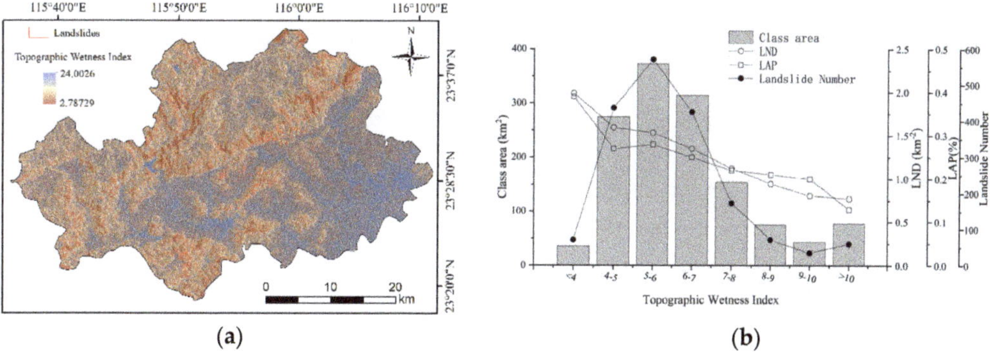

Figure 10. Relationship between TWI and landslide distribution; (**a**) TWI and landslide locations in the study area; (**b**) Statistical analysis results.

The range of topographic relief in the study area is between 0 and 299, and landslides in the range of 100 to 299 account for 10% of the total. Therefore, this study divided the topographic relief into six intervals with a 20-unit interval spacing. According to the analysis in Figure 11b, both LND and LAP initially increase with the increase in topographic relief and then decrease. LND and LAP reach their peak values in the 80–100 interval, at 2.72 km^{-2} and 0.54%, respectively. The maximum landslide count is reached in the 60–80 interval, with 491 landslides. Due to the smallest area and large span of the 100–299 interval, neither LND nor LAP reaches the maximum value. This indicates that larger topographic relief is more likely to trigger landslide occurrences.

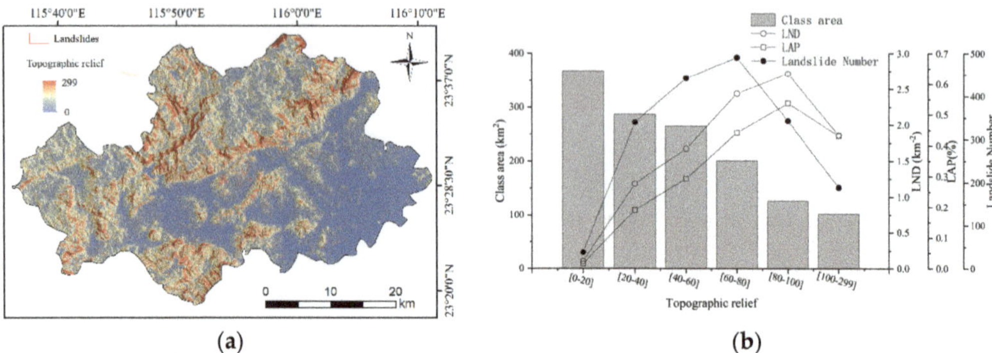

Figure 11. Relationship between topographic relief and landslide distribution; (**a**) Topographic relief and landslide locations in the study area; (**b**) Statistical analysis results.

The earliest strata in Jiexi County is the Triassic period, and the most recent one is the Quaternary period. This study divided it into four intervals based on the strata. Among them, the Cenozoic period includes the Quaternary, and the Mesozoic period includes the Cretaceous, Jurassic, and Triassic. According to the analysis in Figure 12b, LND reaches its maximum value in the Triassic period at 2.06 km^{-2}; LAP is highest in the Jurassic and Triassic periods, at 0.34% and 0.33%, respectively; the Jurassic period has the highest number of landslides, with 1359 occurrences. The Mesozoic period accounts for 82.6% of the study area and is mainly composed of sandstone, mudstone, and weathered granite layers. In terms of stability, these lithologies are more prone to landslides due to erosion and permeation by rainwater.

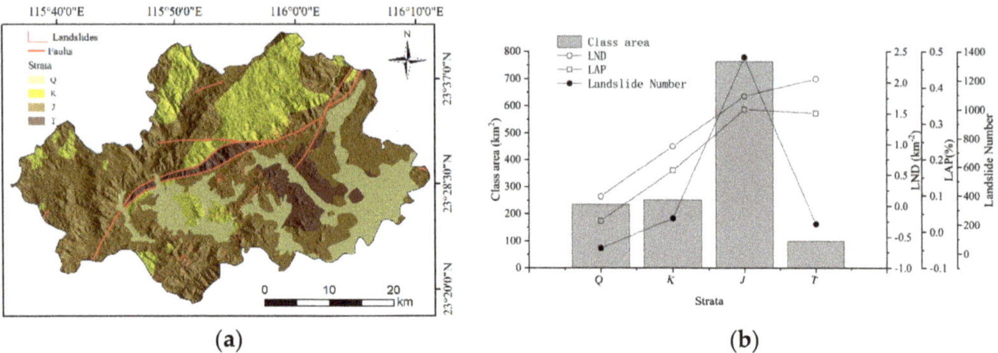

Figure 12. Relationship between strata and landslide distribution; (**a**) Lithologies and landslide locations in the study area; (**b**) Statistical analysis results.

The water system in the study area is highly developed, and there is a close relationship between rainfall-induced landslides and the distance to river. The distance to the river in this area ranges from 0 to 4630 m, and this study divided it into five intervals with a 1 km interval spacing. According to the analysis in Figure 13b, both LND and LAP reach their maximum values in the 3000–4000 m interval, at 3.23 km^{-2} and 0.68%, respectively. The highest landslide counts are found in the ranges of less than 1000 m and 1000–2000 m, with 693 and 759 landslides, respectively. Due to the small area coverage of the 3000–4000 m interval (only 2% of the total), it leads to inflated values of LND and LAP. In summary, this study can conclude that there is a high-density distribution of landslides within the 3000–4000 m range. Places closer to the water system are more prone to landslide occurrences.

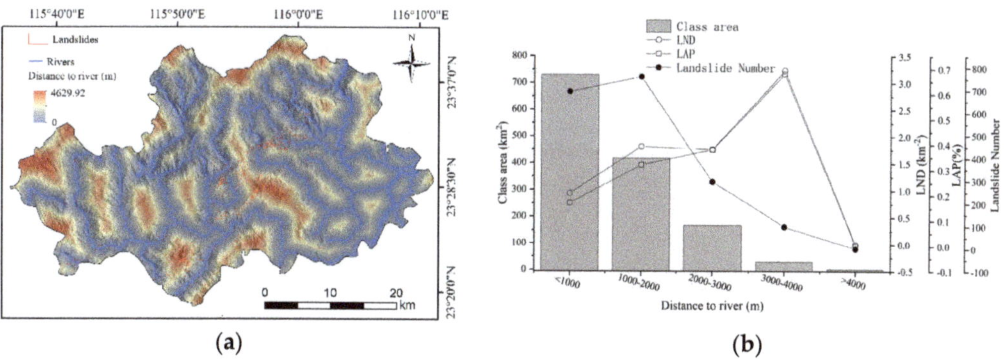

Figure 13. Relationship between distance to river and landslide distribution; (**a**) Distance to river and landslide locations in the study area; (**b**) Statistical analysis results.

Although the development of landslides is influenced by daily rainfall, continuous rainfall is even more conducive to landslide development [50]. This study overlayed the daily rainfall data from 27 to 31 August in Figure 3 to generate a 5-day accumulated rainfall dataset (Figure 14a). As the accumulated rainfall in the study area ranges from 283.03 to 425.1, this study divided it into six intervals with a 25 mm interval spacing. According to the analysis in Figure 14b, as the accumulated rainfall increases, both LND, LAP, and the number of landslides also increase. In the interval range greater than 400 mm, LND, LAP, and the number of landslides reach their maximum values at 2.59 km^{-2}, 0.52%, and 971, respectively. In general, the distribution of accumulated rainfall is proportional to the development law of landslide. An increase in accumulated rainfall leads to an increase in the number and area of landslides, making areas with higher rainfall more favorable for landslide development.

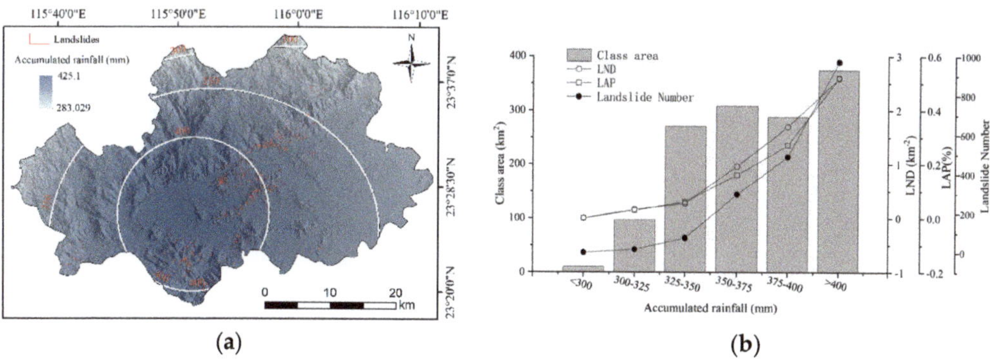

Figure 14. Relationship between accumulated rainfall and landslide distribution; (**a**) Accumulated rainfall and landslide locations in the study area; (**b**) Statistical analysis results.

5. Discussion

5.1. Importance of Landslide Database

At present, the research on landslides has achieved comprehensive and abundant results, including regional landslide interpretation/automatic identification [51,52], disaster susceptibility/hazard/risk analysis [53–55], landslide dynamics mechanism research [56], and so on. Among these, the establishment of a landslide database serves as the foundation for regional landslide research [57,58]. A detailed regional landslide database can provide more effective support for local geological hazard prevention and control [59,60]. On one hand, the spatial distribution of landslides in the study area based on a rich and comprehensive database is the most direct theoretical basis for disaster prevention and mitigation [61–64]. On the other hand, with the development of artificial intelligence, it shows superior performance in both disaster assessment and image automatic identification. However, due to the complex changes in geological environments and diverse characteristics of different types of landslides in different regions, the generality of research results still needs to be improved further. Adequate training samples undoubtedly play a pivotal role in improving model performance and are particularly crucial in areas with insufficient landslide data. In this study, this study has interpreted a total of 1844 rainfall-induced landslides and provided a comprehensive and detailed database. Since the inventory of rainfall-induced landslides in Guangdong is very small [4], this database can add a valuable case study to the study of rainfall landslides in Guangdong.

5.2. Interpretation of the Results of Environmental Factor Analysis

Jiexi County is located in a coastal area, with its southern and eastern parts closer to the sea. It is influenced by the southeast monsoon, making the southeast slopes of the region slope windward. Windward slopes often receive more precipitation, making them more prone to landslides [65]. In this rainfall event, the most landslide-prone slope direction was southeast (Figure 8b). Granite layers are one of the most common geological formations in the Guangdong province. Typically, weathered granite forms a large amount of residual soil with low hardness and poor stability, and widespread rock fractures. In studies of other areas in Guangdong, landslides often occur in wind-weathered granite and residual accumulated soil layers [31,33]. As shown in Figure 12b, landslides mainly concentrated in Jurassic strata, which aligns with the distribution of granite strata in the study area according to Table 1, which is consistent with previous research findings. Slope angle and topographic relief are both important factors affecting landslide development. Figures 9b and 11b indicate that LND and LAP increase with an increase in slope angle and topographic relief, respectively. Most of the landslides induced by this rainfall event were concentrated in areas with steep slopes and substantial topographic relief. The closer to the river, the stronger the water activity, resulting in more pronounced erosion on the slopes [66]. In areas closer to the river, there are more landslides, and they are more likely to occur (Figure 13b). This study referred to other studies analyzing the development law of rainfall-induced landslides in southeast regions [4,20] and found similar trends in the relationship between landslide occurrence and TWI, which shows an initial increase followed by a decrease. When the value is 4–7, it is the most prone to landslide. Rainfall, as the main triggering factor for rainfall-induced landslides, is closely related to the development and distribution of landslides and is also influenced by geological and topographic factors [67–69]. Figure 14b shows that areas with higher accumulated rainfall amounts have more landslides, indicating a positive correlation.

5.3. Possible Points to Be Strengthened

In Figure 4a,d,e and Figure 14a, this study can observe a concentric pattern in the rainfall distribution. This is actually due to the interpolation of data from distant national rain gauge stations. Compared to the distances between the national rain gauge stations, this paper study area is relatively small, and there are few available national rain gauge stations surrounding the study area. This results in a concentric pattern in the rainfall distribution after data interpolation. Such interpolation work leads to inaccuracies in the rainfall data corresponding to each landslide. In future work, this study can use local rain gauge stations for interpolation to make this paper analysis more refined and accurate.

With the development of remote sensing technology, an increasing number of high-resolution remote sensing images are being used for landslide identification. Generating a landslide database through visual interpretation not only saves a considerable amount of external work time but also enables a quick response to related disaster prevention and mitigation efforts [70]. From Figure 7b, it can be observed that the landslides in this event are primarily distributed in lower elevation areas with elevations ranging from 100–200 m.

As indicated in Figure 15, vegetation in lower elevation areas recovers rapidly, and due to the long time elapsed since the landslide events, the morphology of landslides may have undergone significant changes [71]. Additionally, there may be other triggering events for landslides, which could introduce considerable error in field investigation results. Based on these factors, this study did not conduct field investigations. This study has to acknowledge that this paper's research results may have missed some landslides. To minimize errors, this study used high spatial resolution (3 m) Planet remote sensing image data for interpretation. The study area was set as the administrative boundaries of Jiexi County, taking into account the maximum possible impact of rainfall events. The image coverage was complete and had good imaging quality, sufficient to support this paper's interpretation accuracy, although there might be some small-scale landslides that did not exhibit obvious features in the image. The temporal scale of the image data was monthly, ensuring that the interpreted landslides correspond accurately to the specific rainfall event and provide comprehensive and accurate results for the database [43].

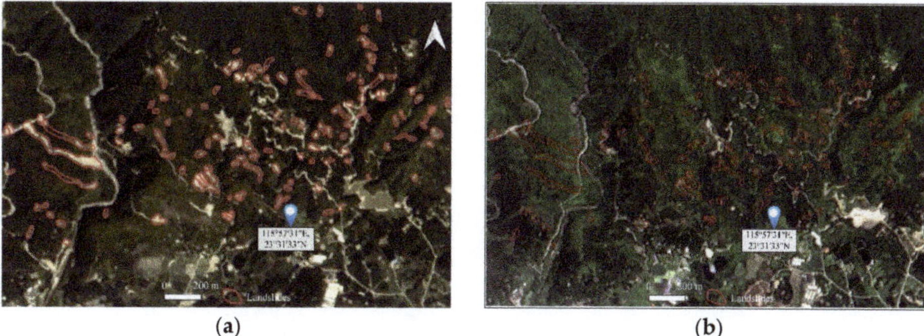

(a)　　　　　　　　　　　**(b)**

Figure 15. Presented is an enlarged view of the area highlighted by the box in Figure 2b. (**a**) Remote sensing image from October 2018; (**b**) Remote sensing image from May 2023.

Furthermore, the analysis in this study focuses on the development law of regional landslides, so the impact of the few observed omissions on the results is not significant. In future research, this study will investigate further various aspects of landslides; for example, the relationship between landslide height and sliding distance [72,73], the position of landslides relative to ridges and rivers [21,74], and the hazard assessment of rainfall-induced landslides [75,76], among others.

6. Conclusions

This study is based on a severe rainstorm event that occurred from 27 to 31 August 2018, and it selected Jiexi County in Guangdong Province as the research area. A total of 1844 rainfall-induced landslides were identified, with a total landslide area of 3.39 km^2. In terms of landslide distribution, landslides were concentrated in the northeast, central, and southwest parts of the research area, which received relatively higher rainfall during the event. To investigate further the influence of regional environment on landslide distribution, this study selected eight influencing factors: elevation, slope aspect, slope angle, topographic wetness index (TWI), topographic relief, lithology, distance to river, and accumulated rainfall. Using landslide number density (LND) and landslide area percentage (LAP) as evaluation indicators, the relationship between rainfall-induced landslide distribution and influencing factors was explored using a data analysis platform. The research results showed that the elevation range of 100–150 m was a high-incidence zone for landslides. Slopes facing east to south were more favorable for landslide development, possibly due to factors such as solar radiation and wind speed. The steeper the slope angle, the greater the impact on landslide development. Landslides exhibited a high-density distribution within the TWI range below 4, with the highest number of landslides occurring in the range of 5–6. Greater topographic relief was associated with a higher likelihood of landslide occurrence. Unstable lithology in the Mesozoic strata was the primary cause of landslides. Landslides exhibited a high-density distribution within the range of 3000–4000 m from the river, and the farther away from the river, the fewer landslides occurred. Larger accumulated rainfall resulted in a higher number and larger area of landslides, possibly due to prolonged infiltration of rainwater, accumulation of water in the lower part of slopes, and weakened shear strength of slope materials, leading to slope instability. The results of this study contribute to understanding the development law of regional rainfall-induced landslides and provide assistance in disaster prevention and mitigation efforts in the area.

Author Contributions: C.X. (Chong Xu) proposed the research concept, organized the landslide interpretation, and provided basic data. C.X. (Chenchen Xie) designed the framework and wrote the manuscript. Y.H. participated in the writing and data analysis. L.L. and T.L. provided basic data and data analysis. All authors have read and agreed to the published version of the manuscript.

Funding: This study was supported by the National Natural Science Foundation of China (42077259) and the National Key Research and Development Program of China (2021YFB3901205).

Acknowledgments: We thank the anonymous reviewer and the editor, whose constructive and helpful comments improved this manuscript.

Conflicts of Interest: The authors declare no conflict of interest.

References

1. Petley, D.N. On the impact of climate change and population growth on the occurrence of fatal landslides in South, East and SE Asia. *Q. J. Eng. Geol. Hydrogeol.* **2010**, *43*, 487–496. [CrossRef]
2. Rong, G.; Li, K.; Han, L.; Alu, S.; Zhang, J.; Zhang, Y. Hazard mapping of the rainfall–landslides disaster Chain based on GeoDetector and Bayesian Network Models in Shuicheng County, China. *Water* **2020**, *12*, 2572. [CrossRef]
3. Hong, Y.; Adler, R.; Huffman, G. Use of satellite remote sensing data in the mapping of global landslide susceptibility. *Nat. Hazards* **2007**, *43*, 245–256. [CrossRef]
4. Ma, S.; Shao, X.; Xu, C. Characterizing the distribution pattern and a physically based susceptibility assessment of shallow landslides triggered by the 2019 heavy rainfall event in Longchuan County, Guangdong Province, China. *Remote Sens.* **2022**, *14*, 4257. [CrossRef]
5. Peruccacci, S.; Brunetti, M.; Luciani, S.; Vennari, C.; Guzzetti, F. Lithological and seasonal control on rainfall thresholds for the possible initiation of landslides in central Italy. *Geomorphology* **2012**, *139–140*, 79–90. [CrossRef]
6. Dadson, S.; Hovius, N.; Chen, H.; Dade, W.B.; Lin, J.; Hsu, M.; Lin, C.; Horng, M.; Chen, T.; Milliman, J. Earthquake-triggered increase in sediment delivery from an active mountain belt. *Geology* **2004**, *32*, 733–736. [CrossRef]
7. Feng, H.; Zhou, A.; Tang, X.; You, S.; Xu, X. Susceptibility analysis of factors controlling rainfall-triggered landslides using certainty factor method. *J. Eng. Geol.* **2017**, *25*, 436–446. [CrossRef]
8. Ma, T.; Li, C.; Lu, Z.; Bao, Q. Rainfall intensity–duration thresholds for the initiation of landslides in Zhejiang Province, China. *Geomorphology* **2015**, *245*, 193–206. [CrossRef]
9. Iverson, R. Landslide triggering by rain infiltration. *Water Resour. Res.* **2000**, *36*, 1897–1910. [CrossRef]
10. Marino, P.; Peres, D.; Cancelliere, A.; Greco, R.; Bogaard, T. Soil moisture information can improve shallow landslide forecasting using the hydrometeorological threshold approach. *Landslides* **2020**, *17*, 2041–2054. [CrossRef]
11. Rahardjo, H.; Li, X.; Toll, D.; Leong, E. The effect of antecedent rainfall on slope stability. *Unsaturated Soil Concepts Their Appl. Geotech. Pract.* **2001**, *19*, 371–399. [CrossRef]
12. Zhang, Y.; Guo, C.; Yang, Z.; Shen, Y.; Wu, R.A.; Ren, S. Study on shear strength of deep-seated sliding zone soil of zhouchangping landslide in Maoxian, Sichuan. *J. Eng. Geol.* **2021**, *29*, 764–776. [CrossRef]
13. Emberson, R.; Kirschbaum, D.; Amatya, P.; Tanyas, H.; Marc, O. Insights from the topographic characteristics of a large global catalog of rainfall-induced landslide event inventories. *Nat. Hazards Earth Syst. Sci.* **2022**, *22*, 1129–1149. [CrossRef]
14. Hong, H.; Liu, J.; Zhu, A.-X.; Shahabi, H.; Pham, B.T.; Chen, W.; Pradhan, B.; Bui, D.T. A novel hybrid integration model using support vector machines and random subspace for weather-triggered landslide susceptibility assessment in the Wuning area (China). *Environ. Earth Sci.* **2017**, *76*, 652. [CrossRef]
15. Youssef, A.; Pourghasemi, H. Landslide susceptibility mapping using machine learning algorithms and comparison of their performance at Abha Basin, Asir Region, Saudi Arabia. *Geosci. Front.* **2021**, *12*, 639–655. [CrossRef]
16. Chi, Y. Regularities of landslides under typhoon rainstorm conditions in Fujian Province. *J. Geol.* **2015**, *39*, 697–701. [CrossRef]
17. Gorsevski, P.V.; Gessler, P.E.; Boll, J.; Elliot, W.J.; Foltz, R.B. Spatially and temporally distributed modeling of landslide susceptibility. *Geomorphology* **2006**, *80*, 178–198. [CrossRef]

18. Feng, H. Rainfall-Triggered Landslide Development Regularity Analysis and Hazard Assessment in Chun'an, West Zhejiang. Ph.D. Thesis, China University of Geosciences, Wuhan, China, 2016.

19. Xu, Y.; Fan, H.; Chen, L.; Zhao, Y.; Li, Y. Relationship between rainfall and landslides in Wuzhou Guangxi. *J. Guangxi Univ.* **2015**, *40*, 949–955. [CrossRef]

20. Ma, S.; Shao, X.; Xu, C. Landslides triggered by the 2016 heavy rainfall event in Sanming, Fujian Province: Distribution pattern analysis and spatio-temporal susceptibility assessment. *Remote Sens.* **2023**, *15*, 2738. [CrossRef]

21. Wu, C.; Chen, S.; Chou, H. Geomorphologic characteristics of catastrophic landslides during typhoon Morakot in the Kaoping Watershed, Taiwan. *Eng. Geol.* **2011**, *123*, 13–21. [CrossRef]

22. Zhong, Y. Landslide related to rainfall and its forecasting. *Chin. J. Geol. Hazard Control.* **1998**, *9*, 81–86.

23. Chen, J.; Chue, Y.; Chen, Y. The application of the genetic adaptive neural network in landslide disaster assessment. *J. Mar. Sci. Technol.* **2013**, *21*, 9. [CrossRef]

24. Chue, Y.; Chen, J.; Chen, Y. Rainfall-induced slope landslide pontential and landslide distribution characteristics assessment. *J. Mar. Sci. Technol.* **2015**, *23*, 14. [CrossRef]

25. Conforti, M.; Ietto, F. Influence of tectonics and morphometric features on the landslide distribution: A case study from the Mesima Basin (Calabria, South Italy). *J. Earth Sci.* **2020**, *31*, 393–409. [CrossRef]

26. Hong, H.; Pourghasemi, H.R.; Pourtaghi, Z.S. Landslide susceptibility assessment in Lianhua County (China): A comparison between a random forest data mining technique and bivariate and multivariate statistical models. *Geomorphology* **2016**, *259*, 105–118. [CrossRef]

27. Li, Q.; Huang, D.; Pei, S.; Qiao, J.; Wang, M. Using physical model experiments for hazards assessment of rainfall-induced debris landslides. *J. Earth Sci.* **2021**, *32*, 1113–1128. [CrossRef]

28. Qin, S.; Jiao, J.J.; Wang, S. The predictable time scale of landslides. *Bull. Eng. Geol. Environ.* **2001**, *59*, 307–312. [CrossRef]

29. Zhu, X.; Xu, Q.; Tang, M.; Nie, W.; Ma, S.; Xu, Z. Comparison of two optimized machine learning models for predicting displacement of rainfall-induced landslide: A case study in Sichuan Province, China. *Eng. Geol.* **2017**, *218*, 213–222. [CrossRef]

30. Qiu, J.; Wen, J.; Gao, W. Stability Analysis and treatment of Guanghui oil depot slope in Shenzhen. *Chin. J. Rock Mech. Eng.* **2009**, *28*, 2201–2207. [CrossRef]

31. Wang, J.; Gong, Q.; Yuan, S.; Chen, J. Combining soil macropore flow with formation mechanism to the development of shallow landslide warning threshold in South China. *Front. Earth Sci.* **2023**, *10*, 1048427. [CrossRef]

32. Wang, Y.; Xia, B. Eco-restoration strategies and measures for the soil and water conservation of Typhoon-hit areas in western Guangdong province-a case study of Magui town, Gaozhou city. *Sci. Soil Water Conserv.* **2012**, *10*, 88–93. [CrossRef]

33. Feng, W.; Bai, H.; Lan, B.; Wu, Y.; Wu, Z.; Yan, L.; Ma, X. Spatial–temporal distribution and failure mechanism of group-occurring landslides in Mibei village, Longchuan County, Guangdong, China. *Landslides* **2022**, *19*, 1957–1970. [CrossRef]

34. Qiu, H.; Cao, M.; Liu, W.; Hao, J.; Hu, S.; Gao, Y.; Liu, Q. Research on variable dimension fractal characteristics of spatial distribution of landslides. *Geoscience* **2014**, *28*, 443. [CrossRef]

35. Xu, C.; Xu, X.; Zheng, W. Compiling inventory of landslides triggered by Minxian-Zhangxian earthquake of July 22, 2013 and their spatial distribution analysis. *J. Eng. Geol.* **2013**, *21*, 736–749. [CrossRef]

36. Zhao, B. Landslides triggered by the 2018 Mw 7.5 Palu supershear earthquake in Indonesia. *Eng. Geol.* **2021**, *294*, 106406. [CrossRef]

37. Li, W. *Records of Jiexi County*; Guangdong People's Publishing House: Guangdong, China, 2005.

38. Shao, X.; Ma, S.; Xu, C.; Xu, Y. Insight into the Characteristics and Triggers of Loess Landslides during the 2013 Heavy Rainfall Event in the Tianshui Area, China. *Remote Sens.* **2023**, *15*, 4304. [CrossRef]

39. Li, C.; Wang, X.; He, C.; Wu, X.; Kong, Z.; Li, X. National 1:200,000 digital geological map (public version) spatial database. *Geol. China* **2019**, *46* (Suppl. 1), 1–10. [CrossRef]

40. Zeng, Z.; Chen, Y.; Wang, D. Observation and mechanism analysis for a record-breaking heavy rainfall event over Southern China in August 2018. *Chin. J. Atmos. Sci.* **2020**, *44*, 695–715. [CrossRef]

41. Cai, J.; Wu, Z.; Chen, X.; Lan, Y.; Guo, Z.; Guo, C. Cause analysis of persistent torrential rain associated with monsoon depression occurred in Guangdong on August 2018. *Torrential Rain Disasters* **2019**, *38*, 576–586. [CrossRef]

42. Zhu, X. *GIS for Environmental Applications: A Practical Approach*; Routledge: London, UK, 2016.

43. Huang, Y.; Xu, C.; Zhang, X. Spatial distribution and influence factors of landslides triggered by the 2019 Ms 6.0 Changning, Sichuan, China Ms6. 0 earthquake: A statistical analysis based on QGIS. In *IOP Conference Series: Earth and Environmental Science*; IOP Publishing: Bristol, UK, 2021; p. 052007.

44. Tian, Y.; Xu, C.; Xu, X.; Chen, J. Detailed inventory mapping and spatial analyses to landslides induced by the 2013 Ms 6.6 Minxian earthquake of China. *J. Earth Sci.* **2016**, *27*, 1016–1026. [CrossRef]

45. Xu, C.; Ma, S.; Tan, Z.; Xie, C.; Toda, S.; Huang, X. Landslides triggered by the 2016 Mj 7.3 Kumamoto, Japan, earthquake. *Landslides* **2018**, *15*, 551–564. [CrossRef]

46. He, X.; Xu, C.; Qi, W.; Huang, Y.; Cheng, J.; Xu, X.; Yao, Q.; Lu, Y.; Dai, B. Landslides Triggered by the 2020 Qiaojia M w5. 1 Earthquake, Yunnan, China: Distribution, Influence Factors and Tectonic Significance. *J. Earth Sci.* **2021**, *32*, 1056–1068. [CrossRef]

47. Huang, Y.; Xu, C.; Zhang, X.; Xue, C.; Wang, S. An Updated Database and Spatial Distribution of Landslides Triggered by the Milin, Tibet M w6. 4 Earthquake of 18 November 2017. *J. Earth Sci.* **2021**, *32*, 1069–1078. [CrossRef]

48. Ma, S.; Xu, C.; Shao, X. Spatial prediction strategy for landslides triggered by large earthquakes oriented to emergency response, mid-term resettlement and later reconstruction. *Int. J. Disaster Risk Reduct.* **2020**, *43*, 101362. [CrossRef]

49. Xu, C.; Xu, X. Spatial distribution difference of landslides triggered by slipping-fault type earthquake on two sides of the fault. *Geol. Bull. China* **2012**, *31*, 532–540.

50. Giannecchini, R.; Galanti, Y.; D'Amato Avanzi, G. Critical rainfall thresholds for triggering shallow landslides in the Serchio River Valley (Tuscany, Italy). *Nat. Hazards Earth Syst. Sci.* **2012**, *12*, 829–842. [CrossRef]

51. Yang, Z.; Xu, C.; Li, L. Landslide detection based on ResU-net with transformer and CBAM embedded: Two examples with geologically different environments. *Remote Sens.* **2022**, *14*, 2885. [CrossRef]

52. Yu, B.; Xu, C.; Chen, F.; Wang, N.; Wang, L. HADeenNet: A hierarchical-attention multi-scale deconvolution network for landslide detection. *Int. J. Appl. Earth Obs. Geoinf.* **2022**, *111*, 102853. [CrossRef]

53. Huang, W.; Ding, M.; Li, Z.; Yu, J.; Ge, D.; Liu, Q.; Yang, J. Landslide susceptibility mapping and dynamic response along the Sichuan-Tibet transportation corridor using deep learning algorithms. *Catena* **2023**, *222*, 106866. [CrossRef]

54. Xu, C.; Xu, X.; Dai, F.; Saraf, A.K. Comparison of different models for susceptibility mapping of earthquake triggered landslides related with the 2008 Wenchuan earthquake in China. *Comput. Geosci.* **2012**, *46*, 317–329. [CrossRef]

55. Yang, Z.; Xu, C.; Shao, X.; Ma, S.; Li, L. Landslide susceptibility mapping based on CNN-3D algorithm with attention module embedded. *Bull. Eng. Geol. Environ.* **2022**, *81*, 412. [CrossRef]

56. Ma, S.; Wei, J.; Xu, C.; Shao, X.; Xu, S.; Chai, S.; Cui, Y. UAV survey and numerical modeling of loess landslides: An example from Zaoling, southern Shanxi Province, China. *Nat. Hazards* **2020**, *104*, 1125–1140. [CrossRef]

57. Huang, Y.; Xu, C.; Zhang, X.; Li, L. Bibliometric analysis of landslide research based on the WOS database. *Nat. Hazards Res.* **2022**, *2*, 49–61. [CrossRef]

58. Li, L.; Xu, C.; Zhang, Z.; Huang, Y. A review of researches on landslide disasters on Loess Plateau. *J. Inst. Disaster Prev.* **2021**, *23*, 1–11. [CrossRef]

59. Shao, X.; Xu, C.; Wang, P.; Li, L.; He, X.; Chen, Z.; Huang, Y.; Xu, X. Two public inventories of landslides induced by the 10 June 2022 Maerkang Earthquake swarm, China and ancient landslides in the affected area. *Nat. Hazards Res.* **2022**, *2*, 269–272. [CrossRef]

60. Wu, W.; Xu, C.; Wang, X.; Tian, Y.; Deng, F. Landslides triggered by the 3 august 2014 Ludian (China) Mw6.2 earthquake: An updated inventory and analysis of their spatial distribution. *J. Earth Sci.* **2020**, *31*, 853–866. [CrossRef]

61. Chen, Z.; Huang, Y.; He, X.; Shao, X.; Li, L.; Xu, C.; Wang, S.; Xu, X.; Xiao, Z. Landslides triggered by the 10 June 2022 Maerkang earthquake swarm, Sichuan, China: Spatial distribution and tectonic significance. *Landslides* **2023**, *20*, 2155–2169. [CrossRef]

62. Huang, Y.; Xu, C.; Li, L.; He, X.; Cheng, J.; Xu, X.; Li, J.; Zhang, X. Inventory and Spatial Distribution of Ancient Landslides in Hualong County, China. *Land* **2022**, *12*, 136. [CrossRef]

63. Medwedeff, W.G.; Clark, M.K.; Zekkos, D.; West, A.J. Characteristic landslide distributions: An investigation of landscape controls on landslide size. *Earth Planet. Sci. Lett.* **2020**, *539*, 116203. [CrossRef]

64. Xiao, Z.; Xu, C.; Huang, Y.; He, X.; Shao, X.; Chen, Z.; Xie, C.; Li, T.; Xu, X. Analysis of spatial distribution of landslides triggered by the Ms 6.8 Luding earthquake in China on September 5, 2022. *Geoenvironmental Disasters* **2023**, *10*, 3. [CrossRef]

65. Chen, Y.; Chang, K.; Wang, S.; Huang, J.; Yu, C.; Tu, J.; Chu, H.; Liu, C. Controls of preferential orientation of earthquake-and rainfall-triggered landslides in Taiwan's orogenic mountain belt. *Earth Surf. Process. Landf.* **2019**, *44*, 1661–1674. [CrossRef]

66. Yuan, X.; Zhao, F.; Duan, Z. Mechanism of loess landslide induced by river action. *Coal Geol. Explor.* **2018**, *46*, 154–160. [CrossRef]

67. Milne, F.; Brown, M.; Knappett, J.; Davies, M. Centrifuge modelling of hillslope debris flow initiation. *Catena* **2012**, *92*, 162–171. [CrossRef]

68. Qi, X.; Tang, C.; Chen, Z.; Shao, C. Coupling analysis of control factors between earthquake-induced landslides and subsequent rainfall-induced landslides in epicenter area of Wenchuan earthquake. *J. Eng. Geol.* **2012**, *20*, 522–531. [CrossRef]

69. Tan, F.; Hu, X.; He, C.; Zhang, Y.; Zhang, H.; Zhou, C.; Wang, Q. Identifying the main control factors for different deformation stages of landslide. *Geotech. Geol. Eng.* **2018**, *36*, 469–482. [CrossRef]

70. Li, L.; Xu, C.; Yang, Z.; Zhang, Z.; Lv, M. An inventory of large-scale landslides in Baoji city, Shaanxi province, China. *Data* **2022**, *7*, 114. [CrossRef]

71. Shao, X.; Ma, S.; Xu, C.; Shen, L.; Lu, Y. Inventory, distribution and geometric characteristics of landslides in Baoshan City, Yunnan Province, China. *Sustainability* **2020**, *12*, 2433. [CrossRef]

72. Hunter, G.; Fell, R. Travel distance angle for" rapid" landslides in constructed and natural soil slopes. *Can. Geotech. J.* **2003**, *40*, 1123–1141. [CrossRef]

73. Tian, Y.; Xu, C.; Chen, J.; Zhou, Q.; Shen, L. Geometrical characteristics of earthquake-induced landslides and correlations with control factors: A case study of the 2013 Minxian, Gansu, China, Mw 5.9 event. *Landslides* **2017**, *14*, 1915–1927. [CrossRef]

74. Meunier, P.; Hovius, N.; Haines, J. Topographic site effects and the location of earthquake induced landslides. *Earth Planet. Sci. Lett.* **2008**, *275*, 221–232. [CrossRef]

75. Lin, W.; Yin, K.; Wang, N.; Xu, Y.; Guo, Z.; Li, Y. Landslide hazard assessment of rainfall-induced landslide based on the CF-SINMAP model: A case study from Wuling Mountain in Hunan Province, China. *Nat. Hazards* **2021**, *106*, 679–700. [CrossRef]

76. Vasu, N.; Lee, S.; Pradhan, A.; Kim, Y.; Kang, S.; Lee, D. A new approach to temporal modelling for landslide hazard assessment using an extreme rainfall induced-landslide index. *Eng. Geol.* **2016**, *215*, 36–49. [CrossRef]

Article

An Infinite Slope Model Considering Unloading Joints for Spatial Evaluation of Coseismic Landslide Hazards Triggered by a Reverse Seismogenic Fault: A Case Study of the 2013 Lushan Earthquake

Gao Li [1,2,3], Mingdong Zang [4,5,*], Shengwen Qi [6,7,8,*], Jingshan Bo [1,2,9], Guoxiang Yang [4,5] and Tianhao Liu [4]

1 Key Laboratory of Earthquake Engineering and Engineering Vibration, Institute of Engineering Mechanics, China Earthquake Administration, Harbin 150080, China; bj_ligao@126.com (G.L.); bojingshan@163.com (J.B.)
2 Key Laboratory of Earthquake Disaster Mitigation, Ministry of Emergency Management, Harbin 150080, China
3 China Nonferrous Metals Resource Geological Survey, Beijing 100012, China
4 School of Engineering and Technology, China University of Geosciences (Beijing), Beijing 100083, China; yanggx@cugb.edu.cn (G.Y.); tianhao.liu@email.cugb.edu.cn (T.L.)
5 Institute of Geosafety, China University of Geosciences (Beijing), Beijing 100083, China
6 Key Laboratory of Shale Gas and Geoengineering, Institute of Geology and Geophysics, Chinese Academy of Sciences, Beijing 100029, China
7 Innovation Academy for Earth Science, Chinese Academy of Sciences, Beijing 100029, China
8 College of Earth and Planetary Science, University of Chinese Academy of Sciences, Beijing 100049, China
9 Geological Engineering Department, Institute of Disaster Prevention, Sanhe 065201, China
* Correspondence: mzang@cugb.edu.cn (M.Z.); qishengwen@mail.iggcas.ac.cn (S.Q.); Tel.: +86-010-8232-2627 (M.Z.); +86-010-8299-8055 (S.Q.)

Citation: Li, G.; Zang, M.; Qi, S.; Bo, J.; Yang, G.; Liu, T. An Infinite Slope Model Considering Unloading Joints for Spatial Evaluation of Coseismic Landslide Hazards Triggered by a Reverse Seismogenic Fault: A Case Study of the 2013 Lushan Earthquake. *Sustainability* 2024, 16, 138. https://doi.org/10.3390/su16010138

Academic Editor: Maurizio Lazzari

Received: 30 September 2023
Revised: 28 November 2023
Accepted: 12 December 2023
Published: 22 December 2023

Abstract: Coseismic landslides pose a significant threat to the sustainability of both the natural environment and the socioeconomic fabric of society. This escalation in earthquake frequency has driven a growing interest in regional-scale assessment techniques for these landslides. The widely adopted infinite slope model, introduced by Newmark, is commonly utilized to assess coseismic landslide hazards. However, this conventional model falls short of capturing the influence of rock mass structure on slope stability. A novel methodology was previously introduced, considering the roughness of potential slide surfaces on the inner slope, offering a fresh perspective on coseismic landslide hazard mapping. In this paper, the proposed method is recalibrated using new datasets from the 2013 Lushan earthquake. The datasets encompass geological units, peak ground acceleration (PGA), and a high-resolution digital elevation model (DEM), rasterized at a grid spacing of 30 m. They are integrated within an infinite slope model, employing Newmark's permanent deformation analysis. This integration enables the estimation of coseismic displacement in each grid area resulting from the 2013 Lushan earthquake. To validate the model, the simulated displacements are compared with the inventory of landslides triggered by the Lushan earthquake, allowing the derivation of a confidence level function that correlates predicted displacement with the spatial variation of coseismic landslides. Ultimately, a hazard map of coseismic landslides is generated based on the values of the certainty factor. The analysis of the area under the curve is utilized to illustrate the improved effectiveness of the proposed method. Comparative studies with the 2014 Ludian earthquake reveal that the coseismic landslides triggered by the 2013 Lushan earthquake predominantly manifest as shallow rock falls and slides. Brittle coseismic fractures are often associated with reverse seismogenic faults, while complaint coseismic fractures are more prevalent in strike–slip seismogenic faults. The mapping procedure stands as a valuable tool for predicting seismic hazard zones, providing essential insights for decision-making in infrastructure development and post-earthquake construction endeavors.

Keywords: Lushan earthquake; coseismic landslide hazard; infinite slope model; unloading joint; natural and social sustainability

1. Introduction

Landslide disasters usually cause significant loss of life and property damage [1,2]. Earthquakes are one of the primary triggers for landslides [3]. In recent years, a rise in strong earthquakes globally has resulted in frequent coseismic landslide hazards, causing great damage in mountainous areas, and posing a serious threat to social sustainability. Consequently, the evaluation of regional coseismic landslide hazards holds paramount importance for pre-earthquake urban planning and post-earthquake reconstruction efforts.

Early attempts to evaluate the coseismic stability of slopes involved the application of the pseudo-static analysis method [4] and the finite-element modeling technique [5]. Accurately predicting slope displacement is important for evaluating seismic slope stability [6,7]. Newmark [8] introduced a straightforward and effective method, still widely used today, for estimating the coseismic permanent displacements of slopes. The models for estimating the coseismic permanent displacements of slopes can be grouped into four categories: rigid-block model, decoupled model, coupled model, and unified model. Newmark's original methodology is generally referred to as rigid-block analysis [9]. To enhance the accuracy of estimating earthquake-induced permanent displacement, a decoupled analysis was developed by Makdisi and Seed [10]. The decoupled analysis accounts for the deformation of the sliding block during the sliding process. Bray and Rathje [11] introduced a streamlined seismic displacement method by utilizing a fully nonlinear decoupled one-dimensional dynamic analysis technique in conjunction with Newmark's rigid-plastic sliding block analysis. Rathje and Bray [12,13] conducted a comparative analysis, contrasting outcomes obtained through rigid-block analysis with both linear and nonlinear coupled as well as decoupled analyses. In the coupled method, deformation and displacement are considered simultaneously, offering a more realistic portrayal of slope response under seismic forces. Bray and Travasarou [14] proposed a simplified approach to predict earthquake-induced permanent displacement based on a nonlinear fully coupled sliding block model, incorporating critical acceleration, dominant sliding block period, and seismic spectral acceleration. Considering the variety in the natural period of the sliding mass during the dynamic response of slopes, a unified simplified model was developed for predicting earthquake-induced sliding displacements of rigid and flexible landslide masses [15]. Recently, Zhang et al. [16,17] introduced a novel model for evaluating seismic slope stability, considering the influence of velocity pulse seismic motion on permanent slope displacement, thereby presenting an advanced framework for assessing seismic slope stability under the specific influence of near-fault pulses.

Jibson et al. [18,19] elucidated the computational process of the Newmark method for the regional hazard mapping of coseismic landslides. This pioneering work significantly advanced the utilization and refinement of the Newmark method within the realm of regional coseismic landslide susceptibility analysis. Subsequently, numerous scholars have adopted and applied the Newmark method in their assessments of regional coseismic landslide susceptibility. Yuan et al. [20] applied the fundamental approach delineated by Jibson [18,19] to calculate the Newmark displacement of slopes using strong ground-motion recordings from the 2013 Lushan earthquake. Various empirical equations with updated fitting coefficients have been formulated for a comparative analysis of Newmark displacement estimation. Ma and Xu [21] introduced an approach that integrates logistic regression and support vector machine models with critical acceleration for the hazard assessment of landslides triggered by the 2013 Lushan earthquake. This method proves to be more effective in coseismic landslide hazard assessment compared to the simplified Newmark model. Jin et al. [22] modified the Newmark model by incorporating attenuation coefficients to the effective internal friction angle and the effective cohesion of geologic units. The coseismic landslide hazard zoning, derived from the modified Newmark model, exhibited a good correlation with the actual distribution of landslides triggered by the 2013 Lushan earthquake. In the context of the 2013 Lushan earthquake, Ma [23] demonstrated that the Newmark method's rapid assessment, even without coseismic landslide data, adequately fulfills the spatial prediction requirements for coseismic landslides during the emergency response phase of disaster reduction.

Valley incisions frequently lead to the formation of shallow unloading joints parallel to the surface in natural slopes [24,25]. Research indicates that during strong earthquakes, rock slopes exhibit collapsing and sliding failures along these shallow unloading joints, with approximately 90% of coseismic landslides being shallow falls and slides [26–29]. Unstable rock blocks are frequently fragmented and mobilized along these joints [30]. However, prior studies have not sufficiently considered the shear strength of these rock joints when evaluating coseismic landslides. In addressing this gap, Zang et al. [31] incorporated the Barton model [32] into the Newmark method to enhance the estimation of slopes' dynamic stability. The method was initially employed and calibrated to assess the hazards associated with coseismic landslides, utilizing data from the 2014 Ludian earthquake in Yunnan Province, southwestern China. The area under the curve analysis revealed that the reliability of the proposed method surpasses that of the conventional Newmark approach. The seismogenic fault associated with the 2014 Ludian earthquake is identified as the Baogunao–Xiaohe fault, characterized by left-lateral strike–slip movement [33]. Field investigations into coseismic landslides have brought to light significant disparities in the spatial distribution patterns between reverse-type and strike–slip-type seismogenic faults [29,34]. The failure modes of slopes exhibit variation in regions affected by different types of seismogenic faults, as do the corresponding calculation parameters. A limitation of the initial calibration lies in its exclusive reliance on coseismic landslides triggered by a strike–slip fault. To address this constraint, our study rigorously reevaluates the initial methodology and enhances its precision through recalibration. This is achieved by incorporating the coseismic landslide dataset from the 2013 Lushan earthquake, identified as a blind reverse-fault event [35]. Following calibration, the methodology becomes applicable for mapping the spatial distribution of coseismic landslides under various seismogenic fault conditions.

This paper provides a concise overview of the features and spatial distribution of landslides triggered by the 2013 Lushan earthquake within the study area, outlines the methodology employed to assess the stability of coseismic slopes, details the procedure for mapping coseismic landslide hazards, and concludes with a discussion of the coseismic hazard assessment results. The paper further compares these findings with those from our prior studies. Finally, it explores the limitations of the method and outlines potential avenues for future research.

2. Study Area

The epicenter of the 2013 M_w 6.6 Lushan earthquake is located at the southern segment of the Longmenshan fault zone, on the eastern margin of the Tibetan Plateau [36,37]. The study area was selected as a rectangular section with high concentrations of coseismic landslides (Figure 1). The study area is located about 12 km northwest of the epicenter and lies on the border of two earthquake-hit counties, i.e., Baoxing County and Lushan County in Sichuan Province, China, covering 138 km^2 (Figure 1). The study area is characterized by deeply incised valleys and high mountains, with an elevation ranging from 1013 to 3308 m above sea level (Figure 1). Exposures of the geological units encompass ages ranging from the Sinian to the Triassic, incorporating a diverse range of formations such as carbonate, diorite, dolomite, limestone, phyllite, sandstone, and shale (Figure 2).

There have been 17 earthquakes with magnitudes $\geq$ 4.7 in the history of the Lushan earthquake zone [38]. Strong tectonic activity provides favorable conditions that make the region prone to landslides. The 2013 Lushan earthquake has triggered thousands of landslides [39]. An inventory of 308 coseismic landslides was compiled by making comparisons between pre-earthquake satellite images and post-earthquake aerial photographs in the study area. The cumulative area of these interpreted landslides spans 5 km^2.

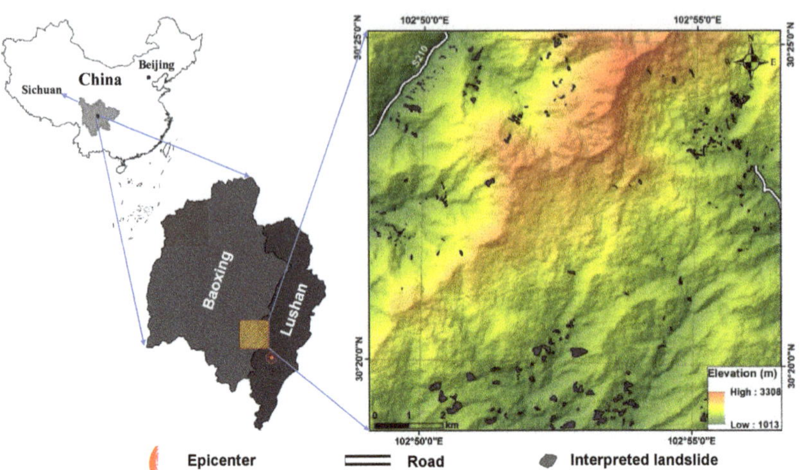

Figure 1. Location of the study area and the interpreted landslides.

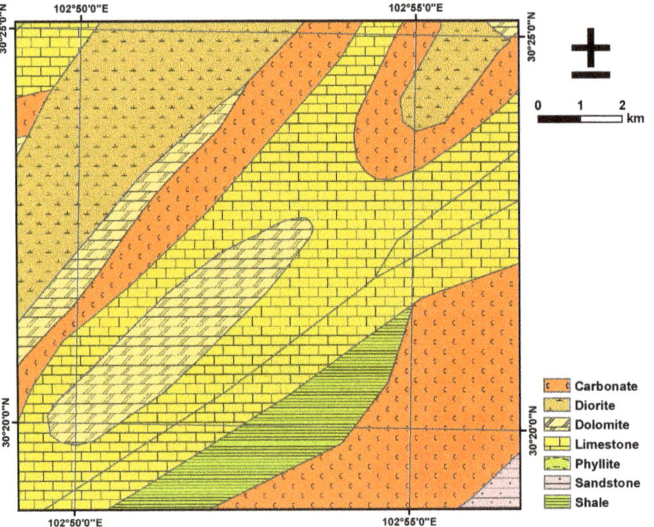

Figure 2. Map depicting the geological units of the study area.

3. Methodology

In 1965, Newmark [8] proposed an infinite slope model for the assessment of the coseismic stability of slopes. The Newmark method [8] simulates a landslide as a rigid friction block situated on an inclined plane with a predetermined critical acceleration (Figure 3). It computes the cumulative permanent displacement of the block in relation to its base, considering the impact of earthquake acceleration–time history [18,19]. The permanent displacement is a valuable index for evaluating the dynamic performance of slopes.

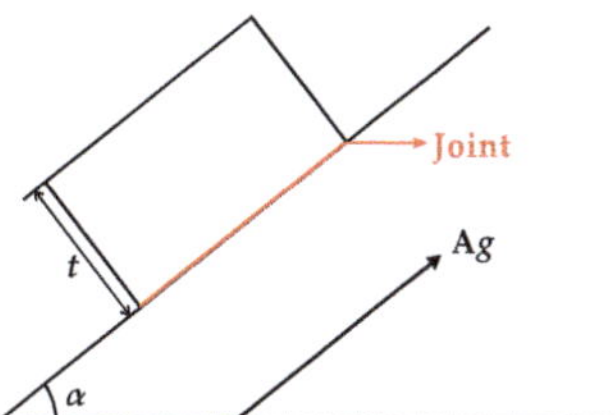

Figure 3. Conceptual sliding-block model of Newmark analysis (adapted from Newmark [8] and Zang et al. [31]). α is the angle of the slope, t is the thickness of the failure rock block, A is a constant, and g is the acceleration due to the Earth's gravity.

Once an acceleration–time history has been selected, segments of the record surpassing the critical acceleration (Figure 4a) are integrated once to establish a velocity profile (Figure 4b). Subsequently, the velocity–time history undergoes a second integration to yield the cumulative displacement of the block (Figure 4c) [18,19,31].

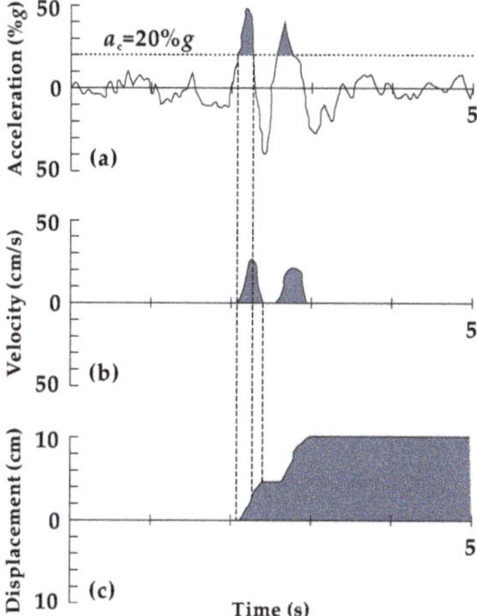

Figure 4. Illustration of the algorithm employed in Newmark analysis (adapted from Jibson et al. [18,19]). a_c is the critical acceleration in terms of g. (**a**) The acceleration–time history of earthquake with a critical acceleration (indicated by the horizontal dashed line) of 20% g superimposed. The color area represents segments of the record that exceed the critical acceleration; (**b**) The velocity–time history of the block. The color area stands the velocity of the block, derived from a single integration of the acceleration–time history corresponding to portions exceeding the critical acceleration; (**c**) The displacement–time history of the block. The color area denotes the cumulative displacement of the block, derived from the double integration of the acceleration–time history corresponding to portions exceeding the critical acceleration.

From Figure 4, we can see that critical acceleration provides a base for predicting the cumulative permanent displacement of a slope. Newmark [8] showed that the critical acceleration of a potential landslide block can be expressed as a simple function of the static factor of safety and the geometry of the slope [18,19]:

$$a_c = (F_S - 1)g\sin\alpha, \tag{1}$$

where a_c is the critical acceleration in terms of g, F_S is the static factor of safety, and α is the angle from the horizontal at which the center of the slide block moves when displacement first occurs [18,19]. In instances of a planar slip surface parallel to the slope, it is typically noted that this angle closely approximates the slope angle.

Natural rock slopes are often cut by a group of shallow unloading joints due to valley incisions, forming an unloading zone on the surface of the slopes [24,25]. Slopes behave as collapsing and sliding failures of shallow unloading joints under intense seismic activity, with approximately 90% of coseismic landslides manifesting as shallow falls and slides [26–29]. The seismic shaking commonly triggers the activation of unstable rock blocks along rock joints (Figure 3). Therefore, the static factor of safety is related to the shear strength of these rock joints.

Zang et al. [31] derived the static factor of safety of a slope based on the Barton [32] shear strength criterion for rock joints in another study:

$$F_S = \frac{\text{Resisting force}}{\text{Driving force}} = \frac{\tau}{\gamma t \sin \alpha} = \frac{\sigma_n \tan\left[\text{JRC}_n \log_{10}\left(\frac{\text{JCS}_n}{\sigma_n}\right) + \phi_b\right]}{\gamma t \sin \alpha} = \frac{\gamma t \cos \alpha \tan\left[\text{JRC}_n \log_{10}\left(\frac{\text{JCS}_n}{\gamma t \cos \alpha}\right) + \phi_b\right]}{\gamma t \sin \alpha} = \frac{\tan\left[\text{JRC}_n \log_{10}\left(\frac{\text{JCS}_n}{\gamma t \cos \alpha}\right) + \phi_b\right]}{\tan \alpha}, \quad (2)$$

where τ is the peak shear strength of the rock joint, γ is the unit weight of the rock mass, t is the thickness of the failure rock block, σ_n is the effective normal stress, JRC_n is the joint roughness coefficient in the in situ scale, JCS_n is the joint wall compressive strength in the in situ scale, and ϕ_b is the basic friction angle.

Considering the impact of joint size, the JRC_n and JCS_n can be defined as [40]:

$$\text{JRC}_n = \text{JRC}_0 \left(\frac{L_n}{L_0}\right)^{-0.02\text{JRC}_0}, \quad (3)$$

$$\text{JCS}_n = \text{JCS}_0 \left(\frac{L_n}{L_0}\right)^{-0.03\text{JRC}_0}, \quad (4)$$

where the adopted nomenclature includes (0) and (n) to represent values corresponding to the laboratory scale and the in situ scale, respectively.

It is impractical to conduct a rigorous Newmark method during the regional analysis. Therefore, researchers have proposed different empirical regressions to estimate Newmark displacement as a function of the critical acceleration and ground motion parameters [14,41–45]. In this study, we chose a vector model developed according to more than 2000 strong motion records [44]:

$$\ln D = 4.89 - 4.85\left(\frac{a_c}{\text{PGA}}\right) - 19.64\left(\frac{a_c}{\text{PGA}}\right)^2 + 42.49\left(\frac{a_c}{\text{PGA}}\right)^3 - 29.06\left(\frac{a_c}{\text{PGA}}\right)^4 + 0.72\ln(\text{PGA}) + 0.89(M_w - 6), \quad (5)$$

where D is the predicted Newmark displacement, PGA is the peak ground acceleration, and M_w is the moment magnitude. D is in centimeters, and a_c and PGA are in units of g.

Figure 5 is a flowchart showing the sequential steps of the spatial evaluation of coseismic landslides. After calculating the Newmark displacement, the predicted displacement was then compared with an inventory of landslides induced by the Lushan earthquake, employing the certainty factor model (CFM) for mapping the spatial distribution of coseismic landslide hazards [46]. The CFM, an inexact reasoning model, initially developed by Shortliffe and Buchanan [46] and refined by Heckerman [47], serves to examine the correlation between the predicted displacement and the occurrence of landslides. Within the CFM framework, the certainty factor (CF) quantifies the net confidence in a hypothesis H based on the evidence E [47], ranging from -1 to 1. A CF of -1 indicates a complete lack of confidence, whereas a CF of 1 signifies absolute confidence. Values exceeding zero lean towards supporting the hypothesis, while those below zero lean towards supporting its negation. The probabilistic interpretation of CF can be articulated as follows [47]:

$$\text{CF} = \begin{cases} \frac{p(H|E) - p(H)}{p(H|E)[1 - p(H)]}, & p(H|E) > p(H) \\ \\ \frac{p(H|E) - p(H)}{p(H)[1 - p(H|E)]}, & p(H|E) < p(H) \end{cases}, \quad (6)$$

where CF stands for the certainty factor, $p(H|E)$ represents the posterior probability denoting the conditional probability for a posterior hypothesis based on evidence, and $p(H)$ denotes the prior probability in the absence of any evidence. For the spatial evaluation of coseismic landslides, $p(H|E)$ was defined as the ratio of the landslide area within a specific displacement area, and $p(H)$ was defined as the proportion of the landslide area within the entire study area [31]. Accordingly, CF values reflected the confidence level regarding the occurrence of coseismic landslides. Positive values indicated an increased confidence in slope failure, whereas negative values indicated a leaning towards its negation [31].

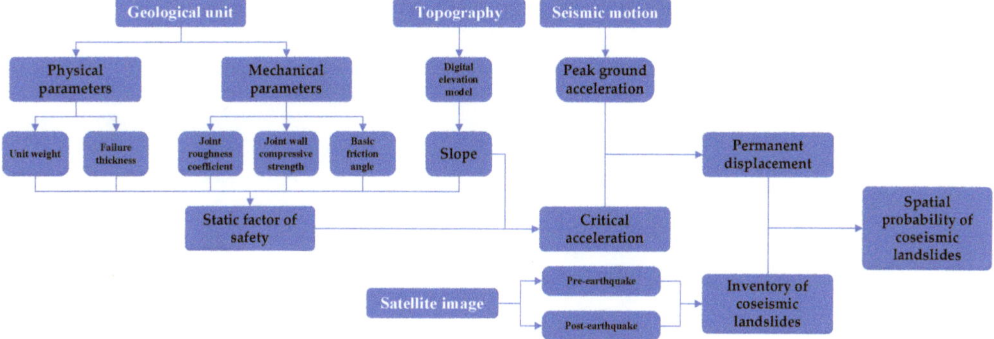

Figure 5. Flowchart depicting the sequential steps involved in the hazard mapping procedure.

4. Mapping of Coseismic Landslide Hazard

The datasets, encompassing geological units, peak ground acceleration (PGA), and a high-resolution digital elevation model (DEM) of topography, were converted into raster format with a grid spacing of 30 m. In this section, each step involved in the mapping of coseismic landslide hazards is discussed in detail.

4.1. Static Factor of Safety Map

We selected a 30 m DEM to produce a slope map (Figure 6) by applying a basic slope algorithm, where the slope was identified as the steepest downhill descent among adjacent cells [48]. The derived slopes ranged from 0° to 75°. The northwestern and northeastern portions of the study area are high mountains and deeply incised valleys with slopes greater than 50°, and the central and southeastern parts of the area are relatively flat (Figure 6).

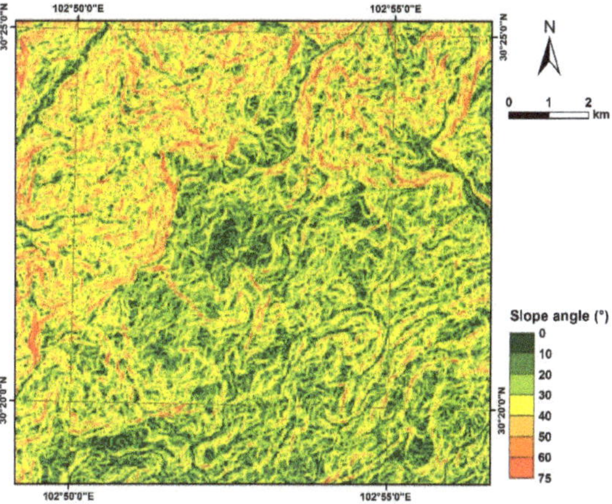

Figure 6. Topographic slope map generated from the DEM of the study area.

We assigned physical and mechanical parameters to the rock types using a digital geological map (Figure 2). The map was rasterized at a 30 m grid spacing for the following calculation after the parameter assignment. Studies have shown that JRC_0 and JCS_0 strongly depend on lithology [49–66]. The values of JRC_0 and JCS_0 assigned to each rock type were estimated based on test data from the references listed in Table 1. These values were obtained from laboratory tests with a sample length of $L_0 = 100$ mm. As we used a projected coordinate system for regional analysis and the grid spacing was 30 m, the engineering dimension, L_n, in a grid cell was equal to 30 m/cosα. We can calculate JRC_n and JCS_n by substituting JRC_0, JCS_0, L_0, and L_n into Equations (3) and (4), respectively. The spatial distribution of JRC_n and JCS_n are shown in Figure 7a,b, respectively. Other parameters such as γ and ϕ_b are listed in Table 1.

Table 1. Shear strengths assigned to rock types in the study area.

Rock Type	γ [1] (kN/m^3)	ϕ_b	JCS_0 (MPa)	JRC_0	φ [1]	c [1] (kPa)	References
Carbonate	23.7	35°	150	9.3	44°	33	Bandis et al. [49] Singh et al. [50] Giusepone and da Silva [51] Yong et al. [52]
Diorite	26.9	30°	200	6.8	50°	40	Sirkiä, et al. [53] Bao et al. [54]
Dolomite	25.9	32°	140	9.5	43°	35	Singh et al. [50] Arzúa et al. [55] Giusepone and da Silva [51]
Limestone	21.5	37°	160	9	45°	30	Bandis et al. [49] Singh et al. [50] Yong et al. [52]
Phyllite	28	28°	130	6	40°	20	Andrade and Saraiva [56] Marques et al. [57] Srivastava [58]
Sandstone	23.5	35°	100	6	42°	24	Coulson [59] Bandis et al. [49] Priest [60] Tang et al. [61] Tang et al. [62]
Shale	24.9	27°	75	8	27°	16	Barton and Choubey [63] Bilgin and Pasamehmetoglu [64] Hashemi and Zoback [65] Qu et al. [66]

[1] Friction angle (φ), cohesion (c), and unit weight (γ) were derived from the Geological Engineering Handbook [67].

For simplicity, the thickness (t) of the rock block in failure (Figure 3) was established at 3 m, based on field investigations of typical slope failures resulting from the Lushan earthquake. We produced a map of the static factor of safety by integrating data layers of α, JRC_n, JCS_n, γ, and ϕ_b in Equation (2). The calculated static factors of safety varied between 0.2 to 133.8. Grid cells displaying static factors of safety below 1 suggested static instability in the slopes, but this did not necessarily imply that they were undergoing sliding under seismic shaking [31]. However, according to Equation (1), to prevent obtaining a negative critical acceleration value, we assigned a minimal static factor of safety of 1.01 to these cells. Keefer [3] showed that the minimum slope angle for earthquake-induced landslides was 5°. In addition, the calculated static factors of safety for the cells with a slope angle of less than 5° were very high, and these cells did not contribute significantly to the statistical analysis due to their limited sample size [18,19]. Therefore, slopes with inclinations less than 5° were excluded from the analysis. The revised static factors of safety now fall within the range of 1 to 10.4, as depicted in Figure 8. Statically unstable slopes, characterized by a static factor of safety below 1.2, are mainly distributed in the northwestern, northeastern, and southeastern parts of the study area, as illustrated in Figure 8.

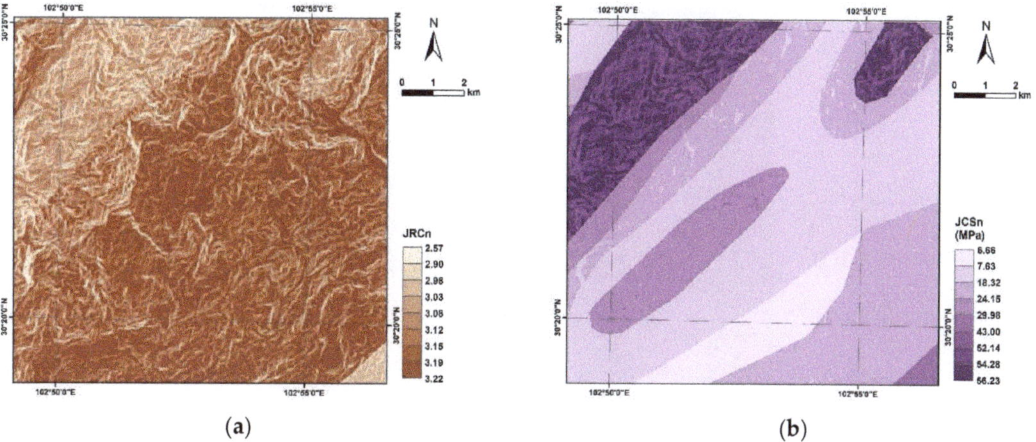

(**a**)　　　　　　　　　　　　　　　　　　　　　　(**b**)

Figure 7. Map showing shear strength components assigned to geological formations in the study area. (**a**) JRC_n component of shear strength; (**b**) JCS_n component of shear strength.

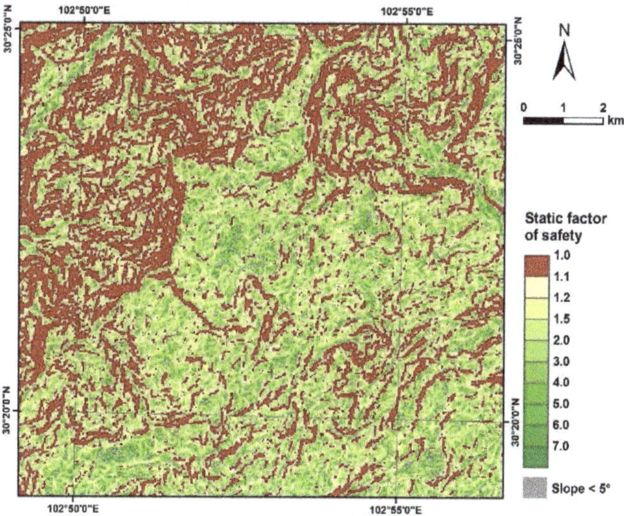

Figure 8. Map depicting the static factor of safety across the study area.

4.2. Critical Acceleration Map

After the adjustment of the static factors of safety, we produced a map of critical acceleration using Equation (1) to integrate the slope angle with the static factors of safety (Figure 9). The critical acceleration represents the intrinsic properties of slopes independent of any shaking scenario; thus, the critical acceleration map serves as an indicator of coseismic landslide susceptibility [18,19]. A smaller value of critical acceleration denotes the lower intensity of shaking needed to overcome the stability of the slope. Therefore, coseismic sliding is more likely to develop on the slope. Figure 9 shows that the dynamically unstable slopes with a critical acceleration of less than 0.3 g are mainly distributed in the northwestern part of the study area, and only a small portion is present in the southern part of the study area.

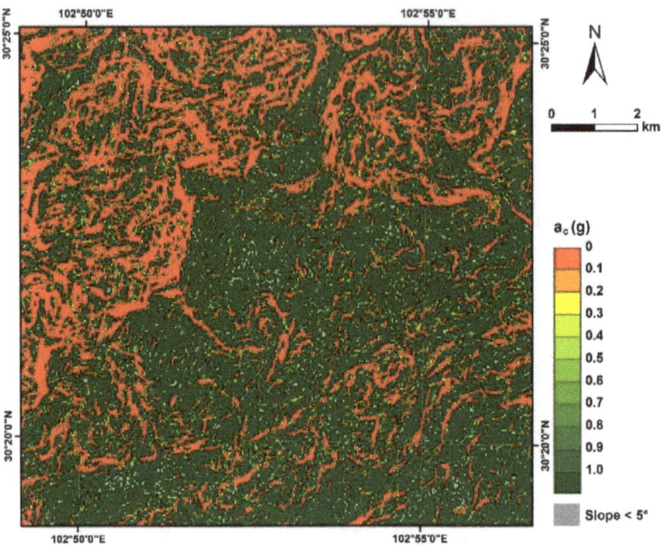

Figure 9. Map illustrating critical accelerations across the study area.

4.3. Predicted Displacement Map

We downloaded the contour map depicting the PGA in the study area (Figure 10) from the United States Geological Survey. The Newmark displacement for each cell was computed using Equation (5), integrating the respective grid values of critical acceleration, PGA, and moment magnitude. The resulting predicted displacements range from 0 to 76 cm (Figure 11). Slopes with a predicted displacement ranging between 0 cm and 4 cm are distributed in the northwestern portion of the study area, slopes with a predicted displacement ranging between 4 cm and 6 cm are distributed in the middle portion of the study area, slopes with a predicted displacement ranging between 6 cm and 15 cm are scattered in the area with a displacement between 4 cm and 6 cm, and slopes with a predicted displacement greater than 15 cm are only distributed in the southeastern portion of the study area. According to the statistical sizes of the areas, displacements less than 4 cm occupy 90.5% of the study area, displacements between 4 cm and 6 cm occupy 5.5% of the study area, and displacements between 6 cm and 15 cm occupy 3.6% of the study area, with displacements exceeding 15 cm constituting a very small area.

4.4. Coseismic Landslide Hazard Map

Jibson et al. [18,19] showed that Newmark displacements serve as an indicator of the coseismic performance of slopes but do not align directly with observable slope movements in the field. Consequently, a comparison of predicted displacements with an inventory of landslides induced by earthquakes was necessary to establish a predictive scenario for coseismic landslide hazards. We employed Equation (6) to generate a coseismic landslide hazard map, expressed in terms of CFs.

The Newmark displacement cells, each 1 cm, were grouped into bins. All cells with displacements between 0 cm and 1 cm were grouped into the first bin, those between 1 cm and 2 cm into the second bin, and so forth [18,19]. We grouped the predicted displacements into 75 bins. For each bin, we calculated the proportion of cells occupied by landslide areas. This proportion represents the posterior probability of the bin according to Equation (6). The proportion of the entire landslide area within the entire study area was calculated to obtain the prior probability, which was the same in each bin. Using Equation (6), we computed the values of CF by combining the corresponding posterior and prior probabilities. The calculated CFs ranged from −1 to 0.99. The CFM provides a necessary linkage between the Newmark displacements and confidence levels of coseismic landslide occurrence in the study area. We produced a coseismic landslide hazard map for the ground-shaking scenario of the Lushan earthquake based on the values of CF, as shown in Figure 12. The inventory of landslides triggered by the Lushan earthquake illustrates the fit of the predicted confidence levels of coseismic landslides.

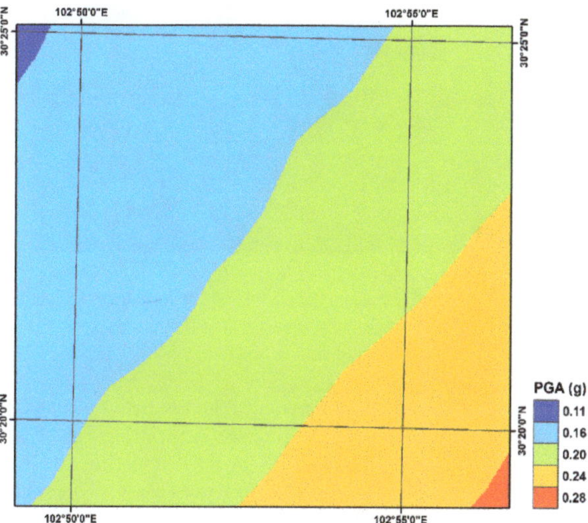

Figure 10. Contour map illustrating peak ground acceleration (PGA). PGA values shown are in *g*.

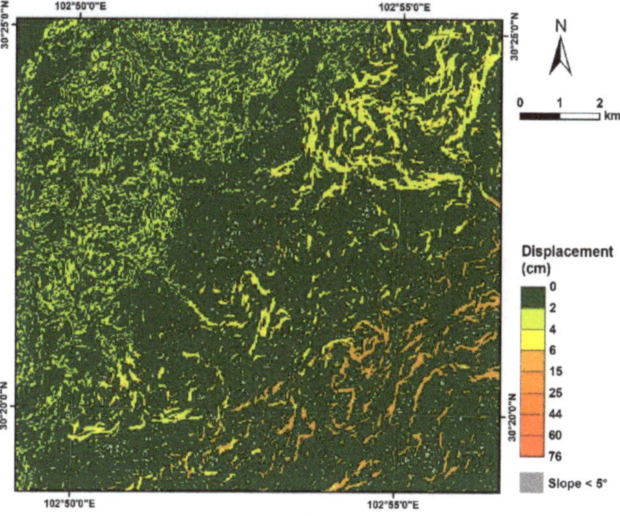

Figure 11. Map illustrating predicted displacements within the study area.

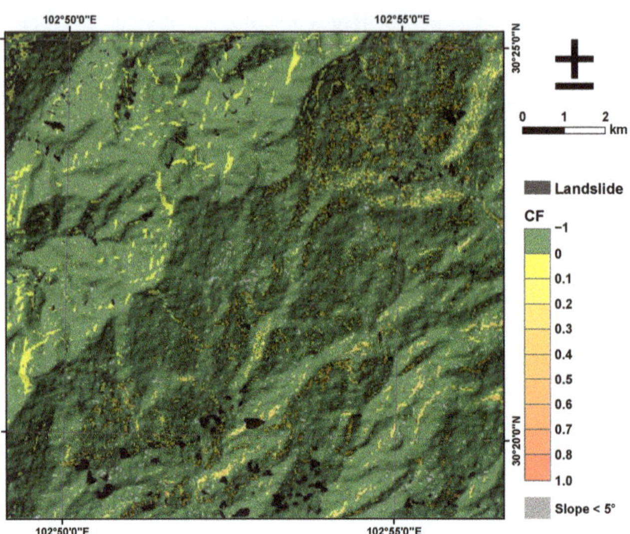

Figure 12. Map showing confidence levels of coseismic landslides. Confidence levels are represented by CF values.

5. Discussion

The predicted displacement refers to the cumulative sliding displacement of slopes for a specific acceleration history. As depicted in Figure 11, these predicted displacements range from 0 to 76 cm. The analysis of the statistical size of each displacement area indicates that 90.5% of the study area experiences displacements less than 4 cm, and 99.6% of the study area experiences displacements less than 15 cm. Only a very small area observes displacements greater than 15 cm. Jibson et al. [18,19] suggested that shallow, disrupted rock falls and rock slides in relatively brittle, poorly cemented sediments tend to collapse at small displacements. Thus, the study area demonstrates higher susceptibility to shallow falls and slides. According to high-resolution aerial photographs and field investigations, most landslides triggered by the Lushan earthquake were rock falls and relatively shallow, disrupted slides, typically 1–5 m in depth [29,39,68]. Therefore, the proposed model is particularly useful for predicting the spatial distribution of the typical coseismic landslides in the study area.

Predicted displacements do not accurately reflect the actual slope movement in the field. Instead, modeled displacements serve as an indicator that can be correlated with field performance [18,19]. By using a function that incorporates Newmark displacement and CF, the confidence levels of slope failure in the field can be estimated, and the spatial variation in coseismic landslides can be predicted under any set of ground-shaking conditions [31]. The number of Newmark displacement cells per 1 cm was uneven. To obtain a more reasonable regression curve correlating the predicted displacement and CF, we grouped these cells into bins based on quantile statistics. The breakpoints were set at 0, 2, 4, 6, 15, 25, 44, 60, and 76, ensuring an equal number of cells in each bin. For each bin, the proportion of cells in landslide areas was calculated, and the CF value of Newmark displacement was plotted as a dot. To fit the data, a modified Weibull [69] curve proposed by Zang et al. [31] was employed, with the following functional form:

$$\mathrm{CF} = m\left[1 - \exp\left(-aD^b\right)\right] - 1,\qquad(7)$$

where CF is the certainty factor, $(m - 1)$ represents the maximum CF value, D stands for the predicted displacement, and a and b are the regression constants. The regression curve derived from the Lushan data is as follows:

$$CF = 1.254\left[1 - \exp\left(-0.669D^{0.682}\right)\right] - 1. \tag{8}$$

The data fit the curve well with an R-squared value of 92%. The CF value serves as a direct indicator for predicting the confidence level of slope failure based on Newmark displacement. Figure 13 shows that the CF value increases monotonically with increasing Newmark displacement. The CF value increases rapidly in the first 10 cm of Newmark displacement, followed by a sudden plateau after the 15 cm range, stabilizing around a CF value of approximately 0.25.

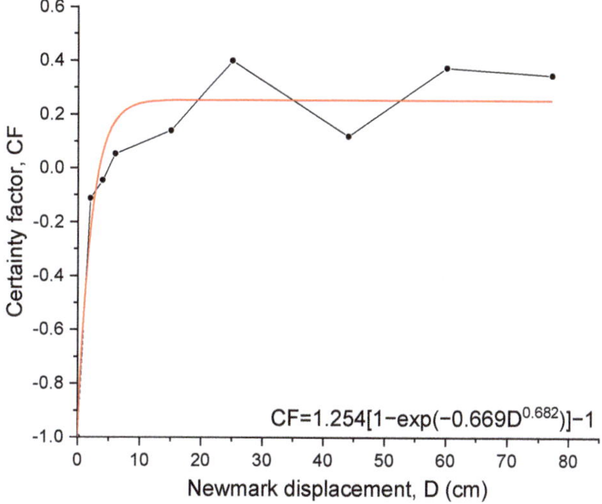

Figure 13. CF as a function of Newmark displacement. Each dot represents the CF value for a specific Newmark displacement bin, while the red line illustrates the fitting curve of the data using a modified Weibull function.

Equation (8) mirrors the form of our previously published equation [31]. However, the equation proposed in this study has different regression constants owing to different data. After calibration, both the curve and its corresponding equation can be applied to any ground-shaking conditions. This allows the prediction of the slope failure's confidence level, based on the predicted Newmark displacement [18,19]. However, a problem exposed by the earlier study should be noted. The maximum CF value is equal to $(m - 1)$, not $m/2$, as indicated by Zang et al. [31]. The values of m, a, and b in Equation (8) may vary in other regions if the lithology, topography, and ground motion conditions significantly differ from those observed in the study area. Additionally, the shape of the fitting curve in Figure 13 might vary in regions where different types of slope failures are predominant. If rock falls and rock slides in more brittle materials were prevalent, the fitting curve may be steeper and could flatten at a lower maximum displacement value [18,19,31]. The seismogenic fault responsible for the 2014 Ludian earthquake is the Baogunao–Xiaohe fault, identified as a left-lateral strike–slip fault [33]. In the case of the 2013 Lushan earthquake, the seismogenic fault is a blind reverse fault [35]. Upon comparing our previous studies [31], we observed that, in contrast to the Ludian earthquake, the predicted displacements caused by the Lushan earthquake exhibit a faster convergence in the fitting curve, with smaller displacement extremes. This suggests that when the seismogenic fault is a reverse fault, slope failures in the seismic region predominantly manifest as brittle fractures. Conversely, if the seismogenic fault is a strike–slip fault, the resulting landslides primarily occur in more compliant materials within the seismic region.

Figure 9 shows the distribution of critical acceleration within the study area. Critical acceleration describes the seismic susceptibility of a slope. Newmark [8] showed that the critical acceleration depends on the static factor of safety and the slope angle (Equation (1)). Comparing Figure 6 with Figures 8 and 9, respectively, we found that the modeling procedure is heavily slope-driven, which is consistent with the findings from Jibson et al. [18,19]. However, in comparing Figure 11 with Figure 10, we observed that the distribution pattern of Newmark displacements bears a resemblance to that of PGA. Areas with displacements between 2 and 4 cm, 4 and 6 cm, and 6 and 15 cm are distributed in bands in the study area, indicating the dominant role of seismic ground motion in triggering coseismic landslides during the Lushan earthquake.

In Section 3, it was mentioned that during seismic activity, unstable rock blocks tend to slide along shallow unloading joints. The static factor of safety is closely related to the shear strength of these rock joints. The conventional Newmark analysis traditionally describes the shear strength of rocks using Coulomb's constants, such as friction angle and cohesion, but these values vary significantly between laboratory conditions with high normal stress and field conditions with low normal stress [32]. Additionally, these values are also dependent on the scale of the analysis [63]. To address these challenges, we introduced the Barton model into the Newmark analysis [31]. This model enables us to predict the shear strength of rock joints and reduce the variability associated with Coulomb's constants. Furthermore, we accounted for scale effects to prevent overestimating the shear strength of geological units in regional analysis by using Equations (3) and (4). A conventional Newmark analysis was conducted using assigned strengths like friction angle and cohesion, outlined in Table 1. The predicted displacements obtained through the conventional Newmark method ranged from 0 to 78 cm, whereas the proposed method yielded values ranging from 0 to 76 cm. Subsequently, the calculation of the CF using the conventional Newmark analysis produced a range of −1 to 0.74, while the proposed method resulted in a range of −1 to 0.99. To evaluate the performance of the two methods, we utilized the area under the curve (AUC) analysis. The AUC plot illustrates the cumulative area of CFs within each interval of calculated values, representing the proportion of the total study area (x-axis), and the proportion of cumulative landslides falling within those CFs (y-axis) [70]. An AUC value of 0.5 suggests a performance equal to random guessing, while a value of 1 indicates perfect performance [70]. Figure 14 depicts the results of the AUC analysis for both methods, with a calculated value of 0.57 for the proposed method and 0.53 for the conventional Newmark method. Therefore, the introduced method yields improved results compared to the conventional Newmark analysis.

However, it is noteworthy that both methods yield relatively small AUC values. This outcome can be attributed to the coseismic landslide interpretation data employed in the analysis. As is well known, landslides can be categorized into three distinct areas: the source area, the transport area, and the deposition area. The interpretation data utilized for the calculations encompass all three areas. The initiation of landslides occurs in the source area, while the deposition area is where the landslide material comes to rest. The source area is typically characterized by steep slopes, leading to relatively large Newmark displacements that align with the model's predictions. Conversely, the deposition area tends to exhibit flatter terrain with smaller slopes, resulting in relatively smaller Newmark displacements. Nevertheless, following the occurrence of a landslide, a substantial amount of material accumulates in the deposition area. Consequently, a significant portion of the landslide area in the interpretation data falls within regions associated with smaller Newmark displacements. This leads to deviations in the predictive outcomes of the model. To achieve more accurate results, it is crucial to precisely delineate the source area of coseismic landslides and employ it as the primary input for hazard assessment. This particular aspect demands significant attention in future evaluations of coseismic landslide hazards. Furthermore, since the proposed method concentrates on assessing the stability of unloading joints based on strength parameters like the joint roughness coefficient and joint wall compressive strength during predictions, its applicability is limited to spatial assessments of rock landslide hazards. The calibrated method proves most effective in predicting the spatial distribution of relatively shallow, disrupted slides and falls in rock and debris. However, it is likely to provide less precise predictions for the distribution of deeper, coherent landslides [18,19].

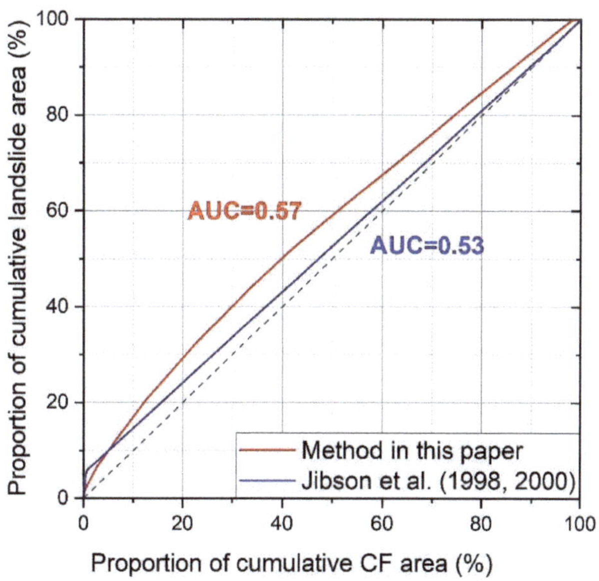

Figure 14. Plots of the area under the curve (AUC) for a comparison between the proposed method and the conventional Newmark method [18,19]. The dot line represents an AUC value of 0.5.

6. Conclusions

This study presents a recalibration of a pioneering method designed to assess coseismic landslide hazards triggered by the 2013 Lushan earthquake. The method incorporates the roughness of potential sliding surfaces on the inner slope. The primary findings are outlined as follows:

(1) The landslides induced by the 2013 Lushan earthquake are primarily concentrated in deeply incised valley regions, consisting mainly of shallow, disrupted rock falls, and rock slides.

(2) For reverse seismogenic faults, slopes within the seismic region predominantly display brittle coseismic fractures. In contrast, strike–slip seismogenic faults tend to result in complaint coseismic fractures on slopes.

(3) The integration of Newmark analysis with Barton's shear strength criterion proves to be practically useful in evaluating regional coseismic landslide hazards. Moreover, the improved method demonstrates higher reliability when compared to the conventional Newmark approach, thereby contributing to the overall sustainability and resilience of communities facing such geological challenges.

Author Contributions: Conceptualization, M.Z. and S.Q.; methodology, G.L. and M.Z.; validation, T.L.; formal analysis, G.L., M.Z. and T.L.; investigation, M.Z.; resources, M.Z.; data curation, M.Z.; writing—original draft preparation, G.L. and M.Z.; writing—review and editing, M.Z., S.Q. and G.Y.; project administration, S.Q. and J.B.; funding acquisition, M.Z., S.Q. and G.Y. All authors have read and agreed to the published version of the manuscript.

Funding: This research was funded by the National Natural Science Foundation of China (grant nos. 41825018, 42207215, and 42277164).

Institutional Review Board Statement: Not applicable.

Informed Consent Statement: Not applicable.

Data Availability Statement: Data will be made available on request.

Acknowledgments: The authors thank the National Natural Science Foundation of China (grant nos. 41825018, 42207215, and 42277164). We appreciate the insightful comments from the reviewing experts and academic editor, which significantly enhanced the quality of this manuscript.

Conflicts of Interest: The authors declare no conflict of interest.

References

1. Mao, Y.; Chen, L.; Nanehkaran, Y.A.; Azarafza, M.; Derakhshani, R. Fuzzy-Based Intelligent Model for Rapid Rock Slope Stability Analysis Using Qslope. *Water* **2023**, *15*, 2949. [CrossRef]
2. Tao, Z.; Wang, Y.; Zhu, C.; Xu, H.; Li, G.; He, M. Mechanical Evolution of Constant Resistance and Large Deformation Anchor Cables and Their Application in Landslide Monitoring. *Bull. Eng. Geol. Environ.* **2019**, *78*, 4787–4803. [CrossRef]
3. Keefer, D.K. Landslides Caused by Earthquakes. *Geol. Soc. Am. Bull. GSA Bull.* **1984**, *95*, 406–421. [CrossRef]
4. Terzaghi, K. Mechanism of Landslides. In *Application of Geology to Engineering Practice*; Paige, S., Ed.; Geological Society of America: New York, NY, USA, 1950; pp. 83–123, ISBN 0-8137-4301-X.
5. Clough, R.W.; Chopra, A.K. Earthquake Stress Analysis in Earth Dams. *J. Eng. Mech. Div.* **1966**, *92*, 197–212. [CrossRef]
6. Yuan, C.; Li, Q.; Nie, W.; Ye, C. A Depth Information-Based Method to Enhance Rainfall-Induced Landslide Deformation Area Identification. *Measurement* **2023**, *219*, 113288. [CrossRef]
7. Zhang, X.; Zhu, C.; He, M.; Dong, M.; Zhang, G.; Zhang, F. Failure Mechanism and Long Short-Term Memory Neural Network Model for Landslide Risk Prediction. *Remote Sens.* **2022**, *14*, 166. [CrossRef]
8. Newmark, N.M. Effects of Earthquakes on Dams and Embankments. *Géotechnique* **1965**, *15*, 139–160. [CrossRef]
9. Jibson, R.W. Methods for Assessing the Stability of Slopes during Earthquakes—A Retrospective. *Eng. Geol.* **2011**, *122*, 43–50. [CrossRef]
10. Makdisi, F.I.; Seed, H.B. Simplified Procedure for Estimating Dam and Embankment Earthquake-Induced Deformations. *J. Geotech. Eng. Div.* **1978**, *104*, 849–867. [CrossRef]
11. Bray, J.D.; Rathje, E.M. Earthquake-Induced Displacements of Solid-Waste Landfills. *J. Geotech. Geoenviron. Eng.* **1998**, *124*, 242–253. [CrossRef]
12. Rathje, E.M.; Bray, J.D. An Examination of Simplified Earthquake-Induced Displacement Procedures for Earth Structures. *Can. Geotech. J.* **1999**, *36*, 72–87. [CrossRef]
13. Rathje, E.M.; Bray, J.D. Nonlinear Coupled Seismic Sliding Analysis of Earth Structures. *J. Geotech. Geoenviron. Eng.* **2000**, *126*, 1002–1014. [CrossRef]
14. Bray, J.D.; Travasarou, T. Simplified Procedure for Estimating Earthquake-Induced Deviatoric Slope Displacements. *J. Geotech. Geoenviron. Eng.* **2007**, *133*, 381–392. [CrossRef]
15. Rathje, E.M.; Antonakos, G. A Unified Model for Predicting Earthquake-Induced Sliding Displacements of Rigid and Flexible Slopes. *Eng. Geol.* **2011**, *122*, 51–60. [CrossRef]
16. Zhang, Y.; Xiang, C.; Chen, Y.; Cheng, Q.; Xiao, L.; Yu, P.; Chang, Z. Permanent Displacement Models of Earthquake-Induced Landslides Considering near-Fault Pulse-like Ground Motions. *J. Mt. Sci.* **2019**, *16*, 1244–1257. [CrossRef]
17. Zhang, Y.; Xiang, C.; Yu, P.; Zhao, L.; Zhao, J.X.; Fu, H. Investigation of Permanent Displacements of Near-Fault Seismic Slopes by a General Sliding Block Model. *Landslides* **2022**, *19*, 187–197. [CrossRef]
18. Jibson, R.W.; Harp, E.L.; Michael, J.A. A Method for Producing Digital Probabilistic Seismic Landslide Hazard Maps; an Example from the Los Angeles, California, Area. *Open-File Rep.* **1998**, 98–113. [CrossRef]
19. Jibson, R.W.; Harp, E.L.; Michael, J.A. A Method for Producing Digital Probabilistic Seismic Landslide Hazard Maps. *Eng. Geol.* **2000**, *58*, 271–289. [CrossRef]
20. Yuan, R.; Deng, Q.; Cunningham, D.; Han, Z.; Zhang, D.; Zhang, B. Newmark Displacement Model for Landslides Induced by the 2013 Ms 7.0 Lushan Earthquake, China. *Front. Earth Sci.* **2016**, *10*, 740–750. [CrossRef]
21. Ma, S.; Xu, C. Assessment of Co-Seismic Landslide Hazard Using the Newmark Model and Statistical Analyses: A Case Study of the 2013 Lushan, China, Mw6.6 Earthquake. *Nat. Hazards* **2019**, *96*, 389–412. [CrossRef]
22. Jin, K.P.; Yao, L.K.; Cheng, Q.G.; Xing, A.G. Seismic Landslides Hazard Zoning Based on the Modified Newmark Model: A Case Study from the Lushan Earthquake, China. *Nat. Hazards* **2019**, *99*, 493–509. [CrossRef]
23. Ma, S.; Xu, C.; Shao, X. Spatial Prediction Strategy for Landslides Triggered by Large Earthquakes Oriented to Emergency Response, Mid-Term Resettlement and Later Reconstruction. *Int. J. Disaster Risk Reduct.* **2020**, *43*, 101362. [CrossRef]
24. Gu, D. *Engineering Geomechanics of Rock Mass*; Science Press: Beijing, China, 1979.
25. Hoek, E.; Bray, J.D. *Rock Slope Engineering*, 3rd ed.; Taylor & Francis: Abingdon, UK, 1981.

26. Harp, E.L.; Jibson, R.W. Landslides Triggered by the 1994 Northridge, California, Earthquake. *Bull. Seismol. Soc. Am.* **1996**, *86*, S319–S332. [CrossRef]
27. Khazai, B.; Sitar, N. Evaluation of Factors Controlling Earthquake-Induced Landslides Caused by Chi-Chi Earthquake and Comparison with the Northridge and Loma Prieta Events. *Eng. Geol.* **2004**, *71*, 79–95. [CrossRef]
28. Dai, F.C.; Xu, C.; Yao, X.; Xu, L.; Tu, X.B.; Gong, Q.M. Spatial Distribution of Landslides Triggered by the 2008 Ms 8.0 Wenchuan Earthquake, China. *J. Asian Earth Sci.* **2011**, *40*, 883–895. [CrossRef]
29. Tang, C.; Ma, G.; Chang, M.; Li, W.; Zhang, D.; Jia, T.; Zhou, Z. Landslides Triggered by the 20 April 2013 Lushan Earthquake, Sichuan Province, China. *Eng. Geol.* **2015**, *187*, 45–55. [CrossRef]
30. Qi, S.; Yan, C.; Liu, C. Two Typical Types of Earthquake Triggered Landslides and Their Mechanisms. In Proceedings of the 11th International and 2nd North American Symposium on Landslides, Banff; CA, USA, 2–8 June 2011; Eberhardt, E., Froese, C., Turner, K., Leroueil, S., Eds.; CRC Press: London, UK, 2012; pp. 1819–1823.
31. Zang, M.; Qi, S.; Zou, Y.; Sheng, Z.; Zamora, B.S. An Improved Method of Newmark Analysis for Mapping Hazards of Coseismic Landslides. *Nat. Hazards Earth Syst. Sci.* **2020**, *20*, 713–726. [CrossRef]
32. Barton, N. Review of a New Shear-Strength Criterion for Rock Joints. *Eng. Geol.* **1973**, *7*, 287–332. [CrossRef]
33. Xu, X.-W.; Jiang, G.-Y.; Yu, G.-H.; Wu, X.-Y.; Zhang, J.-G.; Li, X. Discussion on Seismogenic Fault of the Ludian Ms6.5 Earthquake and Its Tectonic Attribution. *Chin. J. Geophys.* **2014**, *57*, 3060–3068.
34. Qi, S.; Xu, Q.; Lan, H.; Zhang, B.; Liu, J. Spatial Distribution Analysis of Landslides Triggered by 2008.5.12 Wenchuan Earthquake, China. *Eng. Geol.* **2010**, *116*, 95–108. [CrossRef]
35. Xu, X.; Wen, X.; Han, Z.; Chen, G.; Li, C.; Zheng, W.; Zhnag, S.; Ren, Z.; Xu, C.; Tan, X.; et al. Lushan MS7.0 Earthquake: A Blind Reserve-Fault Event. *Chin. Sci. Bull.* **2013**, *58*, 3437–3443. [CrossRef]
36. Han, Z.; Ren, Z.; Wang, H.; Wang, M. The Surface Rupture Signs of the Lushan "4.20" Ms 7.0 Earthquake at Longmen Township, Lushan County and Its Discussion. *Seismol. Geol.* **2013**, *35*, 388–397.
37. Chen, X.L.; Yu, L.; Wang, M.M.; Lin, C.X.; Liu, C.G.; Li, J.Y. Brief Communication: Landslides Triggered by the Ms = 7.0 Lushan Earthquake, China. *Nat. Hazards Earth Syst. Sci.* **2014**, *14*, 1257–1267. [CrossRef]
38. Xu, C.; Xu, X.; Zheng, W.; Wei, Z.; Tan, X.; Han, Z.; Li, C.; Liang, M.; Li, Z.; Wang, H.; et al. Landslides Triggered by the April 20, 2013 Lushan, Sichuan Province Ms 7.0 Strong Earthquake of China. *Seismol. Geol.* **2013**, *35*, 641–660.
39. Xu, C.; Xu, X.; Shyu, J.B.H. Database and Spatial Distribution of Landslides Triggered by the Lushan, China Mw 6.6 Earthquake of 20 April 2013. *Geomorphology* **2015**, *248*, 77–92. [CrossRef]
40. Barton, N.; Bandis, S. Effects of Block Size on the Shear Behavior of Jointed Rock. In Proceedings of the 23rd U.S Symposium on Rock Mechanics (USRMS), Berkeley, CA, USA, 25–27 August 1982; 1982; pp. 739–760.
41. Ambraseys, N.N.; Menu, J.M. Earthquake-Induced Ground Displacements. *Earthq. Eng. Struct. Dyn.* **1988**, *16*, 985–1006. [CrossRef]
42. Jibson, R.W. Regression Models for Estimating Coseismic Landslide Displacement. *Eng. Geol.* **2007**, *91*, 209–218. [CrossRef]
43. Saygili, G.; Rathje, E.M. Empirical Predictive Models for Earthquake-Induced Sliding Displacements of Slopes. *J. Geotech. Geoenviron. Eng.* **2008**, *134*, 790–803. [CrossRef]
44. Rathje, E.M.; Saygili, G. Probabilistic Assessment of Earthquake-Induced Sliding Displacements of Natural Slopes. *Bull. N. Z. Soc. Earthq. Eng.* **2009**, *42*, 18–27. [CrossRef]
45. Hsieh, S.-Y.; Lee, C.-T. Empirical Estimation of the Newmark Displacement from the Arias Intensity and Critical Acceleration. *Eng. Geol.* **2011**, *122*, 34–42. [CrossRef]
46. Shortliffe, E.H.; Buchanan, B.G. A Model of Inexact Reasoning in Medicine. *Math. Biosci.* **1975**, *23*, 351–379. [CrossRef]
47. Heckerman, D. Probabilistic Interpretations for Mycin's Certainty Factors. In *Machine Intelligence and Pattern Recognition: Uncertainty in Artificial Intelligence*; Kanal, L.N., Lemmer, J.F., Eds.; Elsevier Science Publishers B.V.: Amsterdam, The Netherlands, 1986; Volume 4, pp. 167–196. ISBN 0923-0459.
48. Burrough, P.A.; McDonnell, R.A. *Principles of Geographical Information Systems*, 2nd ed.; Oxford University Press: Oxford, UK, 1998; ISBN 0198322663.
49. Bandis, S.C.; Lumsden, A.C.; Barton, N.R. Fundamentals of Rock Joint Deformation. *Int. J. Rock Mech. Min. Sci. Geomech. Abstr.* **1983**, *20*, 249–268. [CrossRef]
50. Singh, T.N.; Kainthola, A.; Venkatesh, A. Correlation Between Point Load Index and Uniaxial Compressive Strength for Different Rock Types. *Rock Mech. Rock Eng.* **2012**, *45*, 259–264. [CrossRef]
51. Giusepone, F.; Da Silva, L.A.A. *Hoek & Brown and Barton & Bandis Criteria Applied to a Planar Sliding at a Dolomite Mine in Gandarela Synclinal*; International Society for Rock Mechanics and Rock Engineering: Goiania, Brazil, 2014.
52. Yong, R.; Ye, J.; Liang, Q.-F.; Huang, M.; Du, S.-G. Estimation of the Joint Roughness Coefficient (JRC) of Rock Joints by Vector Similarity Measures. *Bull. Eng. Geol. Environ.* **2018**, *77*, 735–749. [CrossRef]
53. Sirkiä, J.; Kallio, P.; Iakovlev, D.; Uotinen, L. Photogrammetric Calculation of JRC for Rock Slope Support Design. In *Proceedings of the Eighth International Symposium on Ground Support in Mining and Underground Construction*; Nordlund, E., Jones, T., Eitzenberger, A., Eds.; Luleå University of Technology: Luleå, Sweden, 2016; p. 13.
54. Bao, H.; Zhang, G.; Lan, H.; Yan, C.; Xu, J.; Xu, W. Geometrical Heterogeneity of the Joint Roughness Coefficient Revealed by 3D Laser Scanning. *Eng. Geol.* **2020**, *265*, 105415. [CrossRef]

															

55. Arzúa, J.; Alejano, L.R.; Clérigo, I.; Pons, B.; Méndez, F.; Prada, F. Stability Analysis of a Room & Pillar Hematite Mine and Techniques to Manage Local Instability Problems. In *Proceedings of the Rock Engineering and Rock Mechanics: Structures in and on Rock Masses*; Alejano, R., Perucho, Á., Olalla, C., Jiménez, R., Eds.; CRC Press: Boca Raton, FL, USA, 2014; pp. 1147–1152.
56. Andrade, P.S.; Saraiva, A.A. Estimating the Joint Roughness Coefficient of Discontinuities Found in Metamorphic Rocks. *Bull. Eng. Geol. Environ.* **2008**, *67*, 425–434. [CrossRef]
57. Marques, E.A.G.; Williams, D.J.; Assis, I.R.; Leão, M.F. Effects of Weathering on Characteristics of Rocks in a Subtropical Climate: Weathering Morphology, in Situ, Laboratory and Mineralogical Characterization. *Environ. Earth Sci.* **2017**, *76*, 602. [CrossRef]
58. Srivastava, L.P. Probabilistic Analysis of Borehole Data for Evaluation of Engineering Properties of Rock Mass. In *Proceedings of the Geotechnical Characterization and Modelling*; Latha, G.M., Rao, P.R., Eds.; Springer: Singapore, 2020; pp. 503–510.
59. Coulson, J.H. *Shear Strength of Flat Surfaces in Rock*; Cording, E.J., Ed.; American Society of Civil Engineers: Urbana, IL, USA, 1972; pp. 77–105.
60. Priest, S.D. *Discontinuity Analysis for Rock Engineering*; Chapman & Hall: London, UK, 1993; ISBN 0-412-47600-2.
61. Tang, Z.C.; Zhang, Q.Z.; Peng, J.; Jiao, Y.Y. Experimental Study on the Water-Weakening Shear Behaviors of Sandstone Joints Collected from the Middle Region of Yunnan Province, P.R. China. *Eng. Geol.* **2019**, *258*, 105161. [CrossRef]
62. Tang, J.-Z.; Yang, S.-Q.; Elsworth, D.; Tao, Y. Three-Dimensional Numerical Modeling of Grain-Scale Mechanical Behavior of Sandstone Containing an Inclined Rough Joint. *Rock Mech. Rock Eng.* **2021**, *54*, 905–919. [CrossRef]
63. Barton, N.; Choubey, V. The Shear Strength of Rock Joints in Theory and Practice. *Rock Mech. Felsmech. Mcanique Roches* **1977**, *10*, 1–54. [CrossRef]
64. Bilgin, H.A.; Pasamehmetoğlu, A.G. *Shear Behaviour of Shale Joints under Heat in Direct Shear*; Barton, N., Stephansson, O., Eds.; CRC Press: Loen, Norway, 1990; pp. 179–183.
65. Hashemi, S.S.; Zoback, M.D. Permeability Evolution of Fractures in Shale in the Presence of Supercritical CO_2. *J. Geophys. Res. Solid Earth* **2021**, *126*, e2021JB022266. [CrossRef]
66. Qu, G.; Shi, T.; Zhang, Z.; Su, J.; Wei, H.; Guo, R.; Peng, J.; Zhao, K. Characteristics Description of Shale Fracture Surface Morphology: A Case Study of Shale Samples from Barnett Shale. *Processes* **2022**, *10*, 401. [CrossRef]
67. Geological Engineering Handbook Editorial Committee. *Geological Engineering Handbook*, 5th ed.; China Architecture & Building Press: Beijing, China, 2018.
68. Xu, C.; Xu, X.; Shyu, J.B.H.; Gao, M.; Tan, X.; Ran, Y.; Zheng, W. Landslides Triggered by the 20 April 2013 Lushan, China, Mw 6.6 Earthquake from Field Investigations and Preliminary Analyses. *Landslides* **2015**, *12*, 365–385. [CrossRef]
69. Weibull, W. *A Statistical Theory of the Strength of Materials*; Generalstabens Litografiska Anstalts Förlag: Stockholm, Sweden, 1939.
70. Miles, S.B.; Keefer, D.K. Evaluation of CAMEL—Comprehensive Areal Model of Earthquake-Induced Landslides. *Eng. Geol.* **2009**, *104*, 1–15. [CrossRef]

sustainability

Article

Probabilistic Approach to Transient Unsaturated Slope Stability Associated with Precipitation Event

Katherin Rocio Cano Bezerra da Costa, Ana Paola do Nascimento Dantas, André Luís Brasil Cavalcante * and André Pacheco de Assis

Department of Civil and Environmental Engineering, University of Brasília, Brasília 70910-900, DF, Brazil; katherincanodacosta@gmail.com (K.R.C.B.d.C.); anapaolans@gmail.com (A.P.d.N.D.); aassis.p@gmail.com (A.P.d.A.)
* Correspondence: abrasil@unb.br

Abstract: The massif rupture is not always reached under saturated conditions; therefore, the analysis of the unsaturated phenomenon is necessary in some cases. This study performed a probabilistic approach for unsaturated and transient conditions to understand the contribution of physical and hydraulic parameters involved in slope stability. The proposed slope stability model was based on the infinite slope method and a new unsaturated constitutive shear strength model proposed in 2021 by Cavalcante and Mascarenhas. The first-order second-moment method, which incorporated multiple stochastic variables, was used in the probabilistic analysis, allowing the incorporation of seven independent variables for the probability of failure analysis as well as for quantifying the contribution of the variables to the total variance of a factor of safety at any state of moisture. This implementation allows a more realistic estimative for the probability of failure, showing in a practical way the decrease and increase of the probability of failure during a rain event. The model provided promising results highlighting the need to migrate from deterministic analyses to more robust probabilistic analyses, considering the most significant number of stochastic variables. The proposed model helps to understand the influence of moisture content on slope stability, being a possible tool in natural disaster risk management.

Keywords: slope stability; probability of failure; unsaturated soils

Citation: Costa, K.R.C.B.d.; Dantas, A.P.d.N.; Cavalcante, A.L.B.; Assis, A.P.d. Probabilistic Approach to Transient Unsaturated Slope Stability Associated with Precipitation Event. *Sustainability* **2023**, *15*, 15260. https://doi.org/10.3390/su152115260

Academic Editors: Chong Xu and Jian Chen

Received: 9 August 2023
Revised: 25 September 2023
Accepted: 25 September 2023
Published: 25 October 2023

1. Introduction

The considerable increase in mass movements increased the susceptibility of tropical mountainous regions to landslides owing to the high degree of soil weathering combined with long stationary periods of rain. Along with increased population and expansion to mountainous areas, these natural conditions directly affect the communities through human and economic losses [1].

During rain events, water precipitates to the surface of massifs, infiltrating the slope and negatively affecting their stability. This process often triggers superficial mass movements identified as translational or planar movements in geotechnics. The stability of a slope or embankment is quantified by the factor of safety (FS), which compares the stresses at work with the resistant strains of the massif. In this case, instability is represented when FS is less than or equal to 1.

In geotechnical engineering, FS can be quantified using several models, namely conceptual, numerical, and analytical models, that are developed according to the geometric, topographic, geotechnical, and hydraulic characteristics of the slope. However, in tropical regions, where the unsaturated state of massifs predominates, understanding mass movements due to the negative pore-water pressure response should be improved [2].

Various authors have developed or applied specific numerical solutions under adjusted conditions and simplifications for each case of stability study, highlighting the need for closed mathematical solutions for understanding the unsaturated flow that would facilitate the computational implementation, the validation of numerical models, and a better representation of the physical phenomenon [2–9].

In the context of unsaturated slope stability, the flow analysis in an unsaturated porous medium and its respective attributes, such as soil water retention curve (SWRC) and unsaturated hydraulic conductivity function of the medium (k-function), should be incorporated.

The conventional deterministic FS quantification is not a reliable indicator of the performance of a slope because the physical and mechanical properties of the soils are inherently associated with the natural uncertainties of the environment. This situation is more evident in the unsaturated state owing to the changes in the uncertainties associated with moisture and suction, which directly influence the stability response. Therefore, probabilistic models should be used to understand the transient conditions that lead to slope faults, considering the statistical dispersion of physical, mechanical, hydraulic, and unsaturated material attributes for a more realistic failure indication [10].

This study evaluated the importance of the probabilistic approach in the unsaturated and transient slope stability analysis. The evaluation methodology proposed by Rojas [11], which implemented the analytical model of infiltration defined by Cavalcante and Zornberg in 2017 [12], and analytical model of unsaturated slope stability defined by Cavalcante and Camapum de Carvalho [13], which considered the constitutive model to evaluate the unsaturated shear strength defined by Cavalcante and Mascarenhas [14], was considered in this work.

It is known that the increasing climatic variability and complex geological conditions increase the limitations inherent in deterministic methods for assessing slope stability. Traditional deterministic approaches often rely on fixed values for soil properties and environmental conditions, overlooking the inherent uncertainties and spatial-temporal variations. This study adopts a probabilistic approach to assess transient unsaturated slope stability, particularly under precipitation events. By incorporating stochastic variables and utilizing the first-order second-moment method (FOSM), our model allows for a more nuanced, risk-informed assessment of slope stability. This approach is critical for providing more realistic estimates in situations where soil properties and hydraulic conditions are not uniformly distributed and are subject to change over time. Thus, the probabilistic framework introduced here offers an improved tool for natural disaster risk management and is especially beneficial in engineering applications where uncertainties are often inevitable.

Unlike conventional geotechnical practices, the proposed methodology considered all independent variables associated with the physical stability problem as continuous random variables. Owing to the high computational cost, random variables were limited. The analyses were performed for one rainfall intensity to quantify the slope performance index and contribution of each physical parameter in the stability. The developed computational methods were implemented using Wolfram Mathematica.

2. Unsaturated Slope Stability

The behavior of materials depends directly on the interaction between the grains; thus, slope stability analysis is usually carried out in the state of effective stresses. However, as shown by Taha et al. [15] and Nishimura and Fredlund [16], the negative pore-water pressure in the unsaturated soil regime, which indicates the amount of water in the voids, directly influences the interaction of the grains, as demonstrated in the significant increases in the shear strength associated with the rise of the matrix suction in some materials. Meanwhile, the potential failure surface is not always conditioned to the saturated state of the soil, considering that the balance between the resisting and acting forces is broken under negative pore-water pressures in some situations, thereby triggering the failure in the unsaturated condition [17]. Thus, the stability of unsaturated slopes under precipitation events should be evaluated to analyze the flow phenomenon, as assessed by the Richards equation [18], and the phenomenon of stresses in the unsaturated porous medium, as validated by the principle of effective strains of Terzaghi in the unsaturated condition [19].

2.1. Unsaturated Flow in Porous Medium

Buckingham [20] incorporated the concept of negative pore pressure into the flow, thereby developing the Darcy–Buckingham law, which derives the fluid velocity in an unsaturated porous medium as follows:

$$v = -\frac{k_s(\psi)}{g}\frac{\partial \Phi}{\partial L} \tag{1}$$

where $k_s(\psi)$ is the function of the unsaturated hydraulic conductivity (LT^{-1}), $\partial \Phi / \partial L$ is the rate of change of potential energy at the extent of flow motion (L), and g is the acceleration due to gravity (LT^{-2})

Subsequently, Richards defined the continuity equation for three-dimensional flow in a porous medium, assuming an incompressible fluid without volume change in the soil mass throughout the flow process (i.e., constant ρ_w) [18]. Thus, Richards combined the continuity equation with the Darcy–Buckingham law (Equation (1)), thereby establishing a partial differential equation for three-dimensional flow in an unsaturated porous medium, as follows:

$$\frac{\partial \theta}{\partial t} = \frac{\partial}{\partial x}\left(\frac{k_x(\psi)}{g\rho_w}\frac{\partial \psi}{\partial x}\right) + \frac{\partial}{\partial y}\left(\frac{k_y(\psi)}{g\rho_w}\frac{\partial \psi}{\partial y}\right) + \frac{\partial}{\partial z}\left(k_z(\psi)\left(\frac{\partial \psi}{\partial y}\frac{1}{g\rho_w} - 1\right)\right) \tag{2}$$

The solution of Equation (2) depends on two constitutive relations: (1) the SWRC, which relates the soil volumetric moisture content with the matrix suction potential, and (2) the k-function, which relates the soil hydraulic conductivity with the matrix suction potential. In practice, the most commonly used constitutive models for obtaining unsaturated soil attributes are those obtained by Brooks and Corey [21], Van Genuchten [22], and Fredlund and Xing [23]. These models established highly nonlinear relationships, resulting in the high complexity in obtaining an analytical solution to Richards' equation under transient conditions [24].

Cavalcante and Zornberg [12,25] proposed new constitutive models to represent the SWRC and k-function, thereby obtaining four analytical solutions for the one-dimensional unsaturated flow problem represented by Richards' partial differential equation in the vertical direction. The constitutive models of the SWRC and k-function are defined in Equations (3) and (4), respectively:

$$\psi(\theta) = \frac{1}{\delta}\ln\left(\frac{\theta - \theta_r}{\theta_s - \theta_r}\right) \tag{3}$$

$$k_z(\theta) = k_s\left(\frac{\theta - \theta_r}{\theta_s - \theta_r}\right) \tag{4}$$

where θ is the volumetric soil moisture content (L^3L^{-3}), θ_s is the saturated volumetric soil water content (L^3L^{-3}), θ_r is the residual volumetric soil water content (L^3L^{-3}), $\theta_s - \theta_r$ is the maximum soil wetting capacity (L^3L^{-3}), k_s is the saturated hydraulic conductivity of the soil (LT^{-1}), and δ is the hydraulic fitting parameter of Cavalcante and Zornberg [12] $(M^{-1}LT^2)$.

In the work of Costa and Cavalcante [26], the hydraulic fitting parameter has a quantitative physical meaning related to the entry of the air pressure into the soil pore matrix. In particular, the hydraulic fitting parameter is higher for sandy soils and lower for finer soils. Subsequently, Costa and Cavalcante [27] developed a constitutive model for obtaining the SWRC and k-function of bimodal soils, represented by Equations (5) and (6), respectively:

$$\theta(\psi) = \theta_r + (\theta_s - \theta_r)[\lambda \exp(-\delta_1|\psi|) + (1 - \lambda)\exp(-\delta_2|\psi|)] \tag{5}$$

$$k(\psi) = k_s[\lambda \exp(-\delta_1|\psi|) + (1-\lambda)\exp(-\delta_2|\psi|)] \tag{6}$$

where δ_1 is the hydraulic fitting parameter of the macroporous region ($M^{-1}LT^2$), δ_2 is the hydraulic fitting parameter of the microporous region ($M^{-1}LT^2$), and λ is the weight factor corresponding to the macroporous region.

The flow phenomenon associated with rainfall depends directly on the storage capacity of the soil on the ground surface. This region determines the amount of water available to initiate the percolation process within the massif. These characteristics are well represented by the analytical solution proposed by Cavalcante and Zornberg [12]. In particular, the authors implemented the assumptions adopted in developing the column test, considering a one-dimensional flow modeling, approximately constant soil porosity throughout the test, and a soil column with a finite compression. The initial condition of the problem was defined by a uniform moisture content throughout the domain with the assumptions of an impermeable region for the lower boundary and constant discharge velocity of the upper boundary, thereby determining the transient surface storage capacity or infiltration rate. This maximum discharge velocity is:

$$v_{i,max} = \frac{\theta_s k_s}{(\theta_s - \theta_r)} \tag{7}$$

Thus, the analytical solution that represents the moisture fronts at any depth of the unsaturated transient flow of a slope is:

$$\theta(z,t) = \theta_i + \left[\frac{v_i}{k_s}(\theta_s - \theta_r) - \theta_i\right]D(z,t) \tag{8}$$

where v_i is the infiltration rate (LT^{-1}), θ_i is the initial volumetric soil water content (L^3L^{-3}), and $D(z,t)$ is the auxiliary function of the analytical solution reported by Cavalcante and Zornberg [12], and represented as follows:

$$
\begin{aligned}
D(z,t) = & \tfrac{1}{2}erfc(Z_{-1}) + \sqrt{\frac{\overline{a}_s^2 t}{\pi \overline{D}_z}}\exp\left(-\frac{(z-\overline{a}_s t)^2}{4\overline{D}_z t}\right) - \tfrac{1}{2}\left(-1 + \frac{\overline{a}_s z}{\overline{D}_z} + \frac{\overline{a}_s^2 t}{\overline{D}_z}\right) \\
& \times \exp\left(\frac{\overline{a}_s z}{\overline{D}_z}\right)erfc(Z_{+1}) + \sqrt{4\frac{\overline{a}_s^2 t}{\pi \overline{D}_z}}\left(1 + \frac{\overline{a}_s}{4\overline{D}_z}(2L - z + \overline{a}_s t)\right) \\
& \times \exp\left(\frac{\overline{a}_s L}{\overline{D}_z} - \frac{1}{4\overline{D}_z t}(2L - z + \overline{a}_s t)^2\right) \\
& - \frac{\overline{a}_s}{\overline{D}_z}\left(2L - z + \frac{3\overline{a}_s t}{2} + \frac{\overline{a}_s}{4\overline{D}_z}(2L - z + \overline{a}_s t)\right) \\
& \times \exp\left(-\frac{\overline{a}_s L}{\overline{D}_z}\right)erfc\left(\frac{(2L - z + \overline{a}_s t)}{2\sqrt{\overline{D}_z t}}\right)
\end{aligned}
\tag{9}
$$

The auxiliary terms that constitute $D(z,t)$ are given by:

$$Z_{\pm 1} = \frac{z \pm \overline{a}_s t}{2\sqrt{\overline{D}_z t}} \tag{10}$$

$$\overline{a}_s = \frac{k_s}{(\theta_s - \theta_r)} \tag{11}$$

$$\overline{D}_z = \frac{k_s}{\delta(\theta_s - \theta_r)\rho_w g} \tag{12}$$

where $\overline{a}_s$ is the unsaturated advective seepage velocity (LT^{-1}) and $\overline{D}_z$ is the unsaturated water diffusivity (L^2T^{-1}).

2.2. Soil-Atmosphere Interaction

In the interaction of the slope surface with the atmosphere, it is essential to highlight the function of hydrogeology, which studies the spatial and temporal dynamics of water inside the massifs to evaluate geotechnical problems. In addition, it covers a closed hydrological cycle, i.e., the phenomena of distributing water in the atmosphere, soil surface, and subsoil [28]. Thus, for surface landslides triggered by rains, the concept of the hydrological cycle is simplified to the physical phenomena of precipitation, runoff, evapotranspiration, interception and infiltration, given by:

$$i(t) = e_s(t) + v_i(t) + e_v(t) + I(t) \tag{13}$$

where $i(t)$ is the precipitation intensity (LT^{-1}), $v_i(t)$ is the infiltration rate (LT^{-1}), $e_s(t)$ is the runoff velocity (LT^{-1}), $I(t)$ is the interception by the vegetation (LT^{-1}), and $e_v(t)$ is the evapotranspiration velocity (LT^{-1}).

In practice, evapotranspiration is the most challenging parameter to quantify. The increase in evapotranspiration at the slope surface produces positive effects on the stability of the surface layers, thereby requiring more significant amounts of water to reach instability [29,30]. Similarly, interception modeling is complex as it depends directly on the type of vegetation on the slope surface. Increased rainwater interception by vegetation decreases the impact of precipitation on the ground surface and reduces the water available to infiltrate the soil [31]. Thus, a more conservative analysis that disregards the action of evapotranspiration and the interception can be obtained by defining the hydro–geotechnical cycle for the study of slope stability associated with precipitation events as:

$$i(t) = e_s(t) + v_i(t) \tag{14}$$

In classical hydrology, the rational method determines the runoff, an accepted theoretical approach for determining runoff flows that contribute to a hydrographic basin [32]. This methodology depends on the runoff coefficient c, which is defined as a function of topography, land use, and slope surface cover, represented by:

$$Q(t) = c \cdot i(t) \cdot A \tag{15}$$

where $Q(t)$ is the flow associated with runoff (L^3T^{-1}), A is the area of the contribution basin (L^2), and c is the runoff coefficient (dimensionless).

It is important to note that the rational method has a limitation regarding the precise mapping data of the runoff coefficient c. Considering the rational method for calculating the instantaneous flow of the runoff associated with the topography of the slope surface and maximum wetting capacity defined by Cavalcante and Zornberg [12] ($v_{i,max}$), Rojas described the runoff velocity as [11]:

$$e_s(t) = \begin{cases} c \cdot i(t) & \text{if } v_i \leq v_{i,max} \\ (i(t) - v_{i,mx}) + c \cdot i(t) & \text{if } v_i > v_{i,max} \end{cases} \tag{16}$$

Thus, by applying Equation (16) to Equation (14), the infiltration rate as a function of the precipitated water and amount of runoff water can be quantified as:

$$v_i(t) = i(t) - e_s(t) \tag{17}$$

2.3. Unsaturated Shear Strength

Determining the soil stress state is based on constitutive models and functions that correlate the variables involved in normal stresses, shear stresses, and deformations by selecting the changes in soil stress and strain. In the unsaturated state, these relationships were established as a function of the principle of effective stresses of Terzaghi [19], where the total normal stress is rewritten as the sum of the pore pressure and effective soil pressure.

The Mohr-Coulomb model was extended to the unsaturated medium to understand the mechanical behavior of soils in this state, where the effective normal stress is defined as the combination of two variables independent of the stress state of the mass incorporating the air. The most commonly used constitutive relationships in practice are those of Croney et al. [33], Bishop [34], and Fredlund et al. [35]. Among these, Bishop's model and Fredlund's model are the most widely used in practice and are given, respectively, by:

$$\tau = c' + (\sigma - u_a) \tan \phi' + \chi (u_a - u_w) \tan \phi' \tag{18}$$

$$\tau = c' + (\sigma - u_a) \tan \phi' + (u_a - u_w) \tan \phi^b \tag{19}$$

where c' is the effective cohesion of the soil ($\mathrm{ML^{-1}T^{-2}}$), ϕ' is the effective friction angle of the soil (°), ϕ^b is the angle of friction of the soil relative to the matrix suction (°), σ is the normal total stress applied to the massif ($\mathrm{ML^{-1}T^{-2}}$), τ is the shearing stress acting on the slope ($\mathrm{ML^{-1}T^{-2}}$), u_w is the pore-water pressure in the soil ($\mathrm{ML^{-1}T^{-2}}$), and χ is the unsaturated effective stress parameter or Bishop's parameter (non-dimensional).

Equations (18) and (19) can be essentially treated as equivalent with the assumption of the following equality [14,36]:

$$\chi \tan \phi' = \tan \phi^b \tag{20}$$

Considering the thermodynamic arguments proposed by Lu et al. to express χ as function of the degree of saturation [37], Cavalcante and Mascarenhas [14] developed a constitutive model for the mechanical behavior of unsaturated soils, which incorporated the constitutive model of Cavalcante and Zornberg for the SWRC as showed at the Equation (3) [12]. Thus, an analytical expression as a function of a single fitting hydraulic parameter δ is defined as:

$$\chi = e^{-\delta |u_a - u_w|} \tag{21}$$

Replacing Equation (21) into Equation (20) and then into Equation (19), the authors defined a new constitutive model for the unsaturated shear strength as:

$$\tau = c' + \left[(\sigma - u_a) + e^{-\delta |u_a - u_w|} \cdot (u_a - u_w) \right] \cdot \tan \phi' \tag{22}$$

From Equation (22), the unsaturated cohesion or total cohesion of soil is represented as:

$$c_{\mathrm{unsat}} = c' + e^{-\delta |u_a - u_w|} \cdot (u_a - u_w) \cdot \tan \phi' \tag{23}$$

The infinite slope method is one of the most widely used methods in evaluating translational or planar mass movements, which generally represent the landslides associated with rainfalls [17]. This analytical method satisfies the static balance of forces involved in the massif and requires a few hypotheses for its implementation. Some of its premises are that the method considers the failure plane parallel to the surface (translational movement), and the slope length must be at least ten times greater than the soil thickness, disregarding the slope length in determining FS [38].

Thus, to represent the slope stability more realistically, Cavalcante and Camapum de Carvalho [13] rewrote the infinite slope model for the assessment of stability in the unsaturated condition by incorporating the analytical solution of Richards Equation for the one-dimensional infiltration proposed by Cavalcante and Zornberg [12]. Subsequently, the authors rewrote the constitutive model of shear strength defined by Fredlund et al. [35], assuming that the relative air pressure on the slope surface corresponds to atmospheric pressure ($u_a = 0$).

Following the methodology of Cavalcante and Camapum de Carvalho [13], the current work implemented the constitutive model to evaluate the unsaturated shear strength developed by Cavalcante and Mascarenhas [14]. The rationale for this choice stems from the model's comprehensive treatment of both physical and hydraulic parameters affecting slope stability in unsaturated conditions. Moreover, this model is a recent addition to the field, incorporating state-of-the-art advancements in unsaturated soil mechanics. Its applicability and efficacy have been validated in various studies, thereby adding a layer of credibility to our analytical framework. We recognize that multiple models exist for evaluating unsaturated slope stability; however, the Cavalcante and Zornberg [12] model offers a robust and well-validated approach that aligns with the complexities and uncertainties considered in our probabilistic analysis. Thus, it serves as a suitable foundation for the methodologies and analyses presented in this manuscript. Therefore, the slope safety factor, which is the ratio between the acting forces and the resistant forces of the slope, is defined by:

$$FS(z,t) = \frac{\tan \phi'}{\tan \beta} + \frac{c' - (u_a - u_w)(z,t) \cdot e^{-\delta \cdot |(u_w - u_w)(z,t)|} \cdot \tan \phi'}{(\gamma_d + \gamma_w \cdot \theta(z,t)) \cdot \cos(\beta) \cdot \sin(\beta) \cdot z} \tag{24}$$

where $(u_a - u_w)(z,t)$ is the negative pore-water pressure as a function of depth and time, which is the matric suction ($ML^{-1}T^{-2}$), γ_d is the specific weight of the grains ($MT^{-2}L^{-2}$), γ_w is the specific weight of water ($MT^{-2}L^{-2}$), and β is the slope angle (°).

Equation (24) represents an analytical, unsaturated, and transient model with physical meaning that considers the variations in soil density, the variations in negative pore-water pressure, and the variations of the effective cohesion as a function of the degree of saturation of the porous medium for the indication of the rupture moment.

3. Failure Probability

Disregarding the uncertainties involved in the slope stability analysis is a limitation the probabilistic approach should overcome. It incorporates the systematic analysis of the geotechnical parameters' uncertainties as continuous stochastic variables represented by a probability distribution. Consequently, the FS under specific conditions is also characterized by a probabilistic distribution instead of deterministic and constant values.

In assessing the probability of slope failure, the methods routinely applied are the Monte Carlo method, first-order second-moment method (FOSM), and point estimates method [10,13,31,39–44]. However, due to advances in technology, recent methodologies using artificial intelligence have been developed for predicting slope stability [45–49].

To evaluate the performance of the slope in the unsaturated condition and discover the contribution of each variable in the deterministic slope stability model (Equation (24)), the probabilistic FOSM method, which can be implemented for analytical models, was chosen by the authors. This approach requires only the statistical moments of the mean and standard deviation. Moreover, it can quantify each independent variable's contribution to the FS's total variance.

3.1. First-Order Second-Moment Method

FOSM is an approximate probabilistic method that uses the Taylor series to determine the moment values of the statistical distributions representing the variables involved in the analyses. To determine the dependent variable's probability distribution function (PDF), the method requires the mean values of the independent variables and derivatives of the independent variables as a function of the dependent variable [50].

The statistical moments obtained by FOSM for the construction of PDF are the mean $(E[x] = \overline{x})$ as a measure of the central tendency and variance $(V[x] = \sigma^2)$ as a measure of the variation or dispersion. When there is no correlation between the independent variables, these moments are determined as follows:

$$E[f(x)] = f(\overline{x}) \tag{25}$$

$$V[f(x)] = \sum \left(\frac{\partial f(x)}{\partial x_i} \right)^2 V(x_i) \tag{26}$$

As FOSM does not estimate the type of probability distribution of independent variables, there should be prior knowledge of the probability distribution that best represents the physical variables involved in the problem. In this case, the normal distribution, one of the most used distributions to model natural phenomena, efficiently represents the PDF of the physical and mechanical parameters of the soil involved in the slope stability analysis [51,52].

However, saturated hydraulic conductivity (k_s) is best represented by a lognormal distribution probability density function, as presented by several authors [31,39,53–56]. In this case, a simple transformation is used to represent the statistical moments of a lognormal distribution in terms of a normal distribution as:

$$E[k_s]_{normal} = \log(k_s) \tag{27}$$

$$V[k_s]_{normal} = \ln\left(\frac{V(k_s)}{k_s^2} + 1\right) \tag{28}$$

Considering $k_s = 10^{\log(k_s)}$, the effect of the hydraulic property on the failure probability analysis could be incorporated by rewriting Equation (8) as:

$$\theta(z,t) = \theta_i + \left[\frac{v_i}{10^{E[ks]_{normal}}}(\theta_s - \theta_r) - \theta_i\right]D(z,t) \tag{29}$$

3.2. Transient Failure Probability

In analyzing the stability of unsaturated slopes, the importance of obtaining the transient response of the pore water pressure should be highlighted to understand the physical and hydraulic conditions that lead to slope failure. Thus, the transient response of the probability of failure should be obtained.

In engineering practice, the application of FOSM allows the analysis of the reliability index of FS, defined by:

$$\beta' = \frac{E[FS] - FSc}{\sigma[FS]} \tag{30}$$

where β' is the reliability index of the FS (dimensionless) and FSc is the critical FS. The reliability index represents the number of standard deviations between FSc and FS based on the analysis. This theory was developed for a normal probability distribution, where a more significant β' suggests a lower probability of failure (P_f).

As presented by Husein Malkawi et al. [57], the reliability index of FS is directly related to the probability of failure under a normal distribution, as follows:

$$P_f(\beta') = P(FS \leq FSc) = 1 - \frac{1}{2}\text{erfc}\left(-\frac{\beta'}{\sqrt{2}}\right) \tag{31}$$

Knowing the analytical solution for the transient FS, presented in Equation (24), the total transient variance can be defined as:

$$V[FS(z,t)] = \sum\left(\frac{\partial FS(z,t)}{\partial x_i}\right)^2 V(x_i) \tag{32}$$

Replacing Equation (32) and Equation (24) into Equation (30), the slope transient reliability index is:

$$\beta'(t,z) = \frac{E[FS(t,z)] - FSc}{\sqrt{V[FS(t,z)]}} \tag{33}$$

Consequently, the transient failure probability for the unsaturated stability problem is:

$$P_f(t,z) = P(FS \leq FSc) = 1 - \frac{1}{2}\text{erfc}\left(-\frac{\beta'(t,z)}{\sqrt{2}}\right) \tag{34}$$

4. Case Study

In this study, the authors used the information from Rojas [11], where the infiltration and unsaturated stability analysis was implemented in the *La Arenosa* basin studied by Aristizábal et al. [58]. This basin is located in the mountainous region of Colombia, where multiple landslides were recorded associated with a precipitation event on 21 September 1991.

An unconditionally unstable slope was chosen for the analysis because of its geometry, as validated by Rojas [11], which clearly exhibited rupture during the high-intensity rain event. As presented in the research by Aristizabal et al. [58], the slope failure is translational, which aligns with the model proposed in this work.

The physical, mechanical, and hydraulic parameters of the slope and coefficient of variation (CoV) of the independent variables incorporated into the probability analysis were adopted from the doctoral theses of Gitirana Jr. [39] and are presented in Table 1.

Table 1. Deterministic and probabilistic parameters for analyzing an unsaturated slope's stability and failure probability.

Parameter Type	Parameter	Mean	CoV
Deterministic Parameters	β (°)	62	–
	δ (kPa^{-1})	0.00249	–
	v_i (ms^{-1})	5.73×10^{-6}	–
	c (-)	0.375	
Probabilistic Parameters	ϕ' (°)	24	9
	C' (kPa)	5	30
	γ_d (kNm^{-3})	18	5
	Log $[k_s]$ (ms^{-1})	Log $[5.4 \times 10^{-6}]$	80
	θ_r (m^{3}m^{-3})	0.026	10
	θ_s (m^{3}m^{-3})	0.43	13
	θ_i (m^{3}m^{-3})	0.05	10

The unsaturated soil attributes were first obtained from the bimodal fit to the SWRC model proposed by Costa and Cavalcante [27]. As shown in Figure 1, the soil, in this case study, exhibited an SWRC with a bimodal behavior defined by two hydraulic fitting parameters δ for the micropore and macropore regions. The unsaturated attributes of the exposed soil shown in Figure 2 were obtained from the constitutive models of Cavalcante and Zornberg [12] for the macropores region.

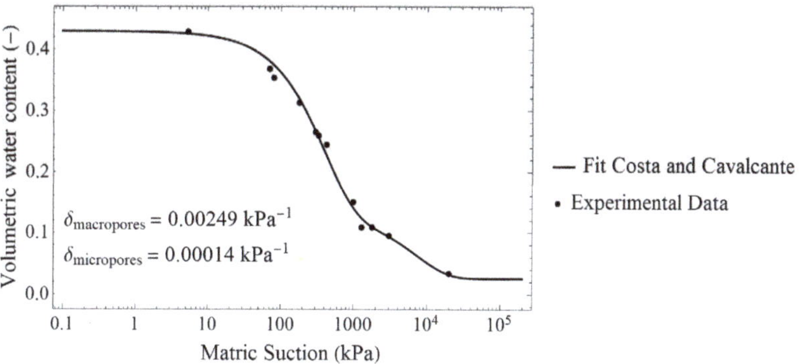

Figure 1. Fitting of the SWRC to a bimodal soil using the approach of Costa and Cavalcante [27].

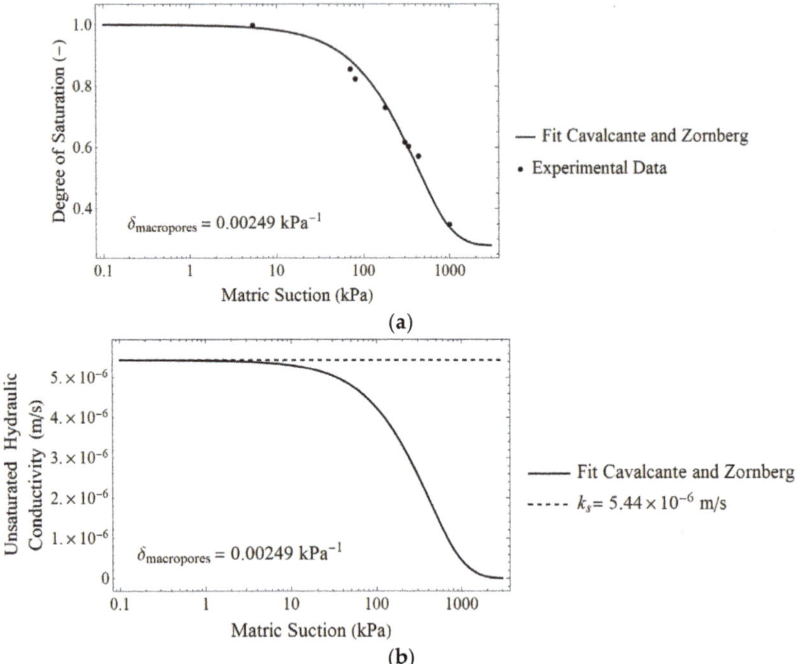

Figure 2. Unsaturated slope attributes: (**a**) SWRC and (**b**) unsaturated hydraulic conductivity function (Cavalcante and Zornberg [12]).

To understand the importance of the probabilistic approach, an event with a high precipitation intensity of 20 mm/h was analyzed. The hydro-geotechnical balance analysis obtained the infiltration rate value v_i using Equation (17).

4.1. Assessment of the Transient and Unsaturated Deterministic Stability

The saturation and slope stability fronts were obtained by applying the model of Cavalcante and Zornberg [12] and Equation (24), respectively. The analyses were conducted for four depths with a distance interval of 0.5 m.

Figure 3a presents the transient saturation profile of the slope during a rainfall of 20 mm/h. This analysis shows that the rainwater percolation inside the slope can fill 90% of voids at a soil depth of 50 cm in approximately 14 h. Thus, during a 50 h analysis, only the first three control depths (0.5, 1.0, and 1.5 m) achieved saturation levels.

Figure 3b presents the transient suction behavior, where the maximum suction generated in the residual soil moisture state is 1134 kPa, and the minimum suction associated with suturing is 6 kPa. As shown in Figure 3c, when the slope stability of the soil is at the residual moisture content (5%) at maximum suction and saturation (12%), the FS performance is lower, which is the opposite of the expected behavior at higher suction values for higher FS. This behavior confirms the promoted stability owing to the water content, as shown in Figure 3c, where the highest stability performance for each depth is given in the time range corresponding to the saturation values of 13–44%. This phenomenon can be explained by the non-linear shear strength enveloping the unsaturated porous media. The increased suction generated by smaller amounts of water in the soil can create the breakage of grains or particle aggregates, thereby impairing the shear strength of soil in the dry state [59].

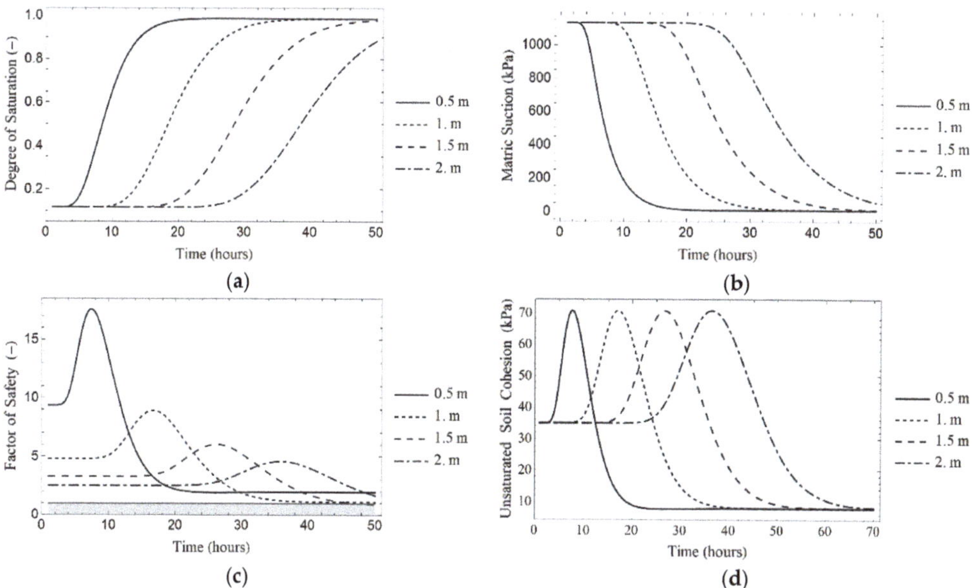

Figure 3. Transient and deterministic analysis during a rainfall of 20 mm/h at depths of 0.5, 1.0, 1.5, and 2.0 m: (**a**) degree of unsaturated saturation, (**b**) matric suction over time, (**c**) slope stability insurance factor under the unsaturated condition, and (**d**) unsaturated soil cohesion.

The transient behavior of unsaturated soil cohesion during a steady rainfall event could be determined by implementing Equation (8) into Equation (23) (Figure 3d). In the scenario with a rainfall of 20 mm/h, the unsaturated cohesion reaches 70 kPa for suction values of approximately 400 kPa and 7 kPa of cohesion for suction values of less than 6 kPa. As expected, slope rupture was inevitable under this scenario. After an analysis of 42 h, the instability, indicated by an FSc value equal to 1, is noted at a depth of 1 m. Consequently, after 2 h (analysis time of 44 h), rupture occurs at a depth of 1.5 m (Figure 3a) when the saturation value reaches 95%, thereby generating a suction of approximately 20 kPa.

4.2. Parametric Analysis of Stochastic Variables

The transient parametric analysis of the stochastic variables defined in Table 1 was performed to identify the response of the FS under different moisture states at a depth of 1 m and is presented in Figure 4.

In Figure 4a, the initial moisture content exhibited important variations for the first 18 h of the analysis, where the saturation varies between 12% and 80%. In Figure 4b, the residual moisture content greatly influences the FS in the same moisture and time range. This response indicates that the proper choice of the initial moisture condition imposed for slope modeling is critical in obtaining more realistic results. Meanwhile, the saturated hydraulic conductivity of the soil (Figure 4g) was shown to be more relevant after the first 25 h of analysis, which demonstrated safety at a lower k_s, a more impermeable soil.

The physical parameter in Figure 4d shows the effect of the specific density of the soil under an unsaturated state, which has higher FS for soils with lower densities. When the mass reaches a saturation of more than 80%, the soil density has lower interference in the FS, resulting in instability for any density and confirming the direct effect of the water in the soil matrix.

In Figure 4e,f, the effect of the soil's effective cohesion and friction angle is directly proportional to the FS, including rupture for cohesion values below 5 kPa and a friction angle of 20° at saturation values higher than 80%. Therefore, these analyses demonstrated the ability of the deterministic analysis to predict ruptures under unsaturated conditions.

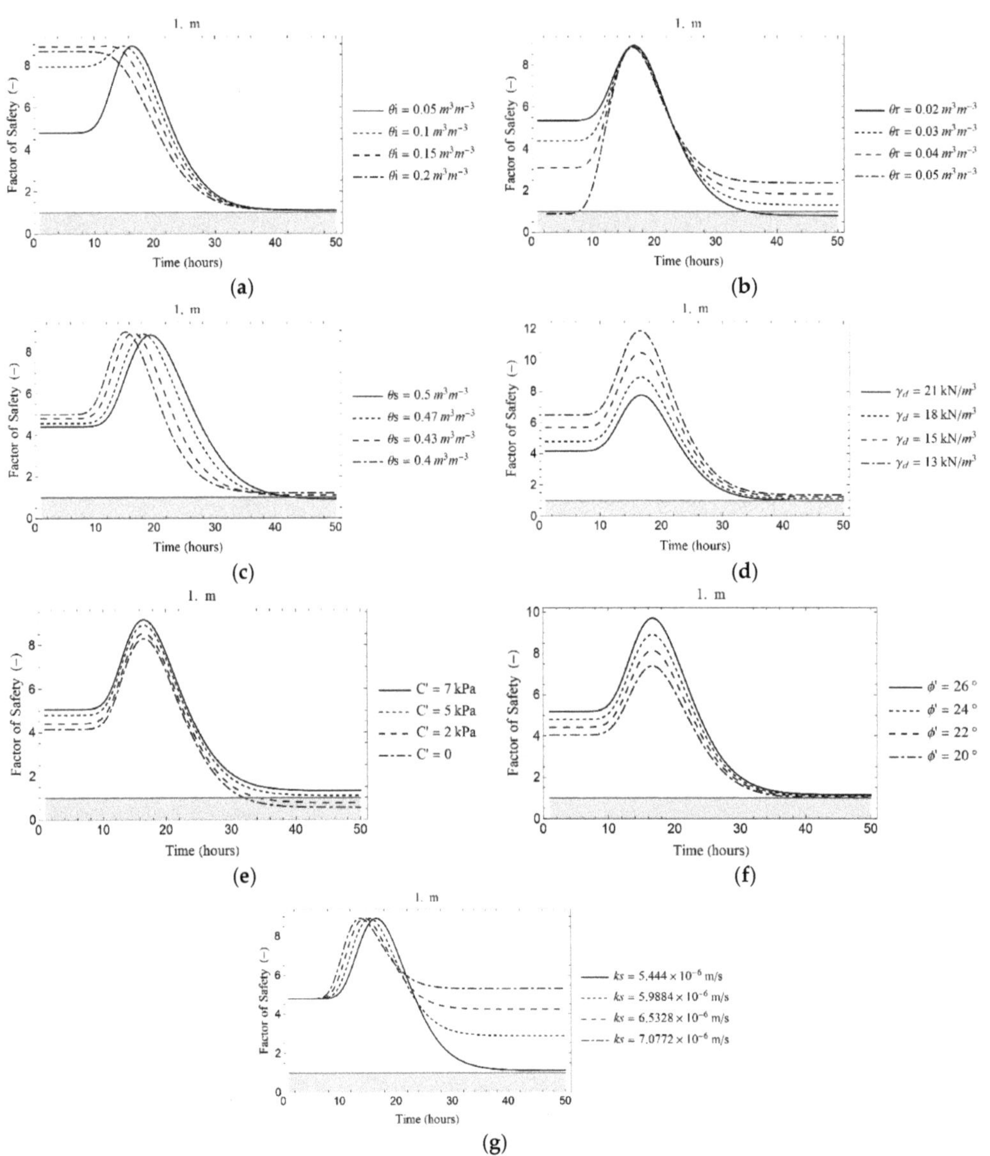

Figure 4. Parametric analysis of the stochastic variables during a 20 mm/h rainfall: (**a**) variation in initial soil moisture content, (**b**) variation in residual soil moisture content, (**c**) variation in saturated soil moisture content, (**d**) variation in density of the soil, (**e**) variation in effective soil cohesion, (**f**) variation in soil friction angle, and (**g**) variation in saturated hydraulic conductivity of the soil.

4.3. Assessment of the Probability of Transient Unsaturated Failure

The seven parameters adopted as continuous random variables among the eleven parameters required in an unsaturated stability model are presented in Table 1. The CoV of these parameters indicates the standard deviation of each variable when multiplied by the statistical mean, which is the second statistical moment needed to construct the normal and log-normal distribution of the relative frequency, as presented in Figure 5.

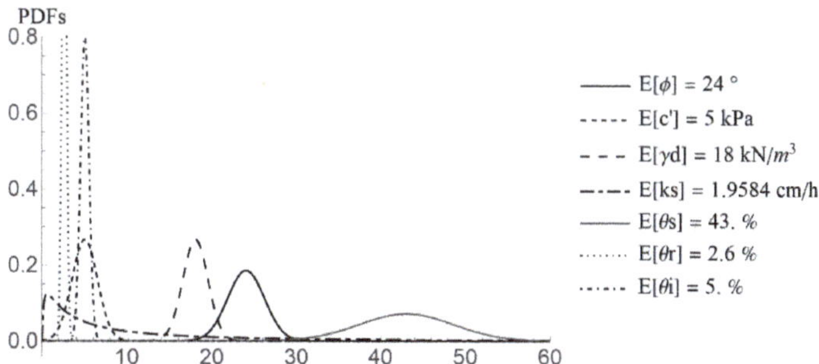

Figure 5. Normal distribution of the parameters and log-normal distribution of the saturated soil hydraulic conductivity.

Similar to the deterministic analysis, four depths were fixed for the probabilistic analysis. The contribution of the statistical parameters was quantified at four different times: 6, 12, 24, and 48 h. This work considered the indexes of the United States Army Corps of Engineers-USACE (Table 2) as preliminary indications, which define the allowable slope failure probability limits considering an FSc value of 1. Thus, the failure probability should be analyzed considering the risk associated with the effects generated by a slope rupture.

Table 2. Permissible failure probabilities [60].

Expected Performance Level	P (FS ≤ FSc)
High	3.0×10^{-7}
Good	3.0×10^{-5}
Above average	3.0×10^{-3}
Below average	6.0×10^{-3}
Poor	2.5×10^{-2}
Unsatisfactory	7.0×10^{-2}
Dangerous	1.6×10^{-1}

The results from the probabilistic analysis implementing FOSM for rainfall of 20 mm/h are presented in Figure 6. Table 3 shows a summary of the statistic moments obtained for different depths concerning time and the volumetric moisture content and suction generated under each scenario.

In Figure 6a,b, the slope stability exhibits good performance at depths of 1.0, 1.5, and 2.0 m, which is analogous to the deterministic analysis results in Figure 3c. However, at a depth of 0.5 m, the probabilistic analysis revealed an average performance in the first 12 h of the study, thereby suggesting rupture at the slope surface.

In Figure 6c,d, a high probability of rupture is observed for all analyzed depths, as represented by the gray area under the curves. Unlike the deterministic analysis shown in Figure 3c, rupture in the entire slope profile after the first 24 h of study is expected in these situations with 12–80% saturation values.

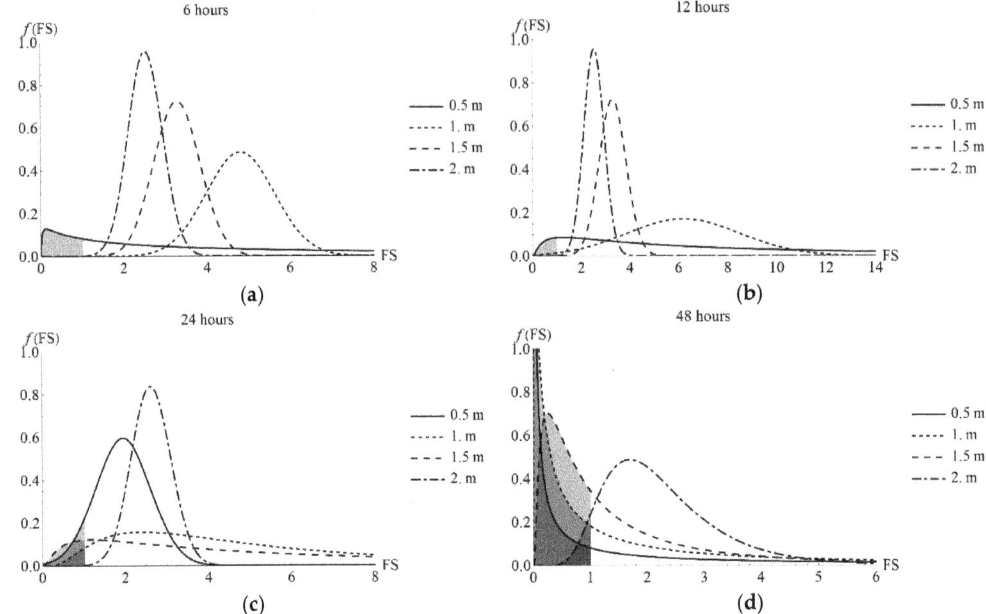

Figure 6. Probability density function during a rainfall of 20 mm/h for different slope depths at (**a**) 6 h, (**b**) 12 h, (**c**) 24 h, and (**d**) 48 h.

Table 3. Summary of the probability of failure by applying the FOSM for a rainfall of 20 mm/h.

Analysis Time (h)	z (m)	θ (m^{3}m^{-3})	S_r (–)	Ψ (kPa)	FS (–)	σ [FS] (–)	p (FS $\leq$ 1) (%)
6	0.5	0.10	0.24	−672.56	15.74	9.03	1.06×10^0
	1.0	0.05	0.12	−1133.86	4.80	0.82	1.85×10^{-4}
	1.5	0.05	0.12	−1133.86	3.28	0.55	1.81×10^{-3}
	2.0	0.05	0.12	−1133.86	2.52	0.42	1.36×10^{-2}
12	0.5	0.35	0.80	−94.66	8.85	4.06	1.20×10^0
	1.0	0.07	0.15	−930.23	6.16	2.32	9.12×10^{-1}
	1.5	0.05	0.12	−1133.74	3.28	0.55	1.82×10^{-3}
	2.0	0.05	0.12	−1133.88	2.52	0.42	1.36×10^{-2}
24	0.5	0.42	0.98	−6.45	1.93	0.67	7.89×10^0
	1.0	0.35	0.81	−90.08	4.39	2.19	3.81×10^0
	1.5	0.12	0.28	−590.39	5.72	3.52	3.79×10^0
	2.0	0.05	0.12	−1103.43	2.61	0.48	3.59×10^{-2}
48	0.5	0.42	0.98	−6.70	1.96	134.90	2.96×10^{-1}
	1.0	0.42	0.98	−6.70	1.10	9.04	4.40×10^0
	1.5	0.42	0.97	−11.56	0.96	3.18	12.35×10^0
	2.0	0.36	0.84	−74.46	2.05	1.55	15.58×10^0

To better understand the contribution of each parameter's variance in the stability problem's total variance, the contributions obtained by FOSM for each analysis time and depth are presented in Figure 7. Contrary to the parametric analysis presented in Figure 4g, the statistical dispersion of the FS represented by the variance (V[FS]) is influenced by the saturated hydraulic conductivity in the first hours of the analysis, as presented in Figure 7a,b. In these scenarios, when k_s accounts for at least 50% of the FS variance, there is a high probability of failure under unsaturated conditions.

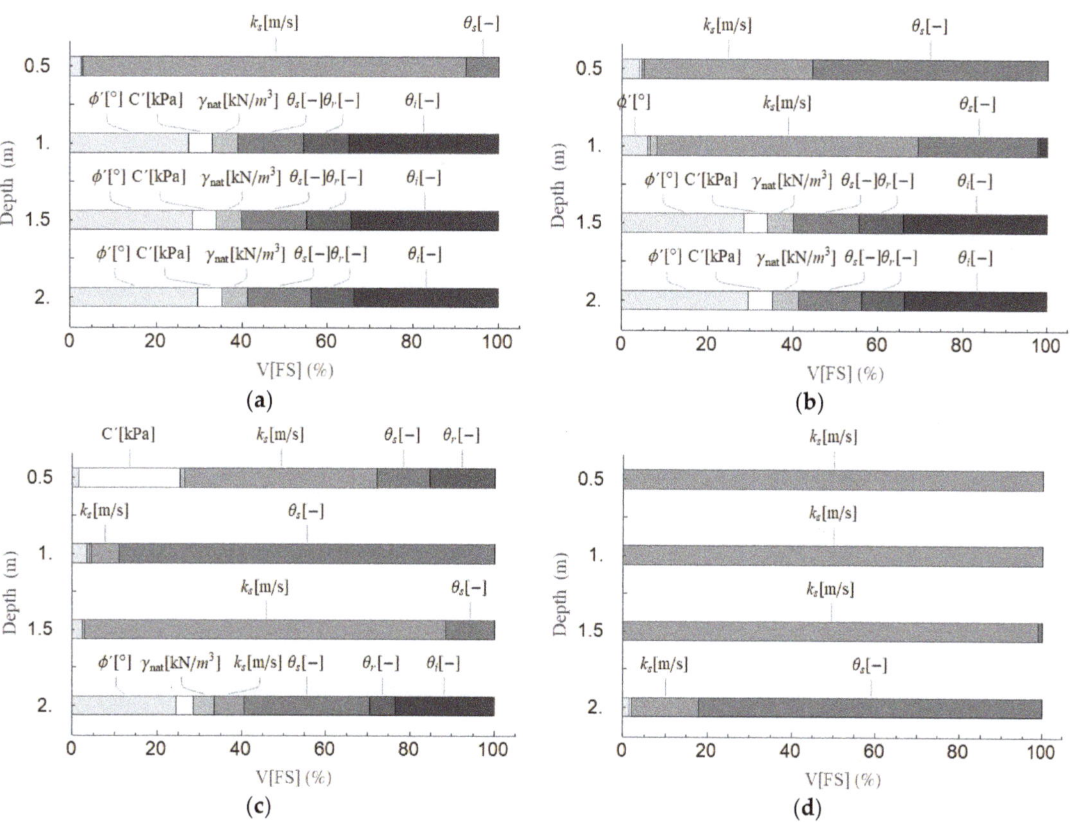

Figure 7. Contribution of the variance of each parameter in the total variance of the evaluation of the unsaturated failure probability during a rainfall of 20 mm/h at (**a**) 6 h, (**b**) 12 h, (**c**) 24 h, and (**d**) 48 h.

The dispersion of the physical parameters (friction and cohesion) affects the total variation of the FS in the first 12 h of the analysis, confirming the importance of these parameters in the unsaturated state. The increased volume of the saturated moisture content in the infiltration process when the soil reaches high saturation values (greater than 80%), checking that this parameter determines the capacity of the soil to receive water.

Figure 8 presents the transient failure probability analysis obtained using Equation (33) to analyze the slope performance profile at different depths and times. Under a rainfall intensity of 20 mm/h, the performance indices oscillated most of the time between the average and unsatisfactory performance regions, demonstrating poor slope surface performance in the first 6 h of the steady rainfall event. Unlike the deterministic analysis, this analysis is more compatible with the field information provided by Aristizábal et al. [58], where the slope was determined to be unconditionally unstable.

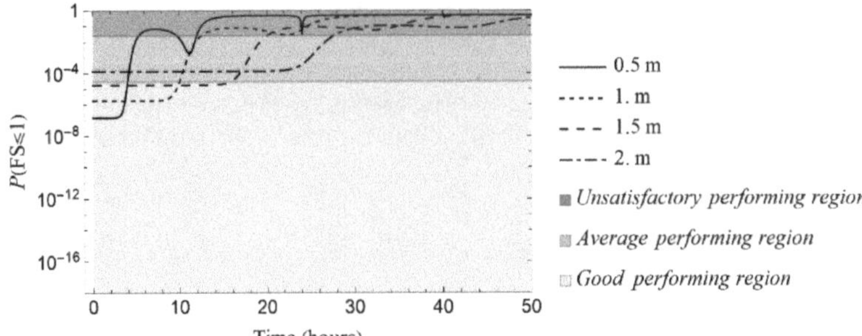

Figure 8. Transient failure probability by applying FOSM for depths of 0.5, 1.0, 1.5, and 2.0 m during a 20 mm/h rainfall.

5. Conclusions

In this study, a probabilistic approach for unsaturated and transient conditions was carried out to understand the contribution of physical and hydraulic parameters on the stability of slopes. The authors implemented a new stability model that implements the Cavalcante and Mascarenhas shear strength model based on the Cavalcante and Zornberg theory, where they proposed a new solution for the unsaturated flow equation as a function of the delta hydraulic adjustment parameter.

For this work, the FOSM demonstrated a satisfactory, quick, and efficient performance as the models used to evaluate the stability under unsaturated conditions were mathematically closed, thereby allowing the implementation of the method in Wolfram Mathematica software 12.0 at a low computational expense. To fully understand the slope stability during a precipitation event, this study demonstrated the importance of using models of unsaturated conditions, where the initial requirements for obtaining the FS are associated with the impact of hydraulic processes.

The probabilistic analysis confirmed its advantages in problems that involve variables with high dispersion and statistics, such as phenomena that depend on the infiltration and percolation processes. For slopes, quantifying the probability of failure becomes an essential tool for the geotechnical aspect, especially in decision-making involving the risk management associated with mass movements. The comparison of the parametric analysis with the analysis of the contribution to the total variance of the FOSM verified that the analyses of the parameter sensitivity were limited to the information of the first statistical moment (E(FS)), which limited the results on the influence of the dispersion of each parameter in the medium to the FS response.

Finally, the probabilistic analysis provided a more efficient approach under transient conditions, thereby allowing the evaluation of a slope under saturated conditions while incorporating the influence of a precipitation event and all parameters involved in the stability problem. Therefore, the probabilistic analysis can efficiently obtain the slope performance indices at any time and depth with a low computational expense. In this way, the authors suggest conducting probabilistic stability analyses, treating all variables as stochastic, to measure the effect of each parameter on the safety factor's total variance. A minimum sample size is needed to determine the variation coefficient for each parameter.

Author Contributions: Conceptualization, K.R.C.B.d.C., A.L.B.C. and A.P.d.A.; methodology, K.R.C.B.d.C., A.L.B.C. and A.P.d.A.; software, K.R.C.B.d.C.; validation, K.R.C.B.d.C. and A.P.d.N.D.; formal analysis, K.R.C.B.d.C.; investigation K.R.C.B.d.C.; writing—original draft preparation, K.R.C.B.d.C.; writing—review and editing, K.R.C.B.d.C. and A.P.d.N.D.; supervision, A.L.B.C. and A.P.d.A. All authors have read and agreed to the published version of the manuscript.

Funding: This study was partly financed by the Coordination for the Improvement of Higher Education Personnel—Brazil (CAPES)—Finance Code 001. The authors also acknowledge the support of the National Council for Scientific and Technological Development (CNPq Grant 305484/2020-6), the National Electric Energy Agency (ANEEL), and its R&D partners Neoenergia/CEB Distribuição S.A. (Grant number PD-05160-1904/2019, contract CEBD782/2019), and the University of Brasília.

Institutional Review Board Statement: Not applicable.

Informed Consent Statement: Not applicable.

Data Availability Statement: All data, models, and codes that support this study's findings are available from the corresponding author upon reasonable request.

Acknowledgments: The authors acknowledge the support the University of Brasilia (UnB) provided.

Conflicts of Interest: The authors declare no conflict of interest.

References

1. Walker, L.R.; Shiels, A.B. Physical Causes and Consequences for Landslide Ecology. *Landslide Ecol.* **2013**, 46–82.
2. Cho, S.E. Stability Analysis of Unsaturated Soil Slopes Considering Water-Air Flow Caused by Rainfall Infiltration. *Eng. Geol.* **2016**, *211*, 184–197. [CrossRef]
3. Tsai, T.L.; Chen, H.E.; Yang, J.C. Numerical Modeling of Rainstorm-Induced Shallow Landslides in Saturated and Unsaturated Soils. *Environ. Geol.* **2008**, *55*, 1269–1277. [CrossRef]
4. Song, Y.S.; Chae, B.G.; Lee, J. A Method for Evaluating the Stability of an Unsaturated Slope in Natural Terrain during Rainfall. *Eng. Geol.* **2016**, *210*, 84–92. [CrossRef]
5. Yang, K.H.; Uzuoka, R.; Thuo, J.N.; Lin, G.L.; Nakai, Y. Coupled Hydro-Mechanical Analysis of Two Unstable Unsaturated Slopes Subject to Rainfall Infiltration. *Eng. Geol.* **2017**, *216*, 13–30. [CrossRef]
6. Cao, W.; Wan, Z.; Li, W. Stability of Unsaturated Soil Slope Considering Stratigraphic Uncertainty. *Sustainability* **2023**, *15*, 717. [CrossRef]
7. Li, S.; Qiu, C.; Huang, J.; Guo, X.; Hu, Y.; Mugahed, A.S.Q.; Tan, J. Stability Analysis of a High-Steep Dump Slope under Different Rainfall Conditions. *Sustainability* **2022**, *14*, 11148. [CrossRef]
8. Mburu, J.W.; Li, A.J.; Lin, H.-D.; Lu, C.W. Investigations of Unsaturated Slopes Subjected to Rainfall Infiltration Using Numerical Approaches—A Parametric Study and Comparative Review. *Sustainability* **2022**, *14*, 14465. [CrossRef]
9. Jeong, S.; Lee, K.; Kim, J.; Kim, Y. Analysis of Rainfall-Induced Landslide on Unsaturated Soil Slopes. *Sustainability* **2017**, *9*, 1280. [CrossRef]
10. Chowdhury, R.; Flentje, P.; Bhattacharya, G. *Geotechnical Slope Analysis*; CRC Press: Boca Raton, FL, USA, 2009; ISBN 9780415469746.
11. Rojas, K.R.C. Estudo Dos Deslizamentos Superficiais Deflagrados por Chuvas Implementando a Análise Não Saturada e Transiente em Escala Regional. Master's Dissertation, University of Brasília, Faculty of Technology, Department of Civil and Environmental Engineering, Brasília, Brazil, 2017.
12. Cavalcante, A.L.B.; Zornberg, J.G. Efficient Approach to Solving Transient Unsaturated Flow Problems. I: Analytical Solutions. *Int. J. Geomech.* **2017**, *17*, 04017013. [CrossRef]
13. Cavalcante, A.L.B.; Camapum de Carvalho, J. *Probabilidade de Ruptura Transiente de Taludes Não Saturados Em Bordas de Reservatórios*; Federal University of Goiás: Goiânia, Brazil, 2017.
14. Cavalcante, A.L.B.; Mascarenhas, P.V.S. Efficient Approach in Modeling the Shear Strength of Unsaturated Soil Using Soil Water Retention Curve. *Acta Geotech.* **2021**, *16*, 3177–3186. [CrossRef]
15. Taha, M.R.; Hossain, M.K.; Mofiz, S.A. Effect of Suction on the Strength of Unsaturated Soils. In Proceedings of the Geo-Denver 2000, Denver, CO, USA, 5–8 August 2000; pp. 521–587.
16. Nishimura, T.; Fredlund, D.G. Relationship between Shear Strength and Matric Suction in an Unsaturated Silty Soil. In *Unsaturated Soils for Asia*; CRC Press: Boca Raton, FL, USA, 2000; pp. 563–568.
17. Fredlund, D.G.; Rahardjo, H. Soil Mechanics for Unsaturated Soils. *Soil Dyn. Earthq. Eng.* **1993**, *12*, 449–450. [CrossRef]
18. Richards, L.A. Capillary Conduction of Liquids through Porous Mediums. *J. Appl. Phys.* **1931**, *1*, 318–333. [CrossRef]
19. Terzaghi, K. *Theoretical Soil Mechanics*, 1st ed.; John Wiley & Sons, Inc.: New York, NY, USA, 1943.
20. Buckingham, E. Studies on the Movement of Soil Moisture. *Bur. Soils Bull.* **1907**, *38*, 61.
21. Brooks, R.H.; Corey, A.T. Hydraulic Properties of Porous Media. *Hydrol. Pap.* **1964**, *3*, 27.
22. Van Genuchten, M.T. A Closed-Form Equation for Predicting the Hydraulic Conductivity of Unsaturated Soils. *Soil Sci. Soc. Am. J.* **1980**, *44*, 892–898. [CrossRef]
23. Fredlund, D.G.; Xing, A. Equations for the Soil-Water Characteristic Curve. *Can. Geotech. J.* **1994**, *31*, 521–532. [CrossRef]
24. Zhang, Z.; Wang, W.; Yeh, T.C.J.; Chen, L.; Wang, Z.; Duan, L.; An, K.; Gong, C. Finite Analytic Method Based on Mixed-Form Richards' Equation for Simulating Water Flow in Vadose Zone. *J. Hydrol.* **2016**, *537*, 146–156. [CrossRef]
25. Cavalcante, A.L.B.; Zornberg, J.G. Efficient Approach to Solving Transient Unsaturated Flow Problems. II: Numerical Solutions. *Int. J. Geomech.* **2017**, *17*, 04017014. [CrossRef]
26. Costa, M.B.A.; Cavalcante, A.L.B. Novel Approach to Determine Soil-Water Retention Surface. *Int. J. Geomech.* **2020**, *20*, 04020054. [CrossRef]

27. Costa, M.B.A.; Cavalcante, A.L.B. Bimodal Soil–Water Retention Curve and k -Function Model Using Linear Superposition. *Int. J. Geomech.* **2021**, *21*, 04021116. [CrossRef]
28. Bordoloi, S.; Ng, C.W.W. The Effects of Vegetation Traits and Their Stability Functions in Bio-Engineered Slopes: A Perspective Review. *Eng. Geol.* **2020**, *275*, 105742. [CrossRef]
29. Chirico, G.B.; Borga, M.; Tarolli, P.; Rigon, R.; Preti, F. Role of Vegetation on Slope Stability under Transient Unsaturated Conditions. *Procedia Environ. Sci.* **2013**, *19*, 932–941. [CrossRef]
30. Gariano, S.L.; Guzzetti, F. Landslides in a Changing Climate. *Earth-Sci. Rev.* **2016**, *162*, 227–252. [CrossRef]
31. Silva, A.P.N. Probabilidade de Ruptura Transiente de Encostas Medida Durante Eventos Significativos de Precipitação. Master's Dissertation, University of Brasília, Faculty of Technology, Department of Civil and Environmental Engineering, Brasília, Brazil, 2018.
32. Terry, J.P.; Wotling, G. Rain-Shadow Hydrology: Influences on River Flows and Flood Magnitudes across the Central Massif Divide of La Grande Terre Island, New Caledonia. *J. Hydrol.* **2011**, *404*, 77–86. [CrossRef]
33. Croney, D.; Coleman, J.D.; Black, W.P. Studies of the Movement and Distribution of Water in Soil in Relation to Highway Design and Performance. *Highw. Res. Board Spec. Rep.* **1958**, *40*, 226–252.
34. Bishop, A.W. The Principle of Effective Stress. *Tek. Ukebl.* **1959**, *39*, 859–863.
35. Fredlund, D.G.; Morgenstern, N.R.; Widger, R.A. The Shear Strength of Unsaturated Soils. *Can. Geotech. J.* **1978**, *15*, 313–321. [CrossRef]
36. Vanapalli, S.K. Shear Strength of Unsaturated Soils and Its Applications in Geotechnical Engineering Practice. In *Keynote Address, Proceedings of the 4th Asia-Pacific Conference Unsaturated Soils, Newcastle, Australia, 23–25 November 2009*; pp. 579–598.
37. Lu, N.; Godt, J.W.; Wu, D.T. A Closed-Form Equation for Effective Stress in Unsaturated Soil. *Water Resour. Res.* **2010**, *46*. [CrossRef]
38. Travis, Q.B.; Houston, S.L.; Marinho, F.M.; Schmeeckle, M. Unsaturated Infinite Slope Stability Considering Surface Flux Conditions. *J. Geotech. Geoenviron. Eng.* **2010**, *136*, 963–974. [CrossRef]
39. Gitirana, G.D.F.N., Jr. Weather-Related Geo-Hazard Assessment Model for Railway Embankment Stability. Ph.D. Thesis, University of Saskatchewan, Saskatoon, SK, Canada, 2005.
40. Azevedo, G.F. Sistema de Análise Quantitativa de Risco Por Escorregamentos Rasos Deflagrados por Chuvas em Regiões Tropicais. Ph.D. Thesis, University of Brasília, Faculty of Technology, Department of Civil and Environmental Engineering, Brasília, Brazil, 2015.
41. Silva Júnior, A.C. Abordagem Rítmica Probabilística Aplicada em Análises de Fluxo e Estabilidade de Taludes. Ph.D. Thesis, University of Brasília, Faculty of Technology, Department of Civil and Environmental Engineering, Brasília, Brazil, 2015.
42. Lee, J.H.; Park, H.J. Assessment of Shallow Landslide Susceptibility Using the Transient Infiltration Flow Model and GIS-Based Probabilistic Approach. *Landslides* **2016**, *13*, 885–903. [CrossRef]
43. Lemos, M.A.C.; Dias, A.D.; Cavalcante, A.L.B. Deterministic and Probabilistic Approach in Slope Stability of a Tropical Soil. *Int. J. Geol. Earth Sci.* **2020**, *6*, 9–14. [CrossRef]
44. Liu, X.; Wang, Y. Probabilistic Simulation of Entire Process of Rainfall-Induced Landslides Using Random Finite Element and Material Point Methods with Hydro-Mechanical Coupling. *Comput. Geotech.* **2021**, *132*, 103989. [CrossRef]
45. Albuquerque, E.A.C.; de Faria Borges, L.P.; Cavalcante, A.L.B.; Machado, S.L. Prediction of Soil Water Retention Curve Based on Physical Characterization Parameters Using Machine Learning. *Soils Rocks* **2022**, *45*, 1–13. [CrossRef]
46. Zhang, W.; Gu, X.; Hong, L.; Han, L.; Wang, L. Comprehensive Review of Machine Learning in Geotechnical Reliability Analysis: Algorithms, Applications and Further Challenges. *Appl. Soft Comput.* **2023**, *136*, 110066. [CrossRef]
47. Aminpour, M.; Alaie, R.; Khosravi, S.; Kardani, N.; Moridpour, S.; Nazem, M. Slope Stability Machine Learning Predictions on Spatially Variable Random Fields with and without Factor of Safety Calculations. *Comput. Geotech.* **2023**, *153*, 105094. [CrossRef]
48. Liu, Y.; Yang, Z.; Li, X. Adaptive Ensemble Learning of Radial Basis Functions for Efficient Geotechnical Reliability Analysis. *Comput. Geotech.* **2022**, *146*, 104753. [CrossRef]
49. Karir, D.; Ray, A.; Kumar Bharati, A.; Chaturvedi, U.; Rai, R.; Khandelwal, M. Stability Prediction of a Natural and Man-Made Slope Using Various Machine Learning Algorithms. *Transp. Geotech.* **2022**, *34*, 100745. [CrossRef]
50. Wong, F.S. First-Order, Second-Moment Methods. *Comput. Struct.* **1985**, *20*, 779–791. [CrossRef]
51. Ering, P.; Babu, G.S. Probabilistic Back Analysis of Rainfall Induced Landslide—A Case Study of Malin Landslide, India. *Eng. Geol.* **2016**, *208*, 154–164. [CrossRef]
52. Marin, R.J.; Velásquez, M.F.; Sánchez, O. Applicability and Performance of Deterministic and Probabilistic Physically Based Landslide Modeling in a Data-Scarce Environment of the Colombian Andes. *J. S. Am. Earth Sci.* **2021**, *108*, 103175. [CrossRef]
53. Cooke, R.A.; Mostaghimi, S.; Woeste, F. Effect of Hydraulic Conductivity Probability Distribution Function on Simulated Solute Leaching. *Water Environ. Res.* **1995**, *67*, 159–168. [CrossRef]
54. Mesquita, M.D.G.B.D.F.; Moraes, S.O.; Corrente, J.E. More Adequate Probability Distributions to Represent the Saturated Soil Hydraulic Conductivity. *Sci. Agric.* **2002**, *59*, 789–793. [CrossRef]
55. Mesquita, M.D.G.B.D.F. Caracterização Estatística da Condutividade Hidráulica Saturada do Solo. Ph.D. Thesis, School of Agriculture Luíz de Queiroz, São Paulo, Brazil, 2001.
56. Bong, T.; Son, Y. Impact of Probability Distribution of Hydraulic Conductivity on Groundwater Contaminant Transport. *KSCE J. Civ. Eng.* **2019**, *23*, 1963–1973. [CrossRef]

57. Malkawi, A.I.H.; Hassan, W.F.; Abdulla, F.A. Uncertainty and Reliability Analysis Applied to Slope Stability. *Struct. Saf.* **2000**, *22*, 161–187. [CrossRef]
58. Aristizábal, E.; Vélez, J.I.; Martínez, H.E.; Jaboyedoff, M. SHIA_Landslide: A Distributed Conceptual and Physically Based Model to Forecast the Temporal and Spatial Occurrence of Shallow Landslides Triggered by Rainfall in Tropical and Mountainous Basins. *Landslides* **2015**, *13*, 497–517. [CrossRef]
59. Campos, T.M.P.; Motta, M.F.B. *Resistência Ao Cisalhamento de Solos Não Saturados*; Brazilian Association of Soil Mechanics and Geotechnical Engineering: São Paulo, Brazil, 2015; ISBN 9788578110796.
60. USACE. *Risk-Based Analysis in Geotechnical Engineering for Support of Planning Studies*; USACE: Washington, DC, USA, 1999.

sustainability

MDPI

Article

Influence of Soluble Salt NaCl on Cracking Characteristics and Mechanism of Loess

Xin Wei *, Li Dong, Xuanyi Chen and Yunru Zhou

School of Human Settlements and Civil Engineering, Xi'an Jiaotong University, Xi'an 710049, China
* Correspondence: weixinstar@xjtu.edu.cn

Abstract: Under the conditions of drought, cracks are likely to appear in loess due to shrinkage, which leads to salt precipitation and accumulation on the surface of loess. Therefore, salinized lands are created in loess areas. Deep study into the influence of soluble salt content on the cracking characteristics and mechanism of loess is of great significance to engineering constructions, geological problems, and disaster prevention for salinized lands in loess regions. In this paper, free desiccation experiments were carried out on the loess samples with different NaCl concentrations (a soluble salt). A high-resolution digital camera was used to obtain the sequence images of loess during the drying process. With the advantage of digital image correlation (DIC) technology and the non-contact full-field strain measurement method, the local displacement and strain on the surface of loess samples were calculated. The microstructure and main elements distribution of loess samples were obtained by scanning electron microscopy (SEM) and energy-dispersive spectrum (EDS) methods. Finally, the influence of NaCl concentrations on cracking characteristics and mechanism of loess was analyzed. The results show that, with the increase in NaCl concentration, the evaporation rate of loess samples decreased and the residual water content increased. The NaCl content can prevent the development of desiccation cracks in loess.

Keywords: clayey loess; soluble salts NaCl; desiccation crack; non-contact full-filed strain measurement method; microstructure

check for
updates

Citation: Wei, X.; Dong, L.; Chen, X.; Zhou, Y. Influence of Soluble Salt NaCl on Cracking Characteristics and Mechanism of Loess. *Sustainability* **2023**, *15*, 5268. https://doi.org/10.3390/su15065268

Academic Editors: Chong Xu and Jian Chen

Received: 30 January 2023
Revised: 3 March 2023
Accepted: 7 March 2023
Published: 16 March 2023

1. Introduction

Loess is a special soil formed since the Quaternary. Defined by various pores and weak cementation, this soil is widely distributed throughout Asia, Europe, North America and South America. Loess in China is unique in the world because of its wide distribution, large thickness, typical sedimentation environment and complete topography [1]. As a special geotechnical material, loess has a sub-stable structure and a large number of pores. Its physical properties, chemical compositions and mechanical behaviors vary under the aqueous conditions [2]. Due to the special physical properties, microstructure and semi-arid climate of the Chinese Loess Plateau, cracks in the loess are likely to be generated. In addition, under drought conditions, salt accumulates in the loess easily on the surface of soil, resulting in different salinization degrees of land [3]. Salt is one of the important components in loess. It exists in the microstructure of loess as cement and skeleton particles and plays an important role in the formation of structural loess and strength. Due to soil salinization or leaching effects, the precipitation of salt from the loess will inevitably affect the integrity of the loess structure, resulting in variations in the mechanical behavior of the loess [4].

With global warming, arid climates are becoming more and more widespread. The phenomenon of cracks in loess due to desiccation is also common. The generation of cracks will have a significant impact on the mechanical and hydraulic properties of soil, leading to various geotechnical problems [5–9]. For example, in landfill projects, cracks in soil may lead to more serious water infiltration and landfill gas emissions, further polluting the surrounding groundwater and air [10,11]. In slope engineering, the drying cracks will increase the risk of water infiltration and further reduce the stability of slopes [12,13]. The cracks formed on the surface of loess slope under the arid climate conditions provide preferential channels for rainfall or evaporation. The salt content in soil is affected by various factors such as water flows in pores and temperature, resulting in the extremely unstable engineering properties of loess. With the effect of leaching, a large amount of salt in the loess is precipitated, forming a white layer on the loess surface. Salinization of loess is often accompanied by the geological problems such as erosion and water and soil loss [14]. Therefore, it is of great significance to carry out research about the impact of salt on the cracks of loess in northwestern China.

Research into the cracks in soil has always been a hot topic. In order to better analyze the evolution characteristics of cracks in soils, Miller et al. introduced the factor of crack strength to quantify the cracking degree of soil [7]. Péron et al. defined the cracking degree of soil with the number of cracks, crack spacing, opening degree of cracks, truncation angles and other parameters [15]. These studies showed that the development of cracks is affected by various factors. The shrinkage cracks of soil are closely related to its mineral compositions and contents, structures, densities, etc. [16,17]. In addition, factors from the natural environment, such as temperature, relative humidity, soil thickness, dry–wet cycle times, boundary conditions and soil size, also have important impacts on the cracks of soil [18,19]. However, there are relatively few studies about the effect of salts on cracks in soils. Waler et al. found that, under drying conditions, the proportions of the cracking area gradually decreased with the increase in soil salt content [20]. DeCarlo et al. found that, with the increase in NaCl concentration, the evaporation rate of the sample under the same conditions decreased. The time of the first crack occurrence was gradually delayed during the drying process. The critical water content corresponding to the first crack gradually decreased. The more salt the sample contains, the lower the degree of cracking on the surface of soil [3,21]. Meanwhile, the crack width, length, fractal dimension, etc., also decrease with the increase in salt content [3]. Some scholars have also studied the cracking mechanism of soils and found different cracking modes, including opening mode, sliding mode and tearing mode [22]. Wei et al. used digital image correlation methods to quantitatively analyze the cracking characteristics and mechanism of the two clays during drying [23].

Although many scholars have conducted numerous field observations [24] and laboratory tests [25–27] on the cracks of different soils, relatively few studies have taken the cracks in loess as a research topic. For example, Corte et al. conducted drying tests on Bloomington clay to explore the influence of soil density, soil sample thickness, etc., on cracks [16]. Wei et al. studied the cracking process, local strains and displacements evolutions of mixtures of kaolinite and montmorillonite [22]. Tang et al. studied the evaporation rate, volumetric shrinkage rate and crack evolution of Nanjing Xiashu clay during drying [28]. Lakshmikantha et al. carried out quantitative research on the cracking parameters of silt [29]. Cordero et al. conducted a study on the sequence of cracks in the mixtures of sand and clay [30]. There are few studies on the dynamics and related mechanisms of cracking mode formation in loess, which is a complex cohesive soil [31]. The spatio-temporal evolution characteristics of the stress field during the drying process of loess have not been quantitatively characterized, leads to a lack of data on the evolution of the stress field during the desiccation of loess.

Several studies showed that the content of Na^+ and Cl^- in the loess is much higher than that of other ions [32,33]. Therefore, in this research, different percentages of NaCl to the loess are chosen for the purpose of analyzing the impact of salt concentration on the cracks in loess during the drying process. The development process of cracks is captured with a high-resolution camera. The parameters of cracks were quantitatively analyzed with an image processing software named PCAS [34–37]. Non-contact full-field strain analyses were conducted using VIC-2D software [18,19,22]. The local displacement and strain on the surface of loess samples were calculated during drying. At the same time, combined with the use pf scanning electron microscopy (SEM) and energy-dispersive spectrum (EDS) element analyses, the effect of salt on the microstructure of loess was interpreted. Through the above experimental methods, the development of cracks in loess was observed. The local displacements and strains on sample surface were quantitatively calculated, and the mechanism of how the soluble salt NaCl impacts on cracks of loess was explained. This study provides insights into factors which influence the evolutions of cracks in loess. In addition, the results are of significance for the understanding of geological disasters related to cracks in saline loess areas.

2. Materials and Methods

2.1. Materials

The soil samples used in this study were taken from Jingyang, Shaanxi Province, China from a depth of approximately 50 m and belonged to clayey loess [38]. The physical properties of the clayey loess are shown in Table 1. In addition, the grain size distribution curve, measured by the laser method, is shown in Figure 1. The clay content is 17% and the silt content is 82.7%. According to the classification standard, the loess belongs to low-plasticity clay (CL).

Table 1. Physical properties of Jingyang loess.

Specific Gravity	Dry Density (g/cm³)	Liquid Limit w_L (%)	Plastic Limit w_P (%)	Plasticity Index I_P	Granulometry (%)		
					Sand Content (4.75–0.075 mm)	Silt Content (0.075–0.002 mm)	Clay Content (≤0.002 mm)
2.69	1.57	35.6	17.1	18.5	0.3	82.7	17

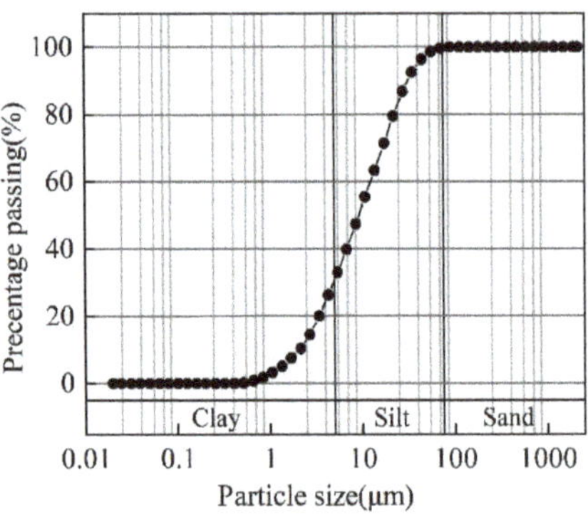

Figure 1. Grain size distributions of Jingyang loess.

2.2. Experimental Methods

2.2.1. Free Desiccation Test

The free desiccation tests of loess with different NaCl concentration were carried out in the laboratory environment with a temperature of around 30 °C and a relative humidity of 35%. The test set-up consists of the soil sample, a balance, a high-resolution camera and a lighting device, as shown in Figure 2. The dimensions of loess samples are 300 mm (length) × 200 mm (width) × 4 mm (thickness), and they are placed on the plexiglass support. The high-resolution camera fixed above the loess sample is connected to a computer. Photos of a loess sample surface are taken at an interval of 10 min to record the whole cracking process. The balance below the sample records the sample weight during drying, which is then used to calculate the water content variations in the sample.

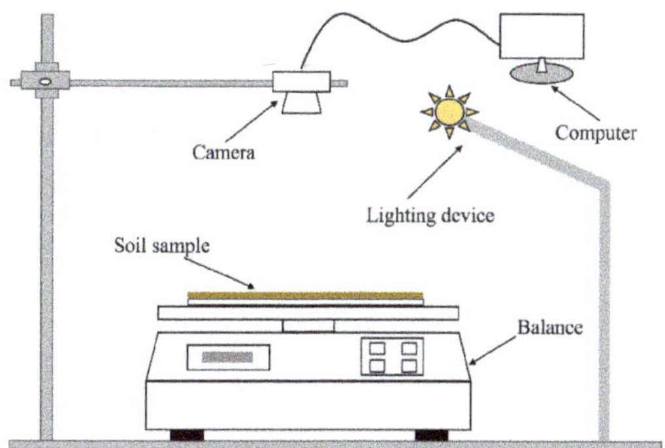

Figure 2. Schematic diagram of the experimental set-up.

2.2.2. Preparation of Loess Samples with Different NaCl Concentration

In this study, the soluble salt NaCl is selected as the influence factor on the cracks in loess [32,33]. Chlorine-salinized and chlorite-salinized soils can be divided into non-saline soil (N < 0.3%), weakly saline soil (0.3% < N < 1.0%), medium saline soil (1.0% < N < 5.0%), strongly saline soil (5.0% < N < 8.0%) and hypersaline soil (N > 8.0%), where N presents the mass percentage of soluble salt content in 100 g dry soil. In order to eliminate the interference of other soluble salts on the test, the soluble salts in the sample were removed with distilled water before the test was performed. The process was carried out by taking a certain amount of disturbed loess and pouring it into the container, injecting distilled water, and then mixing them completely; removing the supernatant with bubble ball, refilling the container with distilled water; repeating the previous steps on approximately 5 occasions; and ensuring that most of the soluble salt is washed out. After the above process, the washed loess samples were put into the oven, ground and passed a 1 mm sieve. The loess sample without soluble salt were prepared [39–45]. In order to study the effect of NaCl concentration on the cracks in loess, four groups of tests were carried out in this research. These corresponded to the four different salinization degree of loess sample. The relevant parameters are shown in Table 2.

Table 2. Different NaCl concentrations of sample in the tests.

Number of the Sample	The Concentration of NaCl (%)	Salinization Degree
1	0	Non-saline soil
2	3.6	Weakly saline soil
3	7.2	Medium saline soil
4	10.8	Strongly saline soil

A certain amount of loess without soluble salts was put into the container. Then, the solutions of NaCl were prepared at different concentrations. The NaCl solutions were poured into the containers with loess and mixed evenly. Finally, slurry loess samples with initial water contents of 36%, equal to the liquid limit, were prepared. The samples were sealed with film for about 24 h to make the mixtures reach a homogeneous state.

2.2.3. Analysis of Crack Parameters with PCAS

PCAS is a software used for particle (pore) and crack recognition and analysis which was developed by Nanjing University. The original photos, taken by the high-resolution camera, are first treated by Photoshop in order to obtain photos with the same surface area and resolution. Then, the pretreated photos are imported into the PCAS software and the photos with cracks are binarized. PCAS software can automatically identify the blocks in the cracking network, repair the crack section, remove the impurities, and identify the crack network [34–37]. Finally, various geometric parameters of the crack network can be obtained, including the number of nodes, number, length, width, area, direction of cracks, etc. Taking the loess sample with 3.6% NaCl concentration at t = 50 h as an example, the image processing and results of PCAS are shown in Figure 3.

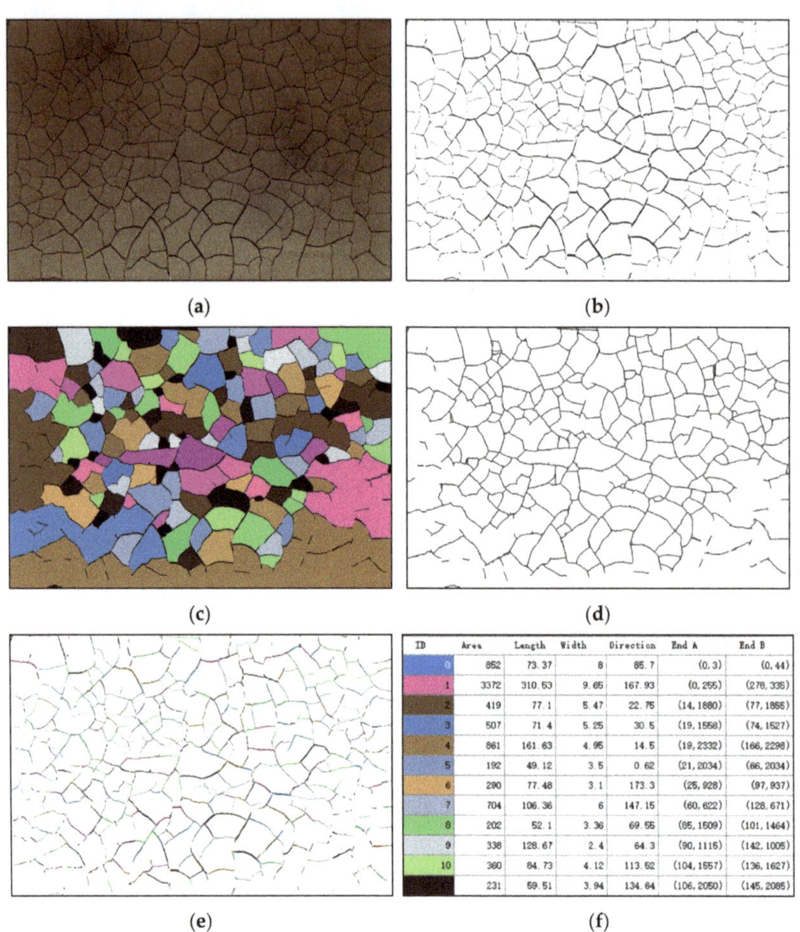

Figure 3. An example of PCAS software results (loess sample with 3.6% NaCl concentration, t = 50 h): (**a**) original photo; (**b**) binarization of photo; (**c**) blockization of cracks; (**d**) skeletonization of cracks; (**e**) analysis of cracks; (**f**) results of crack parameters.

2.2.4. Digital Image Correlation Method

The digital image correlation (DIC) method [18,19,22] was used in this research. Random speckle patterns were created by spraying black paint onto the surface of the soil sample at a distance of about 50 cm from the surface of the loess samples. Speckle patterns are necessary in the DIC method to enable the image processing software to calculate the local surface displacements and strains. The DIC principle is to calculate the displacement and strain fields by correlating the speckle pixel subsets before and after the deformation of loess samples. The paint used in the tests is quick-drying and has good adhesion and compatibility. In this study, the local displacement and strain of loess samples during drying are calculated with commercial software VIC-2D, the accuracy of which is estimated to be better than about ±0.5% [22]. Considering the characteristics of the paint and the accuracy of the VIC-2D software, the drying of the paint barely affects the results of local displacements and strains. The following two displacement components and three strain components can be obtained with VIC-2D:

U (mm)—longitudinal displacement along x axis;
V (mm)—transversal displacement along y axis;
ε_{xx} (%)—longitudinal strain along x axis;
ε_{yy} (%)—transversal strain along y axis;
ε_{xy} (%)—shear strain.

The above five components can be used to accurately analyze the cracking behavior of the loess samples during the drying process and to assess the influence of the sample heterogeneity. Figure 4 shows the longitudinal strains ε_{xx} of the loess sample with 7.2% NaCl concentration at t = 4.67 h. The positive values of ε_{xx} show tensile strains (in red or yellow) and the negative values denote compressive strains (in green, blue to purple). The white vectors in the figure show the displacement directions of the deformed specimen compared with the referenced specimen (0 h), and the length of the vector represents the value of the displacement. It is worth noting that near the boundaries of the cracks, VIC-2D software cannot calculate the displacement and strain because of the large deformations. These areas that cannot be calculated are shown in a gray color in Figure 4.

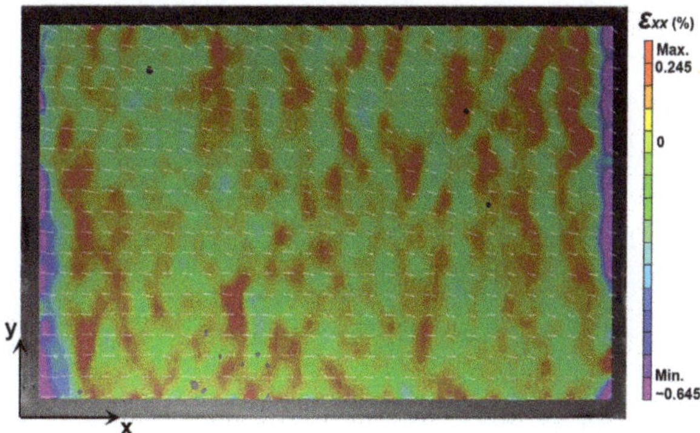

Figure 4. An example of longitudinal strain ε_{xx} (loess sample with 7.2% NaCl concentration, t = 4.67 h).

2.3. Microstructure Analysis with SEM

Different NaCl concentrations will affect the microstructure of loess, especially in the cementation degrees and cementation modes among soil particles. The mechanical properties and macroscopic cracking characteristics of soil will be further affected. Therefore, in this study, the cracking section of soil sample, after the free-drying test has been performed, is selected for use in SEM observation to analyze the variations in microstructures and cementations with different NaCl concentrations. Combined with the EDS element analysis method, the distribution of elements can be measured qualitatively and quantitatively to better analyze the cementation degree of NaCl.

3. Results and Discussion

3.1. Influence of NaCl to the Characteristics of Cracks in Loess

During the free-drying process, the concentration of NaCl influences the water content variations in loess samples. The curve of water content variations versus time of loess samples with different NaCl concentrations is shown in Figure 5. The initial water content of these four loess samples is 36%. The water content of all loess samples decreases in the beginning of desiccation and then samples become stable after a certain time. However, with the increase in NaCl concentration, the residual water content of loess samples increases gradually. The maximum and minimum residual water contents are 21.5% and 1.3%, respectively, corresponding to the sample with the maximum NaCl concentration of 10.8% and the minimum NaCl concentration of 0%. This is because the ion concentration in the pore water of the loess sample increases with the increase in NaCl concentration, resulting in the increase in the osmotic suction of water molecules. Therefore, the ability of water evaporation decreases during drying process in the samples with larger NaCl concentration.

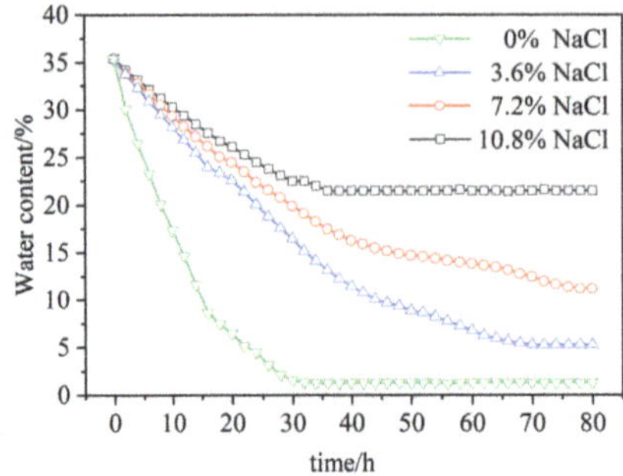

Figure 5. Variations in water content versus time of loess samples with different NaCl concentration.

NaCl has a great influence on the evolution characteristics of cracking in loess. By comparing the final crack patterns at the end of desiccation in these four loess samples, it can be found that, for the loess samples with 0%, 3.6% and 7.6% NaCl, crack networks are generated. However, for loess samples with 10.8% NaCl, there is no crack (Figure 6). With the increase in NaCl concentration, the cracks of loess samples become less developed.

Figure 6. Different crack patterns in loess samples with different NaCl concentration at the end of free desiccation tests: (**a**) 0% NaCl; (**b**) 3.6% NaCl; (**c**) 7.2% NaCl; (**d**) 10.8% NaCl.

In order to better analyze the influence of NaCl on the evolution characteristics of cracks in loess, the final crack patterns of different samples were analyzed with PCAS software (Figure 7). In addition, the quantitative analysis on the crack parameters of loess samples was carried out with PCAS, such as crack ratio, crack number, total crack length, etc. Figure 8 shows the evolution of crack parameters versus water contents. R_{sc} is the surface crack ratio, that is, the ratio of the crack area to the total surface area of the loess sample. N_c and L_c represent the number and the total length of cracks, respectively.

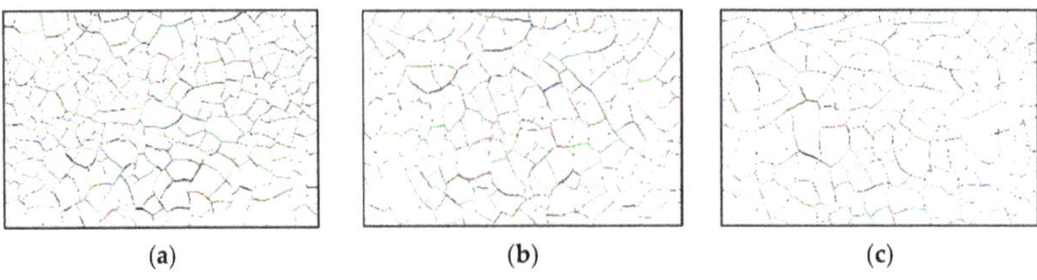

Figure 7. Images of cracks in loess samples with different NaCl concentration after PCAS treatment at the end of free desiccation tests: (**a**) 0% NaCl; (**b**) 3.6% NaCl; (**c**) 7.2% NaCl.

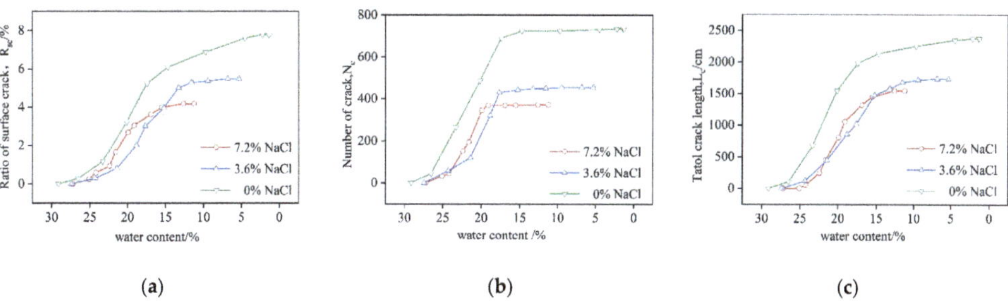

(a)   **(b)**   **(c)**

Figure 8. Variations in cracking parameters on the surface of loess samples versus water content: (**a**) Variations in R_{SC}; (**b**) Variations in N_C; (**c**) Variations in L_C.

It can be seen from Figure 8 that the crack parameters of loess samples, R_{sc}, N_c and L_c, firstly increase and then stabilize with the decrease in the water content during the drying process. However, with the increase in NaCl concentration, the above three parameters all gradually decrease. The increase in NaCl concentration restrains the development of cracks in loess samples.

3.2. Influence of NaCl on Cracking Mechanisms in Loess

The appearance of cracks will result in the failure of loess. For drying cracks in loess, matrix suction and tensile strength are the two key mechanical parameters to controlling the initiation, evolution and stability of cracks. During the drying process of loess, the water in the soil begins to evaporate from the surface. The free water in the pores of loess is firstly evaporated. With continuous drying, the water in the interieur of loess will be continuously transferred to the upper water–air surface in order to maintain the evaporation. During this process, matrix suction and the surface tension of pore water will be generated. Thus, tensile stress will appear in the loess. When the tensile stress exceeds the tensile strength of loess, cracks will be generated and the tensile stress around the cracks will be released [28].

In order to understand the cracking mechanism of loess during the drying and shrinkage process of soil, the evolution characteristics of the stress, especially the tensile stress, is very significant. In this research, the local displacements and strains on the surface of the loess sample can be obtained by the DIC method. The longitudinal strain ε_{xx} results of the loess sample with 0% NaCl at different times are taken as examples (Figure 9). At t = 2 h, the ε_{xx} in zone A is relatively larger than that in other areas, equals to about 0.26% (Figure 9a), which indicates that there is strain energy accumulation in this zone. It can be predicted that more cracks will appear in zone A during the following drying process. When t = 8 h, a new crack was developed in zone A. From Figure 9b it can be observed that the energy accumulated in zone A has been released around the new crack and the strain field is readjusted. In the process of crack evolution, when the tensile stress exceeds the tensile strength of the soil, the crack is generated. The energy accumulated in related zones is released, and the strain field is readjusted [18,22].

For the purpose of analyzing the effect of NaCl on the cracking mechanism of loess, the loess samples with different NaCl concentrations were selected at the same time of t = 5 h. The strains ε_{xx} of these four samples are presented in Figure 10. The maximum longitudinal strains ε_{xxmax} for these samples are 0.36%, 0.355%, 0.245% and 0.0011%, respectively (Figure 10a–d). With the increase in NaCl concentration in the loess sample, the maximum strain ε_{xxmax} of the loess sample at the same time gradually decreases. It is concluded that the increase in NaCl concentration reduces the energy accumulation capacity in the loess sample, thus slowing down the rate of crack evolution.

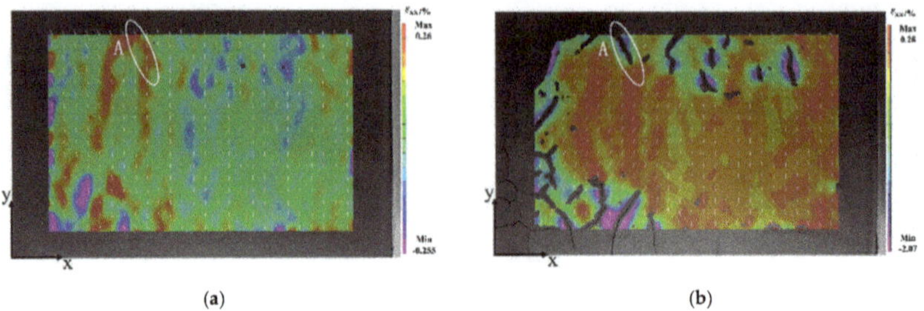

Figure 9. Longitudinal strains results of 0% NaCl loess samples at different times: (**a**) t = 2 h; (**b**) t = 8 h (Zone A presents area of interest).

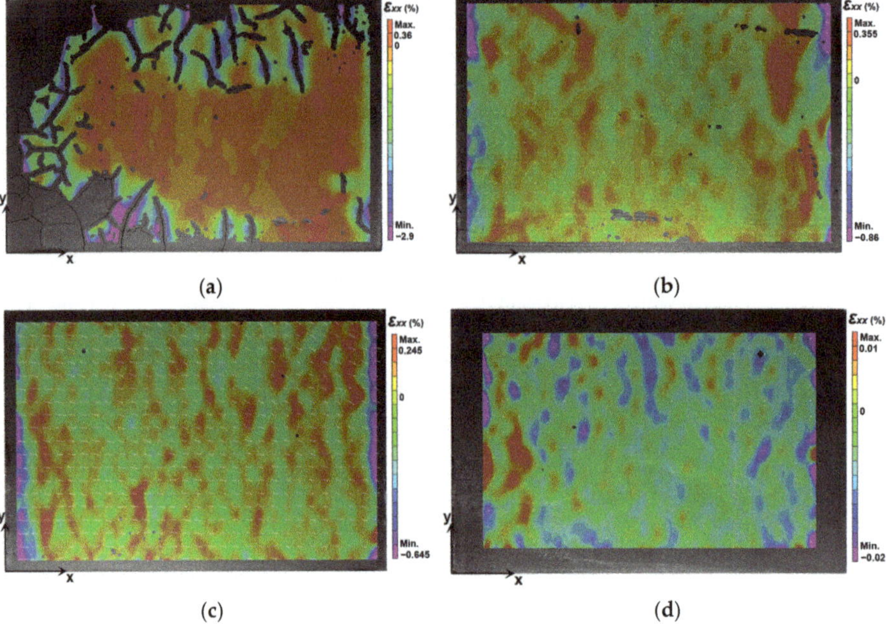

Figure 10. Longitudinal strain results of loess samples with different NaCl concentration at t = 5 h: (**a**) 0% NaCl; (**b**) 3.6% NaCl; (**c**) 7.2% NaCl; (**d**) 10.8% NaCl.

Some researchers have analyzed the distribution of strain and displacement near the cracks and concluded that there are three modes in the process of crack evolution, as shown in Figure 11: (1) opening mode; (2) sliding mode; (3) tearing mode [21]. The results of strain ε_{xx} of loess samples with 0% NaCl and 7.2% NaCl are taken as examples to further explore the influence of NaCl concentration on cracking mode (Figure 12). The local displacement D on the surface of the loess sample can be decomposed into the components D_{PA} and D_{PE}, which are parallel and perpendicular to the direction of crack, respectively. When the directions of the displacements on both sides of the crack are perpendicular to the developing direction of crack, it indicates that the strains acting on both sides of the crack are mainly tensile. The crack is developed in the opening mode. When the directions of displacements of both sides of the crack are parallel to the developing direction of the crack, the crack is mainly formed by shear strain. It can be concluded that the crack corresponds to a shearing mechanism. As shown in Figure 12a, there are components D_{PA} and D_{PE} on both sides of the crack during drying in loess sample with 0% NaCl. It shows that there are tensile strains perpendicular to the developing direction of crack and shear strains parallel to the developing direction of crack on both sides. Therefore, in this crack, a mixed opening–sliding mechanism is activated (Figure 12a). In Figure 12b, for loess sample with 7.2% NaCl, the displacement directions of both sides of the crack are perpendicular to the extension direction of crack, which indicates an opening mode. As shown in Figure 12c, for a loess sample with 7.2% NaCl, a crack is developed in a mixed opening–sliding mode. It can be concluded that the opening mode and sliding modes exist in the drying process of loess samples with different NaCl concentrations. The NaCl concentration has no significant effect on the cracking modes of loess.

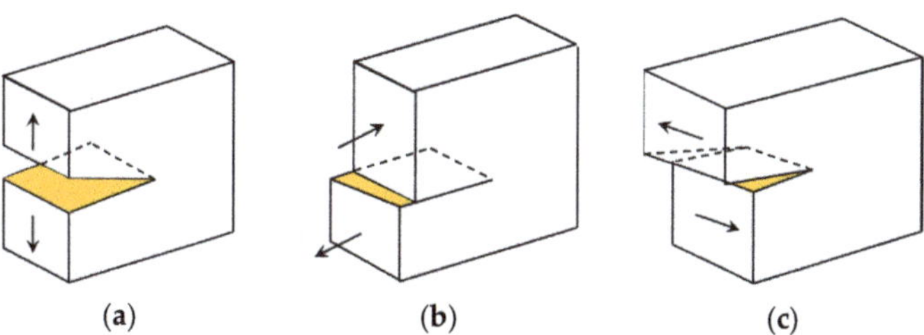

(a)   (b)   (c)

Figure 11. Three cracking modes: (**a**) opening mode; (**b**) shearing mode; (**c**) tearing mode.

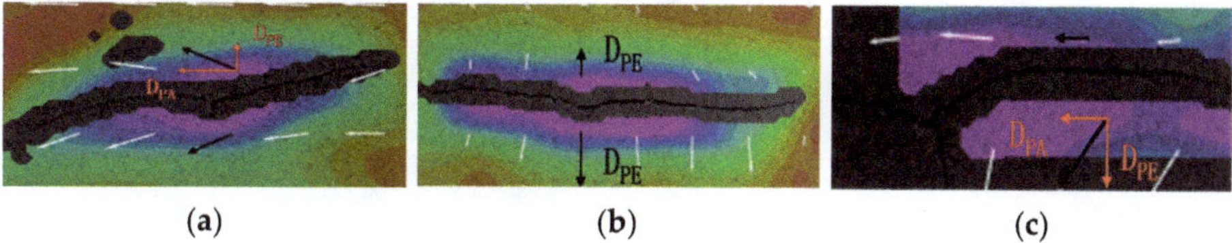

(a)   (b)   (c)

Figure 12. Different cracking modes in loess samples with different NaCl concentration during free desiccation tests: (**a**) Crack with mixed opening–sliding mode in loess sample with 0% NaCl; (**b**) Crack with opening mode in loess sample with 7.2% NaCl; (**c**) Crack with mixed opening–sliding mode in loess sample with 7.2% NaCl.

3.3. Microscopic Cracking Mechanism of Loess with Different NaCl Concentration

In order to better analyze the influence of NaCl on the cracking characteristics and mechanisms of loess from a microscopic perspective, this study uses scanning electron microscopy (SEM) methods to observe the microstructure of the cracking surface of loess samples with different NaCl concentrations at the end of the free desiccation test with the same magnification (5000 times), as shown in Figure 13.

Figure 13a shows SEM photos of loess sample with 0% NaCl. It can be observed that flaky clay particles are attached to the surface of the larger silt particles. There are many pores among the soil particles. Figure 13b shows the microstructure of the loess sample with 3.6% NaCl. Compared with the loess sample with 0% NaCl, the microstructure of loess samples with a small amount of NaCl has changed significantly at the end of the free-drying test. A layer of smooth cementation is produced on the surface of the loess particles. Some flaky clay particles and silt particles with larger particle size are flocculated and agglomerated with a layer of smooth film-like cementation, which significantly reduces the dimensions of pores among soil particles. Therefore, the microstructures of loess become more compacted. This can be explained by the interaction between NaCl and clay particles. The thickness of electric double layers can also be used for the variation of microstructures. Na^+ in NaCl reacts with clay particles by cation exchange, which results in the flocculation and agglomeration of clay particles. Figure 13c shows the microstructure of loess sample with 7.2% NaCl. With the increase in NaCl concentration in loess samples, almost all flaky clay particles and larger silt particles are tightly cemented by the smooth layer. The cementation between soil particles is significantly enhanced. When the NaCl content increases to 10.8%, a large number of NaCl crystals precipitate from the loess sample during the free desiccation. Some of the NaCl crystals are with large volumes and are embedded between the soil particles or adhere to the structural surface, which plays an important role in cementing the soil particles, as shown in Figure 13d.

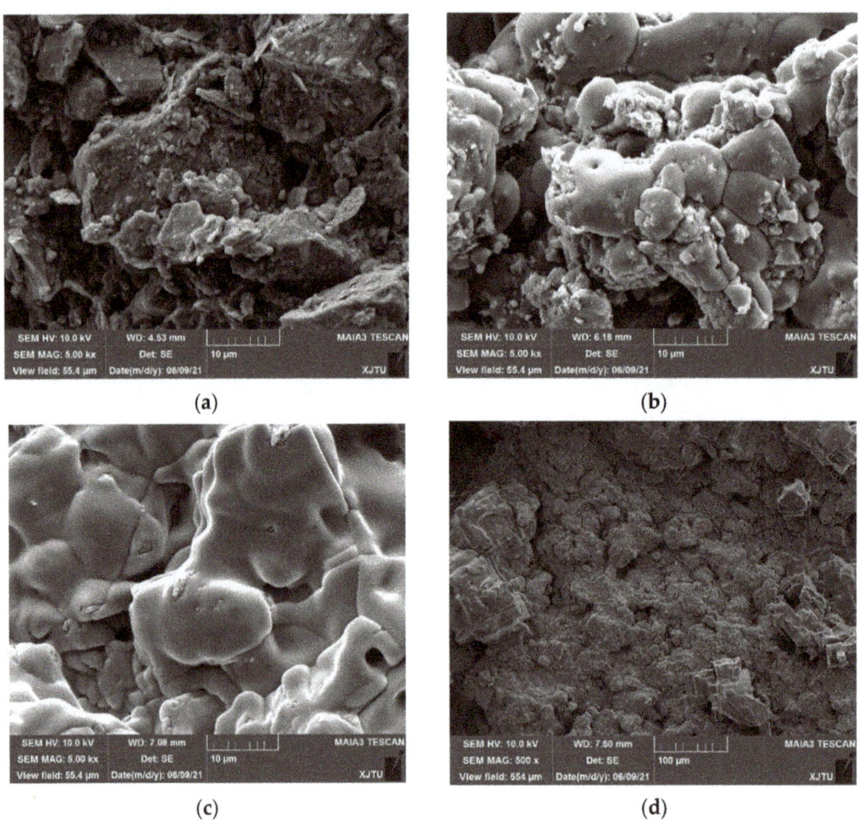

Figure 13. SEM images of loess samples with different NaCl concentration: (**a**) 0% NaCl; (**b**) 3.6% NaCl; (**c**) 7.2% NaCl; (**d**) 10.8% NaCl.

In order to further verify that the smooth layer on the surface of soil particles in the loess sample is caused by NaCl, the energy-dispersive spectrum (EDS) element analysis was combined with the SEM observation in different zones of the loess sample. Figure 14a shows the SEM image of the loess sample with 3.6% NaCl. Two zones, A and B, were selected on loess sample, which is on the soil particles and smooth layer, respectively. The element distributions diagrams in zones A and B, obtained from the EDS analysis are presented in Figure 14b,c. The related contents of elements are shown in Table 3. It is verified that for the smooth layer which cements the particles in zone B, the elements are mainly O, Si, Na and Cl, etc. The content of Na and Cl are larger than the other elements, which are 31.45% and 64.65%, respectively. In zone A, except the elements like Na and Cl, there are Al and Ca compared with the element in zone B. This is due to the existence of moderately soluble salt and insoluble salts, such as SiO_2, $CaCO_3$, Al_2O_3, in the salt content of loess [14]. The moderately soluble salt and insoluble salt were not eliminated during the sample preparation. Figures 15 and 16 show the EDS analysis results of loess samples with 7.2% NaCl and 10.8% NaCl, respectively. The correspondent contents of element are presented in Tables 4 and 5. The results are similar for these loess samples: for smooth layers, the main elements are Na and Cl. Therefore, the SEM observation combined with EDS analysis on loess samples with different concentration of NaCl identified that the smooth layer formed during the drying process is the cementation of NaCl, which increases the cementation of loess particles.

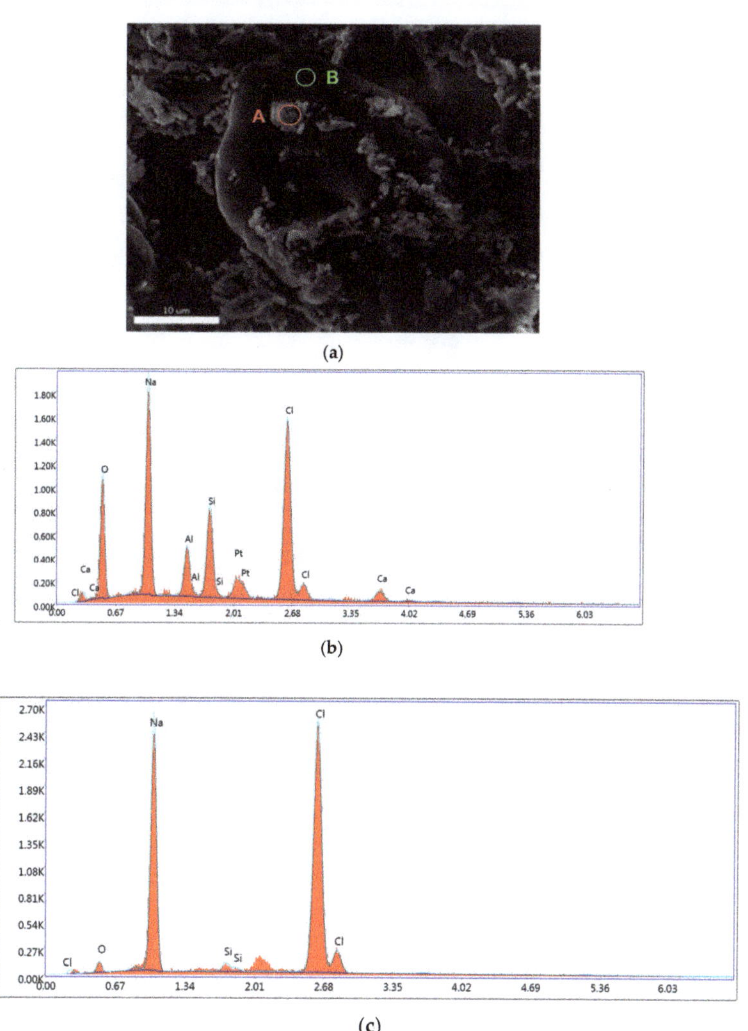

Figure 14. Results of EDS element analysis: (**a**) SEM photo of loess sample with 3.6% NaCl; (**b**) Element analysis in zone A; (**c**) Element analysis in zone B.

Table 3. Comparisons of element contents in zones A and B on the surface of loess sample with 3.6% NaCl.

Element Content (%)	Na	Cl	Si	O	Ca
Zone A	21.37	35.52	9.65	19.29	0
Zone B	31.45	64.65	1.20	2.71	3.63

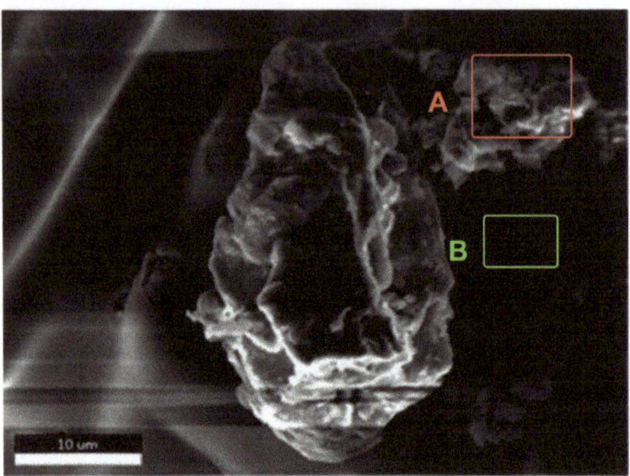

Figure 15. SEM photo of loess sample with 7.2% NaCl (zone A and B represent areas of soil particles and smooth layer, respectively).

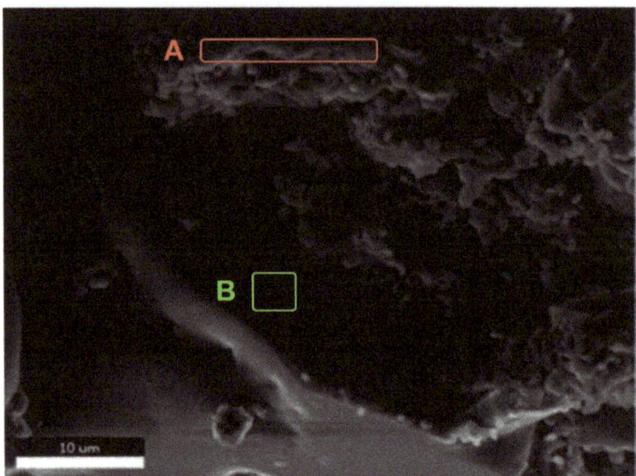

Figure 16. SEM photo of loess sample with 10.8% NaCl (zone A and B represent areas of soil particles and smooth layer, respectively).

Table 4. Comparison of element contents in zones A and B on the surface of loess sample with 7.2% NaCl.

Element Content (%)	Na	Cl	Si	O	Ca
Zone A	27.37	40.84	8.86	19.57	0
Zone B	45.78	52.08	0	2.13	0

Table 5. Comparison of element contents in zones A and B on the surface of loess sample with 10.8% NaCl.

Element Content (%)	Na	Cl	Si	O	Ca
Zone A	9.75	15.88	14.87	43.62	2.53
Zone B	44.57	54.94	0.04	0.45	0

The above microscopic observations and analysis have confirmed the cementation degree in loess samples with different NaCl concentrations. For the loess samples with NaCl concentration of 3.6%, 7.2% and 10.8%, respectively, the flocculation and agglomeration of soil particles and smooth layer caused by Na^+ make the cementation degree between soil particles become larger. Thus, the soil structure becomes denser, leading to the reduction in crack evolution of loess.

4. Conclusions

In this paper, the free desiccation tests were carried out on loess samples with different NaCl concentrations. The whole process of crack initiation and propagation was captured with a high-resolution camera. With the help of the image processing software named PCAS, the variations in the crack parameters of loess samples were assessed. With the DIC method, the local displacement and strain on the surface of the loess sample were calculated. The microstructure and the element distribution of loess samples were obtained by SEM and EDS. Finally, the influence of different NaCl concentration on the cracking characteristics and mechanism in loess was interpreted. The conclusions drawn are as follows:

(1) With the increase in NaCl concentration, the evaporation rate of loess samples gradually decreases and the residual water content gradually increases during the free-drying process.
(2) NaCl in the loess sample will restrain the development of cracks in loess. With the increase in NaCl concentration in the loess sample, some parameters of cracks in the loess sample, such as crack ratio, numbers, and total length, decrease.
(3) Based on the local displacements and strains measurement results of the loess sample, the evolution characteristics of cracks can be predicted. The cracking modes in loess can be classified as opening, sliding and mixed opening–sliding modes. The concentration of NaCl does not affect the cracking modes of loess. The increase in NaCl reduces the accumulation capacity of energy in loess samples and decreases the rate of crack evolution.
(4) During the free desiccation process, the NaCl in the loess sample will react with the mineral components in the clay to form a smooth layer, which has the function of cementing clay particles and making the soil structure compact. The cementation increases with the increase in NaCl concentration, thus restraining the crack evolution.

Author Contributions: Conceptualization, X.W.; methodology, X.W.; software, L.D.; validation, X.C. and Y.Z.; investigation, L.D. and X.C.; resources, X.W.; data curation, L.D.; writing—original draft preparation, X.W.; writing—review and editing, X.W. and L.D.; supervision, X.W.; project administration, X.W.; funding acquisition, X.W. All authors have read and agreed to the published version of the manuscript.

Funding: This research was funded by the National Natural Science Foundation of Youth, grant number 42007278 and the Fundamental Research Funds for the Central Universities, grant number xhj032021017-02.

Institutional Review Board Statement: Not applicable.

Informed Consent Statement: Not applicable.

Data Availability Statement: The data that support the findings of this study are available within the article.

Acknowledgments: Ling. X. of School of Human Settlements and Civil Engineering, Xi'an Jiaotong University provided the experimental devices and aided with the field investigation. The authors would like to extend their deepest gratitude to him.

Conflicts of Interest: The authors declare no conflict of interest.

References

1. Xu, L.; Coop, M.R.; Zhang, M.S.; Wang, G.L. The mechanics of a saturated silty loess and implications for landslides. *Eng. Geol.* **2018**, *236*, 29–42. [CrossRef]
2. Rogers, C.D.F.; Dijkstra, T.A.; Smalley, I.J. Hydroconsolidation and subsidence of loess: Studies from China, Russia, North America and Europe. *Eng. Geol.* **1994**, *37*, 83–113. [CrossRef]
3. DeCarlo, K.F.; Shokri, N. Slinity effects on cracking morphology and dynamics in 3-D desiccating clays. *Water Resour. Res.* **2013**, *50*, 3052–3072. [CrossRef]
4. Zuo, L.; Xu, L.; Baudet, B.A.; Gao, C.Y.; Huang, C. The structure degradation of a silty loess induced by long-term water seepage. *Eng. Geol.* **2020**, *272*, 105634. [CrossRef]
5. Yesiller, N.; Miller, C.J.; Inci, G.; Yaldo, K. Desiccation and cracking behavior of three compacted landfill liner soils. *Eng. Geol.* **2000**, *57*, 105–121. [CrossRef]
6. Morris, P.H.; Graham, J.; Williams, D.J. Cracking in drying soils. *Can. Geotech. J.* **1992**, *29*, 263–277. [CrossRef]
7. Miller, C.J.; Mi, H.; Yesiller, N. Experimental analysis of desiccation cracking propagation in clay liners. *J. Am. Water Resour. Assoc.* **1998**, *34*, 677–686. [CrossRef]
8. Tang, C.S.; Shi, B.; Liu, C.; Zhao, L.Z.; Wang, B.J. Influencing factors of geometrical structure of surface shrinkage cracks in clayey soils. *Eng. Geol.* **2008**, *101*, 204–217. [CrossRef]
9. An, N.; Tang, C.S.; Xu, S.K.; Gong, X.P.; Shi, B.; Inyang, H.I. Effects of soil characteristics on moisture evaporation. *Eng. Geol.* **2018**, *239*, 126–135. [CrossRef]
10. Li, H.D.; Tang, C.S.; Cheng, Q.; Li, S.J.; Gong, X.P.; Shi, B. Tensile strength of clayey soil and the strain analysis based on image processing techniques. *Eng. Geol.* **2019**, *253*, 137–148. [CrossRef]
11. Tang, C.S.; Zhu, C.; Leng, T.; Shi, B.; Cheng, Q.; Zeng, H. Three-dimensional characterization of desiccation cracking behavior of compacted clayey soil using X-ray computed tomography. *Eng. Geol.* **2019**, *255*, 1–10. [CrossRef]
12. Wang, L.L.; Tang, C.S.; Shi, B.; Cui, Y.J.; Zhang, G.Q.; Hilary, I. Nucleation and propagation mechanisms of soil desiccation cracks. *Eng. Geol.* **2018**, *238*, 27–35. [CrossRef]
13. Jiang, N.J.; Tang, C.S.; Yin, L.Y.; Xie, Y.H.; Shi, B. Applicability of Microbial Calcification Method for Sandy-Slope Surface Erosion Control. *J. Mater. Civil Eng.* **2019**, *31*, 04019250. [CrossRef]
14. Zuo, L.; Lv, B.R.; Xu, L.; Li, L.W. The influence of salt contents on the compressibility of remolded loess soils. *Bull. Eng. Geol. Environ.* **2022**, *81*, 185. [CrossRef]
15. Péron, H.; Herchel, T.; Laloui, L. Fundamentals of desiccation cracking of fine-grained soils experimental characterization and mechanisms identification. *Can. Geotech. J.* **2009**, *46*, 1177–1201. [CrossRef]
16. Corte, A.; Higashi, A. *Experimental Research on Desiccation Cracks in Soils*; U.S. Army Snow Ice and Permafrost Research Establishment, Research Report No. 66; Corps of Engineers: Wilmette, IL, USA, 1960.
17. Xu, L.; Coop, M.R. Influence of structure on the behaviour of a saturated clayey loess. *J. Can. Geotech.* **2016**, *53*, 1026–1037. [CrossRef]
18. Wei, X.; Hattab, M.; Taibi, S.; Bicalho, K.V.; Xu, L.; Fleureau, J.M. Crack development and coalescence process in drying clayey loess. *Geomech. Eng.* **2021**, *25*, 535–552.
19. Wei, X.; Bicalho, K.V.; Hajjar, A.E.; Taibi, S.; Hattab, M.; Fleureau, J.M. Experimental techniques for the study of the cracking mechanisms in drying clays. *Geotech. Test. J.* **2020**, *44*, 323–338. [CrossRef]
20. Waler, P. Irrigation on Swelling Clays Soils. Master's Thesis, Agricultural Engineering, Central Region, Moldova, 1989.
21. Shokri, N.; Zhou, P.; Keshmiri, A. Patterns of desiccation cracks in saline bentonite layers. *Transp. Porous Med.* **2015**, *110*, 333–344. [CrossRef]
22. Wei, X.; Hattab, M.; Bompard, P. Highlighting some mechanisms of crack formation and propagation in clays on drying path. *Géotechnique* **2016**, *66*, 287–300. [CrossRef]
23. Wei, X.; Hattab, M.; Fleureau, J.M. Micro-macro-experimental study of two clayey materials on drying paths. *Bull. Eng. Geol. Environ.* **2013**, *72*, 495–508. [CrossRef]
24. Kodikara, J.; Costa, S. Desiccation cracking in clayey Soils: Mechanisms and Modelling. In *Multiphysical testing of Soils and Shales*; Springer: Berlin/Heidelberg, Germany, 2013; pp. 21–32.
25. Li, J.H.; Li, L.; Chen, R. Cracking and vertical preferential flow through landfill clay liners. *Eng. Geol.* **2016**, *206*, 33–41. [CrossRef]
26. Wang, D.Y.; Tang, C.S.; Shi, B. Studying the effect of drying on soil hydromechanical properties using micro-penetration method. *Environ. Earth Sci.* **2016**, *75*, 1–13.
27. Zhang, X.D.; Chen, Y.G.; Ye, W.M. Effect of salt concentration on desiccation cracking behavior of GMZ bentonite. *Environ. Earth Sci.* **2017**, *76*, 531. [CrossRef]
28. Tang, C.S.; Shi, B.; Liu, C.; Suo, W.B.; Gao, L. Experimental characterization of shrinkage and desiccation cracking in thin clay layer. *Appl. Clay Sci.* **2011**, *52*, 69–77. [CrossRef]
29. Lakshmikantha, M.R.; Prat, P.C.; Ledesma, A. *An Experimental Study of Cracking Mechanisms in Drying Soils*; Thomas Telford Publishing: London, UK, 2006; pp. 533–540.
30. Cordero, J.A.; Useche, G.; Prat, P.C. Soil desiccation cracks as a suction–contraction process. *Géotech. Lett.* **2017**, *7*, 279–285. [CrossRef]

31. Lu, H.; Li, J.; Wang, W. Cracking and water seepage of Xiashu loess used as landfill cover under wetting–drying cycles. *Environ. Earth Sci.* **2015**, *74*, 7441–7450. [CrossRef]
32. Zhang, Y.S.; Qu, Y.X. Cements of sand loess and their cementation in North Shaan Xi and West Shan Xi. *J. Eng. Geol.* **2005**, *13*, 18–28, (In Chinese with English abstract).
33. Liu, D.S. (Ed.) *Loess and the Environment*; Science Press: Beijing, China, 1985.
34. Liu, C.; Shi, B.; Zhou, J.; Tang, C.S. Quantification and characterization of microporosity by image processing, geometric measurement and statistical methods: Application on SEM images of clay materials. *Appl. Clay Sci.* **2011**, *54*, 97–106. [CrossRef]
35. Liu, C.; Tang, C.S.; Shi, B. Automatic quantification of crack patterns by image processing. *Comput. Geosci.* **2013**, *57*, 77–80. [CrossRef]
36. Zeng, H.; Tang, C.S.; Cheng, Q.; Zhu, C.; Yin, L.Y.; Shi, B. Drought induced soil desiccation cracking behavior with consideration of basal friction and layer thickness. *Water Resour. Res.* **2019**, *56*, e2019WR026948. [CrossRef]
37. Fang, H.Q.; Ding, X.M.; Jiang, C.Y.; Peng, Y.; Wang, C.Y. Effects of layer thickness and temperature on desiccation cracking characteristics of coral clay. *Bull. Eng. Geol. Environ.* **2022**, *81*, 391. [CrossRef]
38. Wei, X.; Yang, Z.T.; Fleureau, J.M.; Hattab, M.; Taibi, S.; Xu, L. Tensile strength identification of remolded clayey soils. *Bull. Eng. Geol. Environ.* **2022**, *81*, 405. [CrossRef]
39. Lebron, I.; Suarez, D.I.; Schaap, M.G. Soil pore size and geometry as a result of aggregate-size distribution and chemical composition. *Soil Sci.* **2002**, *167*, 165–172. [CrossRef]
40. Malik, M.; Mustafa, M.A.; Letey, J. Effect of mixed Na/Ca solutions on swelling, dispersion and transient water flow in unsaturated montmorillonitic soils. *Geoderma* **1992**, *52*, 17–28. [CrossRef]
41. Shokri, N.; Lehmann, P.; Vontobel, P.; Or, D. Drying front and water content dynamics during evaporation from sand delineated by neutron radiography. *Water Resour. Res.* **2008**, *44*, 1–11. [CrossRef]
42. Ding, J.; Yu, D. Monitoring and evaluating spatial variability of soil salinity in dry and wet seasons in the Werigan-Kuqa Oasis, China, using remote sensing and electromagnetic induction instruments. *Geoderma* **2014**, *235–236*, 316–322. [CrossRef]
43. Feikema, P.M.; Baker, T.G. Effect of soil salinity on growth of irrigated plantation eucalyptus in south-eastern Australia. *Agric. Water Manag.* **2011**, *98*, 1180–1188. [CrossRef]
44. Zhang, G.H.; Yu, Q.C.; Wei, G.Q.; Chen, B.; Yang, L.S.; Hu, C.Y.; Li, J.P.; Chen, H.H. Study on the basic properties of the soda-saline soils in Songnen plain. *Hydrogeol. Eng. Geol.* **2007**, *2*, 37–40.
45. Zhang, Y.; Hu, Z.; Li, L. Improving the structure and mechanical properties of loess by acid solutions—An experimental study. *Eng. Geol.* **2018**, *244*, 132–145. [CrossRef]

sustainability

Article

Spatial and Temporal Analysis of Global Landslide Reporting Using a Decade of the Global Landslide Catalog

Chelsea Dandridge [1,*], Thomas A. Stanley [2,3], Dalia B. Kirschbaum [4] and Venkataraman Lakshmi [1]

1   Engineering Systems and Environment, University of Virginia, Charlottesville, VA 22903, USA
2   GESTAR II, University of Maryland Baltimore County, Baltimore, MD 21250, USA
3   Hydrological Sciences Laboratory, National Aeronautics and Space Administration Goddard Space Flight Center, Greenbelt, MD 20771, USA
4   Earth Science Division, National Aeronautics and Space Administration Goddard Space Flight Center, Greenbelt, MD 20771, USA
*   Correspondence: cld9mt@virginia.edu

Abstract: Rainfall-triggered landslides can result in devastating loss of life and property damage and are a growing concern from a local to global scale. NASA's global landslide catalog (GLC) compiles a record of rainfall-triggered landslide events from media reports, academic articles, and existing databases at global scale. The database consists of all types of mass movement events that are triggered by rainfall and represents a minimum number of events occurring between 2007 and 2018. The GLC collection is no longer being compiled, and the dataset will not be updated past 2018. The research presented here evaluates global patterns in landslide reporting from events in the GLC. The evaluation includes an analysis of the spatial and temporal distribution of global landslide events and associated casualties and comparisons with other landslide inventories. This database has been used to estimate landslide hotspots, evaluate geographic patterns in landslides, and train and validate landslide models from local to global scales. The most notable landslide hotspots are in the Pacific Northwest of North America, High Mountain Asia, and the Philippines. Additionally, the relationship between country GDP and income status with landslide occurrence was determined to have a positive correlation between economic status and landslide reporting. The GLC also indicates a reporting bias towards English-speaking countries. The general goal of this research is to assess the decade of global landslide reports from the GLC and show how this database can be used for rainfall-triggered landslide research.

Keywords: global landslide catalog; rainfall-triggered landslides; landslide reporting; landslide inventory; landslide database

Citation: Dandridge, C.; Stanley, T.A.; Kirschbaum, D.B.; Lakshmi, V. Spatial and Temporal Analysis of Global Landslide Reporting Using a Decade of the Global Landslide Catalog. *Sustainability* **2023**, *15*, 3323. https://doi.org/10.3390/su15043323

Academic Editors: Chong Xu and Jian Chen

Received: 9 December 2022
Revised: 29 January 2023
Accepted: 2 February 2023
Published: 11 February 2023

1. Introduction

Rainfall-triggered landslides are a mounting global concern due to increased frequency of extreme precipitation due to climate change [1–3]. Quality landslide inventories are necessary for assessing landslide risk and hazard [4,5]. NASA's global landslide catalog (GLC) compiles a record of rainfall triggered landslides globally from news reports, academic papers, and pre-existing databases at NASA Goddard Space Flight Center. Landslides are not recorded consistently across the globe, with that being said the GLC represents a minimum number of landslide reports from 2007 to 2018. The original framework and collection methods for the GLC are detailed in Kirschbaum et al. (2010) [6]. Comprehensive analyses of the GLC database were also undertaken in Kirschbaum et al. (2012) and Kirschbaum et al. (2015) [7,8]. Single event entries in the GLC may consist of multiple landslides near one another triggered by the same rainfall event. The GLC is published along with other landslide inventories in the Cooperative Open Online Landslide Repository (COOLR), which contains landslides reported by citizen scientists and other inventories from the broader research community [9]. Even though the GLC is no longer being compiled, landslides can still be reported using the Landslide Reporter application and added to the COOLR collection. The GLC represents all landslides triggered by rainfall including mudslides, rockslides, and debris flows. Each event includes location information (nominal and geographic), time of event, triggering mechanism, type of landslide, relative size, location accuracy, impacts such as estimated economic damage, casualties, and fatalities. The location accuracy is based on a qualitative radius of confidence in kilometers. Location accuracy provided in the GLC is affected by the capability of the news reports to convey the location information and preciseness. Additionally, the GLC may not be complete for non-English-speaking countries due to the collection method and most reports originating from English-language media [8]. The year 2010 yielded the highest number of annually recorded events in the GLC, which can be attributed to unusually high precipitation and increased landslide events globally compared to other years. A detailed analysis of the anomalies in 2010 has been performed and discussed by [7].

Historically, there have been few efforts to compile landslides at global scale [10]. Detailed landslide inventories are more commonly found at regional or national scale [11–15] or are event-based after a significant rainfall event [16–18]. However, Froude and Petley (2018) present a global database of fatal landslides from 2004 to 2016, but it does not include non-fatal events [4]. There are several methods to produce landslide inventories and most are event-based or region-specific. For example, Bessette-Kirton et al. (2019) present landslides triggered by Hurricane Maria in 2017 in Puerto Rico that were mapped using post-storm satellite and aerial imagery [19]. From their mapping, an estimated 40,000 landslides occurred as a result of the hurricane event. The national landslide inventory for Cuba only represents landslides that cause major damage and does not reveal qualitative information for most of the landslides in the inventory or represent the entire country [11]. The AVI project is the most extensive database for landslides and floods in Italy with events manually gathered from historical news articles and scientific reports from 1918 to 1990 [20]. Alternatively, Amatya et al. (2019) utilized high-resolution imagery from 2012 and 2018 with object-based image analysis to map a landslide inventory along the Karnali Highway in Nepal, and then several inventories in Southeast Asia in Amatya et al. (2021) [17,21]. This method was able to identify almost 60% of the landslides identified manually. The Fatal Landslide Event Inventory of China (FLEIC) is a record of 1911 landslides from 1950 to 2016. Each entry represents a single landslide gathered from geological records, media reports, and literature [22]. The GLC can be used to supplement existing inventories that may be limited temporally or spatially.

Several landslide studies have used the GLC in their creation of landslide databases and to assess global patterns of landslides. Lin and Wang (2018) [22] used events in the GLC to analyze fatal landslides in China and create the Fatal Landslide Event Inventory of China (FLEIC). They also used the same methodology to determine the location and radius of confidence for additional landslides in China. This study noted that the short time period available in the GLC (less than ten years) limited temporal trend analysis. Chandrasekaran et al. (2013) investigated damages to infrastructure in Nilgiris, India caused by rainfall-induced landslides using the GLC [23]. They found that roughly ten percent of global landslide fatalities occurred in India. Benz and Blum (2019) proposed an algorithm to detect global landslide clusters triggered by the same rainfall event and applies it to events reported in the GLC [24]. They found that more than 40% of events can be related to another event, and 14% of events are part of a cluster with more than ten landslides triggered by the same rainfall event with results varying greatly geographically. Culler et al. (2021) evaluates post-fire landslide susceptibility in different regions across the globe using the GLC for landslide events and comparing antecedent precipitation at burned and unburned locations from MODIS [25]. They found that wildfires increase landslide susceptibility, but post-fire landslides are not uniform and vary geographically. This study noted that the GLC was chosen because it offered the largest spatial and temporal range of any landslide inventory. Whiteley et al. (2019) included the GLC in their review of geophysical monitoring of global rainfall-triggered landslides to draw the conclusion that landslide distribution is not uniform across the globe [26]. Additionally, Froude and Petley (2018) used the GLC in their analysis of global fatal landslides from 2004 to 2016 [4]. They found that most fatal landslide clusters occur around cities in countries with lower gross national income.

Furthermore, the GLC has been used in several studies to train and validate landslide models. Lin et al. (2017) used landslides from the GLC and the World Geological Hazard Inventory as training and validation data for a logistic regression model for global landslide susceptibility [27]. Farahmand and Aghakouchak (2013) used 581 events from 2003, 2007, 2008, and 2009 in the GLC as training and validation data for a support vector machines machine learning algorithm to predict global landslides [28]. This study noted that the GLC is composed of major landslides and therefore, the model is not calibrated for small landslides. The model reliably predicted historical landslides. Liao et al. (2010) proposed an early warning system for rainfall-triggered landslides over Java Island, Indonesia using events in the GLC from 2003 and 2007 and the SLIDE (slope infiltration distributed equilibrium) model with NASA's Tropical Rainfall Measurement Mission (TRMM) precipitation estimates [29]. Kirschbaum et al. (2015a) used the GLC within the LHASA framework for Central America and Hispaniola [30]. This study related GLC events to long-term precipitation from TMPA from 2001 to 2013. The LHASA model was able to correctly identify the potential for most GLC events. This study noted that the GLC is the only event-based database for landslides across all countries within Central America and the Caribbean region. Furthermore, Kirschbaum et al. (2012) compared GLC events from Hurricane Mitch in Central America in 1998 to the susceptibility maps globally and regionally, and all the nine landslides occurred in high susceptibility zones in the regional map, and eight landslides occurred in high susceptibility zones in the global map [31].

Following the analysis of the GLC by Kirschbaum et al. (2015b), this research further examines various attributes and limitations associated with the landslide information reported by the GLC using a longer time period of recorded events [6]. Global patterns of landslides and associated fatalities are examined both geographically and temporally. Fatality and landslide hotspots are assessed at global and continental scale. The GLC has not previously been compared against other landslide inventories. Here, we evaluate twenty-seven region-specific event inventory databases and their similarities and differences against events reported in the GLC. A quantitative spoken language analysis has yet to be performed with the GLC reports. To determine if there is any bias regarding the language of reported events in the GLC, the relationship between spoken language and landslide reporting is investigated. The economic status of varying regions could affect the amount and method of landslide reporting, thus economic variables are compared against landslide activity reported in the GLC to determine the effect of economic status on landslide event reporting. The results of this study give a deeper understanding of the global patterns of landslide reporting, and are also useful for applying the GLC in rainfall-triggered landslide analysis.

2. Data

This section lists and describes the datasets used for analysis in this research and organizes the information in Table 1.

Table 1. Data descriptions and source information.

Name	Description	Source
Global Landslide Catalog	Global compilation of rainfall-triggered landslides from 2007 to 2018	[6]
Area (km^2)	A country's total area	[32]
Population Density (people/km^2)	Midyear population divided by land area in square kilometers	[33]
GDP per Capita (USD)	Gross Domestic Product (GDP) divided by the midyear population	[34]
Landslide Hazard	Global landslide susceptibility map	[5,35]
Landslide Inventories	Landslide event inventories from 2008 to 2018	[3,17,36–38]
	Region-specific landslide inventory for the Apulian region of southern Italy collected from 2008 to 2016	[39]

2.1. GLC

The Global Landslide Catalog is a record of landslides collected from news reports, academic articles, and pre-existing inventories. There are a total of 11,334 landslides in the GLC between 2007 and 2018, which are publicly available as geospatial point or tabular data from [6].

2.2. Land Area

The land area is a country's total area, excluding area under inland water bodies, national claims to continental shelf, and exclusive economic zones. The land area estimates are reported in km^2. A full data description and download information are available from [32].

2.3. Population Density

The population density is the midyear population divided by land area in square kilometers. A full data description and download information are available from [33].

2.4. Gross Domestic Product per Capita

Gross Domestic Product (GDP) is the sum of gross value added by all resident producers in the economy plus any product taxes and minus any subsidies not included in the value of the products. The GDP per capita is the GDP divided by the midyear population. Values are collected from the latest year available for each country. A full data description and download information are available from [34].

2.5. Landslide Susceptibility

The global landslide susceptibility map has a spatial resolution of roughly 1 km and covers most of the world's land surface, but not Antarctica. The susceptibility model combines slope, geology, road networks, faults, and forest loss using a heuristic fuzzy methodology [5,35]. Susceptibility is classified by increasing severity as very low, low, medium, high, and very high. The receiver operating characteristic (ROC) curve and area under the curve (AUC) is used to assess the performance of the global susceptibility map. The AUC ranges from 0.61 to 0.85 when compared against local landslide inventories from varying locations across the globe, and the AUC when compared against the GLC is 0.82. Further explanation on the quality of the map can be found in Stanley and Kirschbaum (2017) [35]. In this research, the global susceptibility rating is extracted for each point representing an event in the GLC.

2.6. Landslide Inventories

Twenty-seven landslide event inventories from 2008 to 2018 were created via object-based image analysis, developed by Amatya et al. (2022); Chang et al. (2014); Chen et al. (2013); Marc et al. (2018); and Van Westen and Zhang (2018) [17,18,36–38]. These inventories consist of landslides mapped after considerable rainfall events in various locations. Most landslides are able to be detected using object-based image analysis, but the areas are not always accurate and manual correction is often necessary [17]. Additionally, a landslide database for the Apulian region of southern Italy was collected from 2008 to 2016 by Vennari et al. (2022) and used for comparison in this study [39]. The event inventories are provided as point or polygon shapefiles which are compared directly to the points provided in the GLC using GIS to determine how many GLC points lie within the inventories mapped extent.

3. Results

3.1. Geographic and Temporal Distribution

A global landslide susceptibility map produced by Emberson et al. (2020) and Stanley and Kirschbaum (2017) is overlaid with landslide events and hotspot estimation and shown in Figure 1A [5,35]. The susceptibility map is used to allocate a susceptibility rating for each point in the GLC by extracting the susceptibility map pixel value at each event location. The most notable landslide hotspots, which are highlighted in red, are in the Pacific Northwest,

High Mountain Asia, and the Philippines. Figure 1B shows the number and percentage of events as well as the number of fatalities for each susceptibility class. Regarding all GLC events from 2007 to 2018, 77% of reports occur in medium or higher susceptibility areas, and 32% of reports occur in very high susceptibility areas. Furthermore, areas described as having very low susceptibility are reported to have the least number of events with only 6.8% of all landslides and have the least number of fatalities associated with landslides. Overall, the number of fatalities increases with susceptibility as shown by the blue line. Figure 1C shows the average number of events in black and the average number of fatalities in blue by month. On average, most events are reported in the months July, August, and September, and the least reports occur in February and November. Most fatalities are reported in June and August, and like the number of reports, the least number of fatalities are reported in February and November.

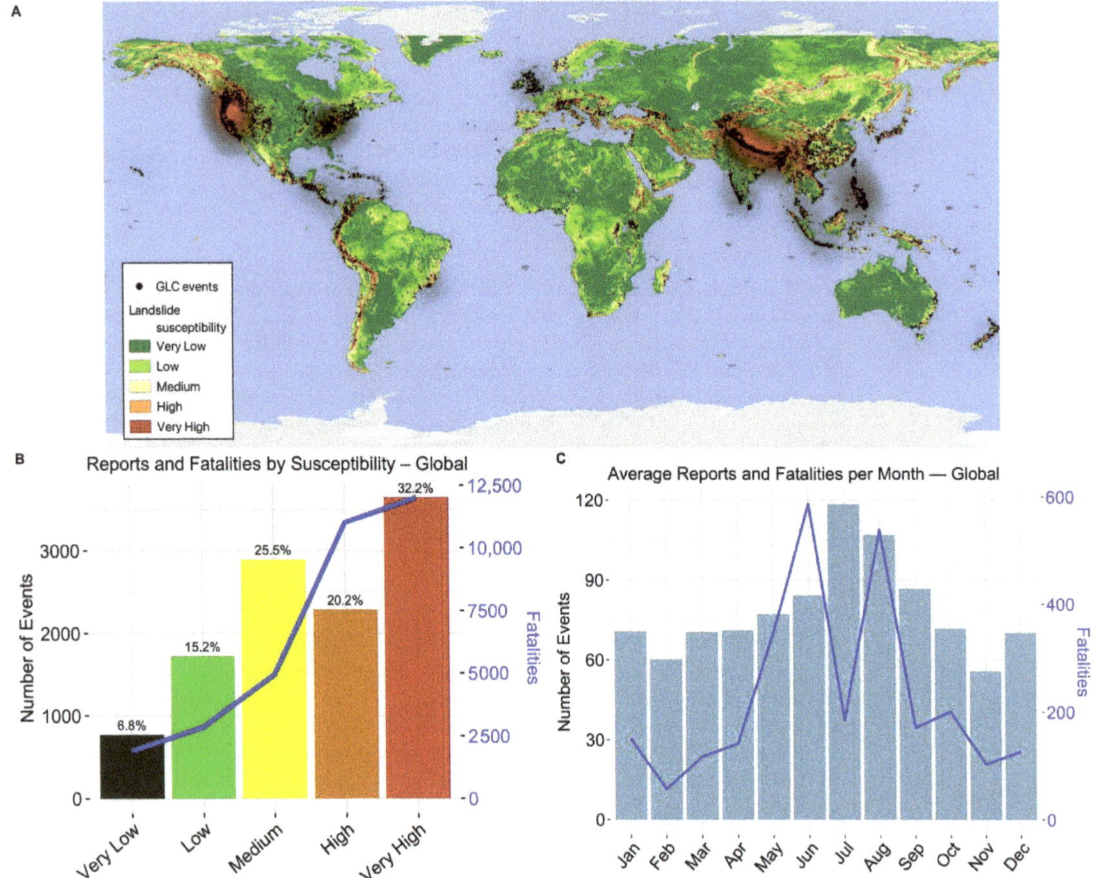

Figure 1. (**A**) Global susceptibility map overlaid with landslide events and hotspot estimation (red); (**B**) Total and percentage of reports and number of fatalities for each landslide susceptibility class; (**C**) Average number of reports and fatalities for each month.

Landslide reports in the GLC are not widespread in Africa as shown in Figure 2A, which displays the landslide susceptibility map overlaid with landslide events and fatalities. Parts of Northern and Central Africa do not have many or any reported events in the GLC despite there being an indication of high landslide susceptibility. The absence of events in this area can be explained by the lack of precipitation over the Sahara Desert. The largest landslide clusters are located near the East African Rift. However, the most fatal landslide reported in Africa occurred on the west coast in Sierra Leone in August of 2017 and had 1141 associated fatalities and resulted in over 3000 displaced persons. Interestingly, the highest percentage of events, approximately 31%, are reported in low susceptibility areas as shown in Figure 2B, which also reports the number of fatalities associated with each susceptibility class, which is most likely due to poor location accuracy in the African region. In Africa, only 12 events are known within 1 km, and 69 events have a location accuracy greater than 5 km. The most fatalities, however, occur in highly susceptible areas. Figure 2C reports the monthly average number of events and average fatalities per month. While May on average has the most landslide reports, the most reported fatalities occur in August due to the substantial fatalities associated with the 2017 Sierra Leone event. In addition, March and August show significantly more reported fatalities than other months when analyzed over 2007–2018.

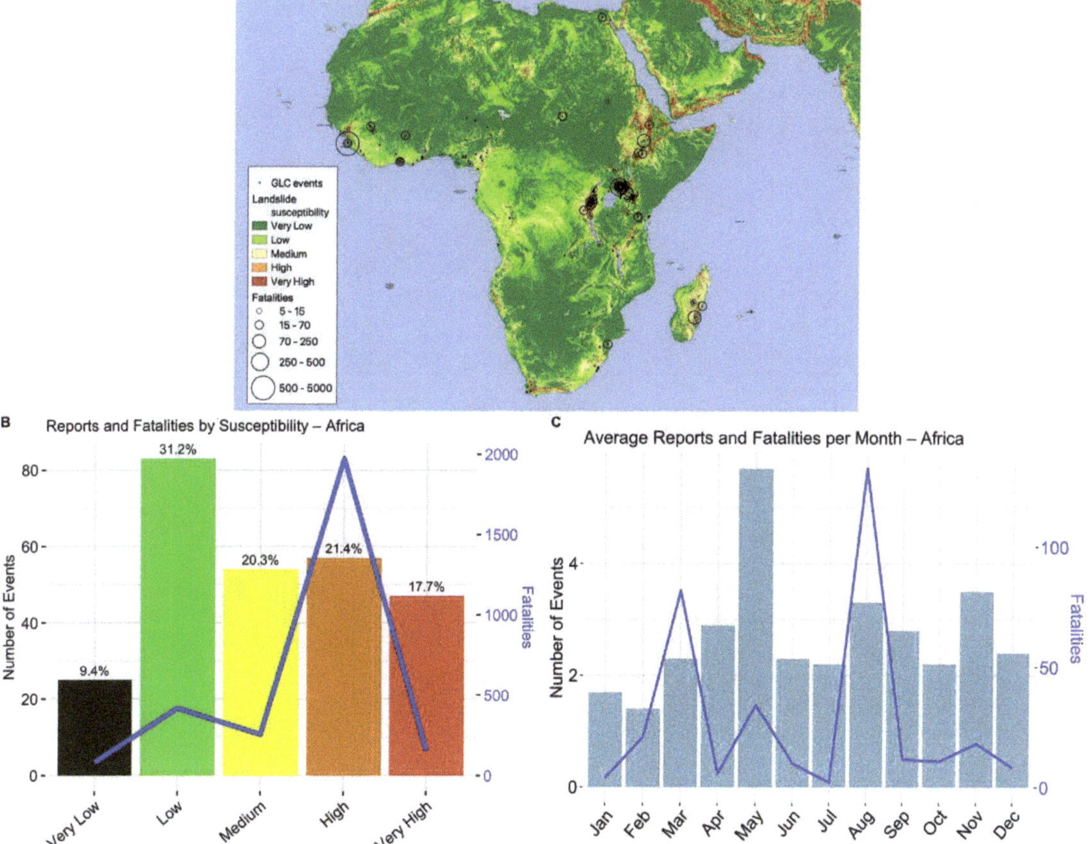

Figure 2. (**A**) Susceptibility map of Africa overlaid with landslide events and fatalities; (**B**) Total and percentage of reports and number of fatalities for each landslide susceptibility class over Africa; (**C**) Average number of reports and fatalities for each month over Africa.

The landslide susceptibility map and reported events as well as the distribution of fatalities throughout Asia are shown in Figure 3A. Many reported events in Asia and fatalities in the GLC occur near the Himalayan Arc. Another landslide hotspot in Asia is the Philippines, reporting 769 events between 2007 and 2018. The largest fatal event in the GLC was reported to have a death toll of 5000 people and occurred after the highest 24 h rainfall in city history triggered a very large landslide near Kedarnath, India in 2013. Oddly, only 22 total events were reported in the Russian Federation from 2007 to 2018 despite the presence of large areas susceptible to landslides. This may indicate that landslide events in the Russian Federation are underrepresented. Figure 3B shows the largest portion (38%) of reports and the most fatalities occurred in very high susceptibility areas. Generally, the amount of fatalities reported increases with susceptibility. Looking at the average number of reports and fatalities by month in Figure 3C, most landslides were reported in June–October with most fatalities being reported in June and August. This trend correlates with the monsoon season experienced in Southeast Asia which lasts approximately May to October [40,41].

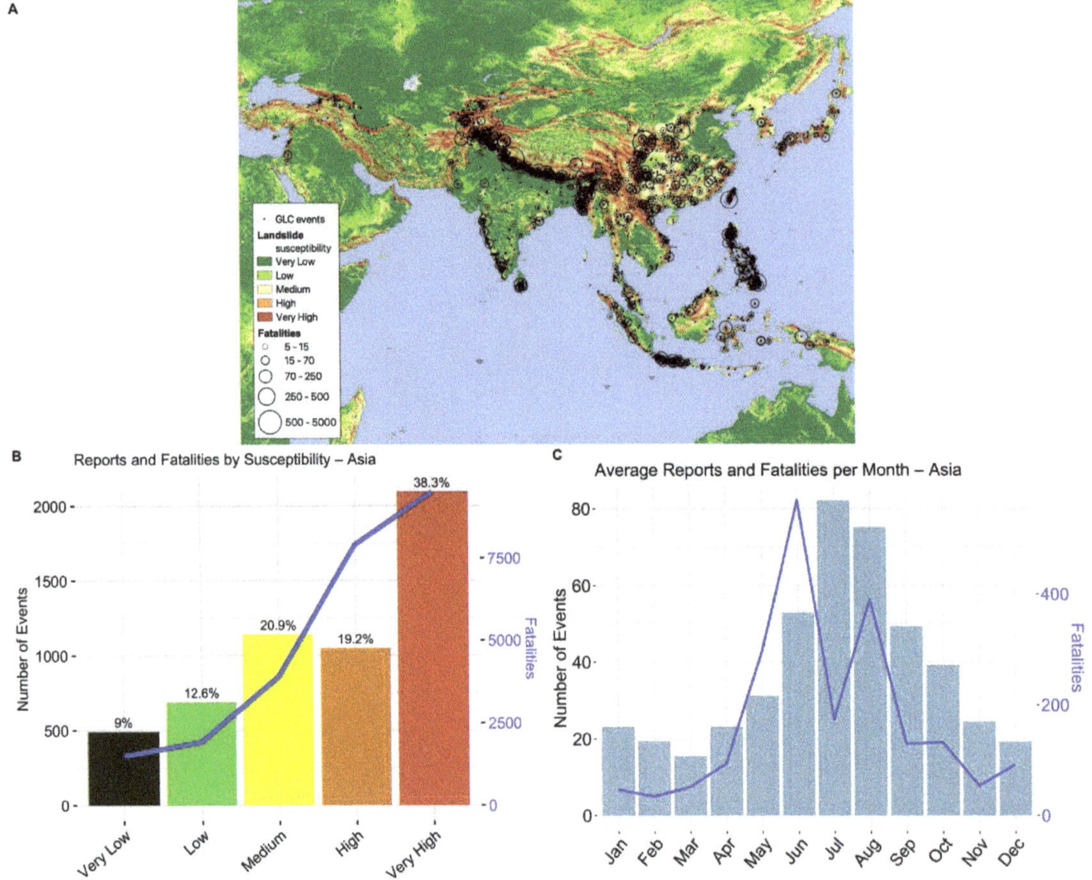

Figure 3. (**A**) Susceptibility map of Asia overlaid with landslide events and fatalities; (**B**) Total and percentage of reports and number of fatalities for each landslide susceptibility class over Asia; (**C**) Average number and percentage of reports and fatalities for each month over Asia.

There are fewer fatal landslides reported in Europe compared to other regions in the world as observed from Figure 4A, which shows the landslide susceptibility and GLC events with associated fatalities over Europe. The highest percentages of events were reported in low (28%) and medium (23%) susceptibility areas, while 22% of events were reported in very high susceptibility areas. However, the highest number of fatalities occurred in locations with very high susceptibility, and only eight fatalities were reported in areas considered to have very low landslide susceptibility from 2007 to 2018 (Figure 4B). Figure 4C shows the average number of landslide events and associated fatalities by month for Europe. The months of February and June have the most reported landslide events in the GLC, while February and October have the most reported fatalities on average from events in the GLC (Figure 4C).

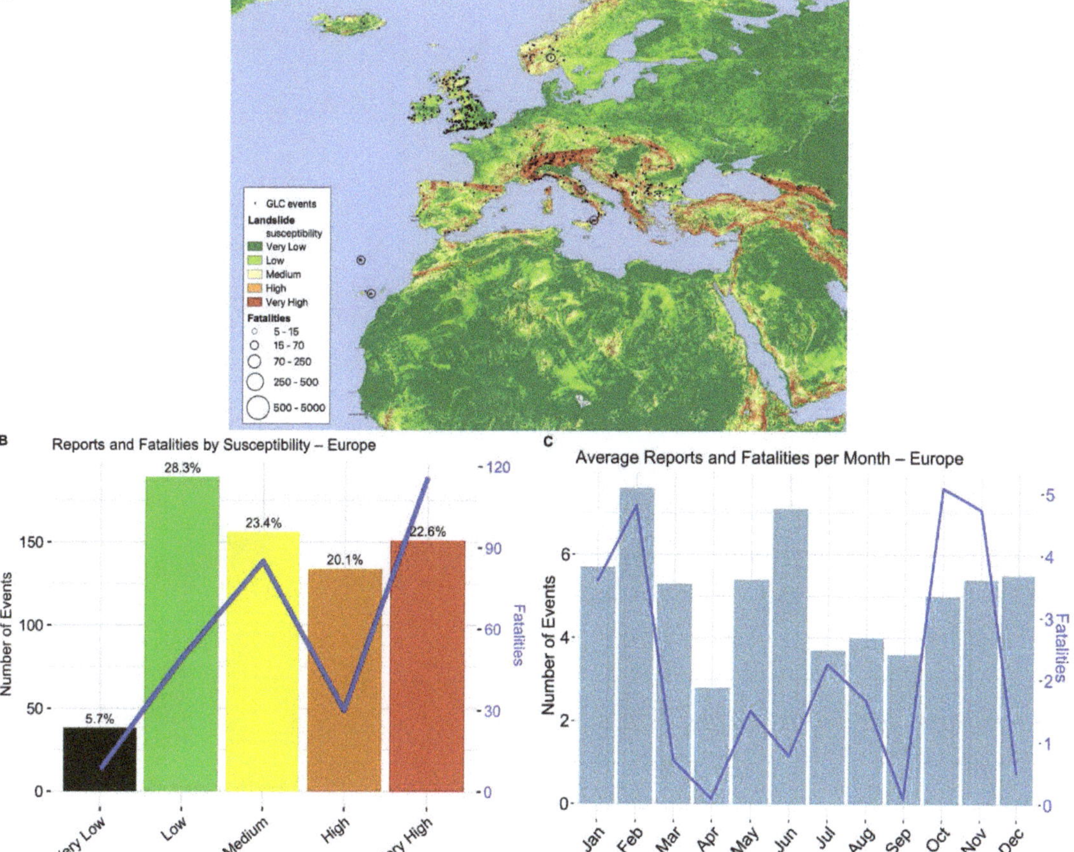

Figure 4. (**A**) Susceptibility map of Europe overlaid with landslide events and fatalities; (**B**) Total and percentage of reports and number of fatalities for each landslide susceptibility class over Europe; (**C**) Average number and percentage of reports and fatalities for each month over Europe.

Most fatal landslides in North America are reported in Mexico, Central America, and the Caribbean islands compared to events in the United States and Canada. This can be observed in Figure 5A, which displays the landslide susceptibility map overlaid with GLC events and reported fatalities in North and Central America. While most events (31.9% of reports) occurred in medium susceptibility zones, 26.8% of reports occur in very high susceptibility zones, and the number of fatalities reported in very high susceptibility areas amounts to almost 1000 casualties, which is significantly higher than the lesser susceptibility classifications (Figure 5B). Figure 5C shows the average number of reported landslide events are highest in March and December and less reports occur from June to November. The associated fatalities by month for North America are noticeably higher from August through October, which could be due to two specific outlier events that occurred in Guatemala in 2015. The August event reported 253 fatalities, and the October event reported 280 fatalities. In the months from December through April, less than five fatalities are reported on average which is significantly less compared to other months.

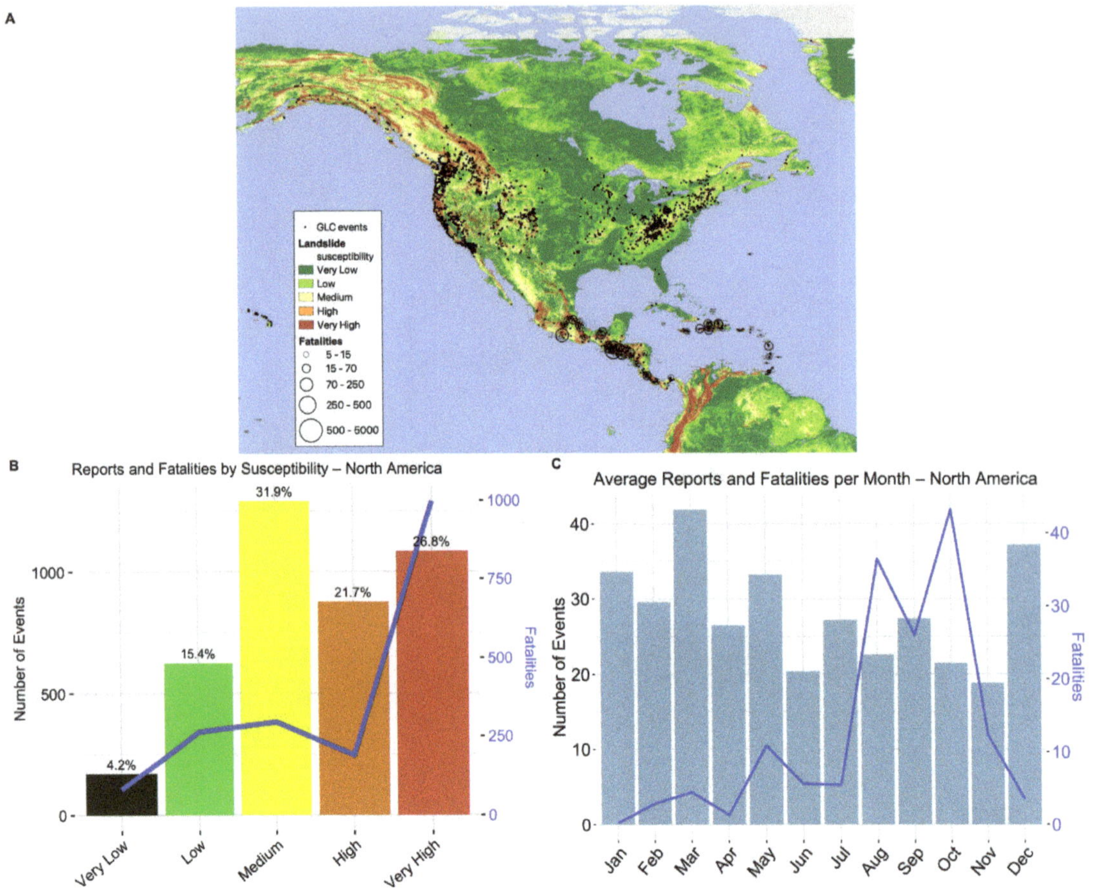

Figure 5. (**A**) Susceptibility map of North America overlaid with landslide events and fatalities; (**B**) Total and percentage of reports and number of fatalities for each landslide susceptibility class over North America; (**C**) Average number and percentage of reports and fatalities for each month over North America.

Regarding Oceania, most landslide events occur on the East Coast of Australia, Papua New Guinea, and New Zealand, which can be seen in Figure 6A, which shows the landslide susceptibility map overlaid with GLC events and corresponding fatalities. There are a total of 98 fatalities from 21 reports from seven countries represented in the GLC from Oceania. Forty-eight percent of fatalities in Oceania, which totals 47 fatalities, occurred in low susceptibility areas which is significantly more than the higher susceptibility categories (Figure 6B). The two most fatal landslides were reported in Papua New Guinea in low susceptibility areas: one in Hela and one in the Eastern Highlands, which report 25 and 10 fatalities, respectively. There may be underrepresentation in the reporting of fatalities due to landslides in Australia, having only one fatality and no injuries reported out of all the events reported in the country, even though there are references of rainfall-triggered landslides that led to additional fatalities and injuries. For example, five people were injured after a landslide blocked a portion of railway tracks and caused the derailment of the train on 26 March 2010 as reported by Leiba (2013) [42]. Looking at the average events and fatalities by month in Figure 6C, there appears to be a trend of higher fatalities from October to January throughout Oceania. However, no clear pattern appears in the number of reported events by month in this region.

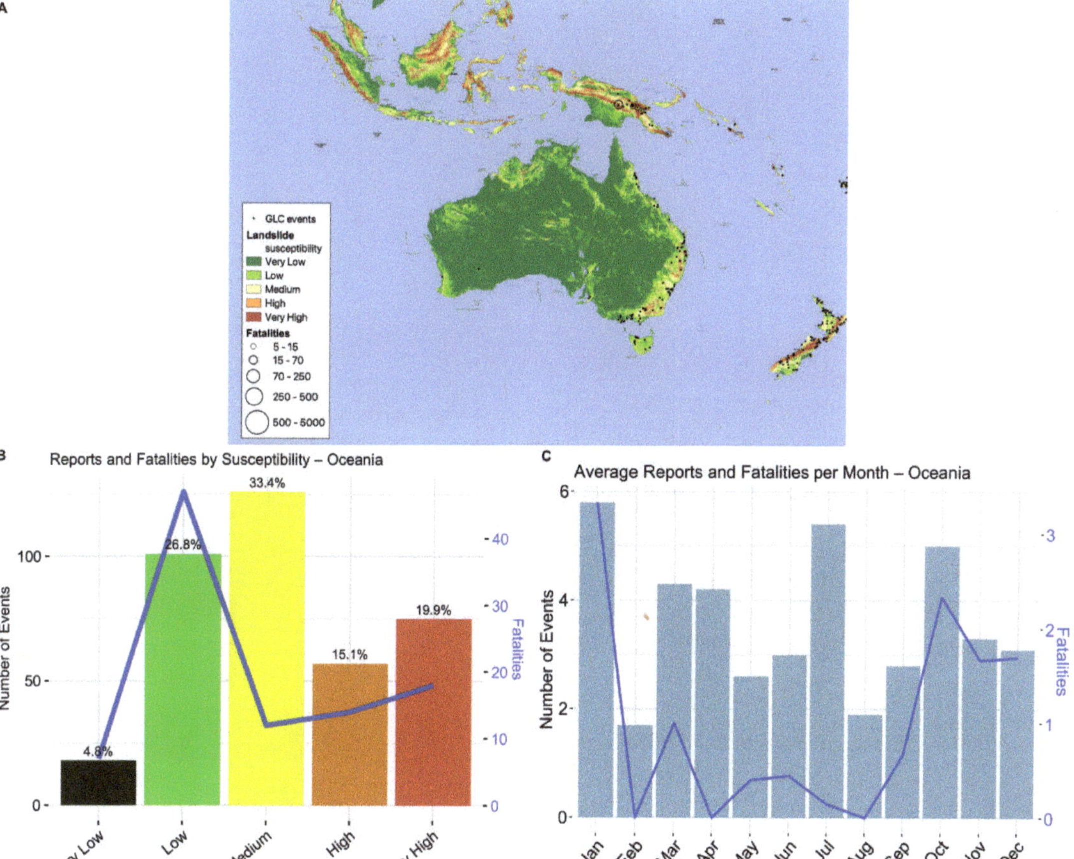

Figure 6. (**A**) Susceptibility map of Oceania overlaid with landslide events and fatalities; (**B**) Total and percentage of reports and number of fatalities for each landslide susceptibility class over Oceania; (**C**) Average number and percentage of reports and fatalities for each month over Oceania.

Landslides in South America are largely reported on the West coast and around Rio de Janeiro. This pattern is shown in Figure 7A which also shows the landslide susceptibility of South America overlaid with reported events and fatalities from the GLC. A large cluster consisting of 209 landslides and numerous fatalities are clustered near Rio de Janeiro, a large and densely populated seaside city in Brazil. Throughout South America, most landslide events and fatalities were reported in very high susceptibility locations compared to lower susceptibility classifications, which is shown in Figure 7B. Two events that reported 378 and 424 fatalities, respectively, both occurred in January

2011 in Rio de Janeiro. These represent the two highest numbers of fatalities associated with a landslide in South America in the GLC. Looking at the monthly average reported events and fatalities in South America in Figure 7C, January reports significantly more fatalities on average, which is likely due to the two events with outlier fatality counts. Alternatively, the period from May through October shows a pattern of less reported events and fatalities compared to other months (Figure 7C).

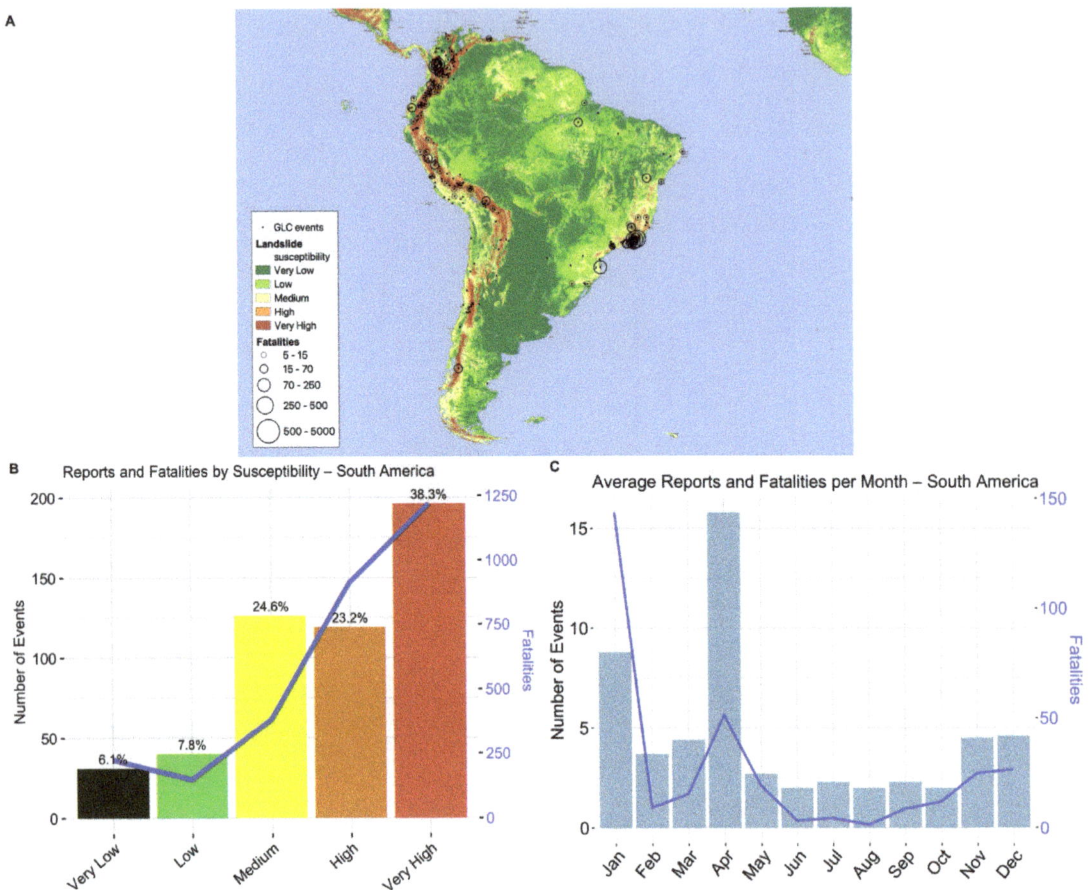

Figure 7. (**A**) Susceptibility map of South America overlaid with landslide events and fatalities; (**B**) Total and percentage of reports and number of fatalities for each landslide susceptibility class over South America; (**C**) Average number and percentage of reports and fatalities for each month over South America.

3.2. Spoken Language Analysis

The reporting language of GLC events was assessed to determine the extent of bias regarding the language of reported events in the GLC. The relationship between spoken language and landslide reporting is investigated and reported in Figure 8. The spoken language for each event was designated by the official language of the country reported in the GLC. The total number and percentage of reports by language are shown in Figure 8A, and the number of fatal events and associated fatalities are shown in Figure 8B. While the plurality (41%) of total events reported in the GLC are from English-speaking countries, the number of fatal events and total fatalities are highest for Hindi-speaking countries. The large number of fatalities reported in Hindi is a result of the 5000 fatalities associated with a single event in India. When excluding this outlier event, Chinese followed by English and Spanish have more total fatalities reported than Hindi. Hindi, Spanish, Chinese, and Nepali-speaking countries report more fatal events than English. Furthermore, India, China, Nepal, and the Philippines make up the countries with the highest number of fatal landslide events in the GLC, reporting 321, 258, 220, and 209 fatal events, respectively. This shows a reporting bias towards English-speaking countries and bias in the collection methods of the reports. More fatal landslides are reported more often than non-fatal landslides regardless of the language spoken in that location.

3.3. Attribute Assessment

A general assessment of the GLC event attribute information is performed to observe trends in the event data. The known versus unknown values of several important attributes that are reported for each event are compared. The landslide category, location accuracy, size, and triggering mechanism attributes were mostly complete with known values for 90% or greater of reported events. However, the event time and landslide setting attributes were reported with more unknown values than known. Over 50% of events in the GLC are known within a five kilometer radius of confidence as shown in Table 2. Only 6.8% of events did not report a radius of confidence, and less than 20% of events have a radius of confidence greater than 25 km. This shows that generally, the location of the landslide reported from GLC events is relatively known within a small radius. The location accuracy is beneficial when using the GLC events for application in model training or validation. Furthermore, the distribution of the event size reported in the GLC is evaluated and the results are displayed in Table 3, which shows 72% of GLC events as medium size, and less than 9% of events are classified as large or very large. As the GLC is a collection of rainfall-triggered landslide events, we evaluated the triggering mechanism reported for each event. The distribution of triggering mechanisms are shown in Table 4, and 83% of events from 2007 to 2018 are in response to some type of precipitation. This attribute can be used to filter events for rainfall-triggered landslide studies that implement the use of the GLC. The reported events in the GLC per year are assessed and a cumulative sum per year is shown in Figure 9. The first three years on record, 2007–2009, consist of less reported events compared to later years, which could indicate that the first several years of landslide event collection are incomplete. The cumulative sum was also assessed by continent and similarly showed fewer reports for the first few years of record. No other regional patterns were discernible from the annual analysis. Furthermore, the reporting of events in the GLC by day of the week per year was evaluated in this study, and no distinct pattern was observed. The year 2010 holds the highest number of landslide events, and as mentioned, this anomaly year has been investigated and justified by extreme and prolonged precipitation in several landslide prone regions by Kirschbaum et al. (2012) [8]. This year also experienced greater numbers of fatal landslides and fatalities associated with landslides.

Table 2. Distribution of the radius of confidence of the reported location for all events.

Location Accuracy	Number of Events	Percentage of Total Events (%)
Exact	1386	12.7
1 km	2185	19.8
5 km	3178	28.8
10 km	1435	13.0
25 km	1470	13.3
50 km	794	7.2
100 km	25	0.2
250 km	16	0.1
Unknown	542	4.8

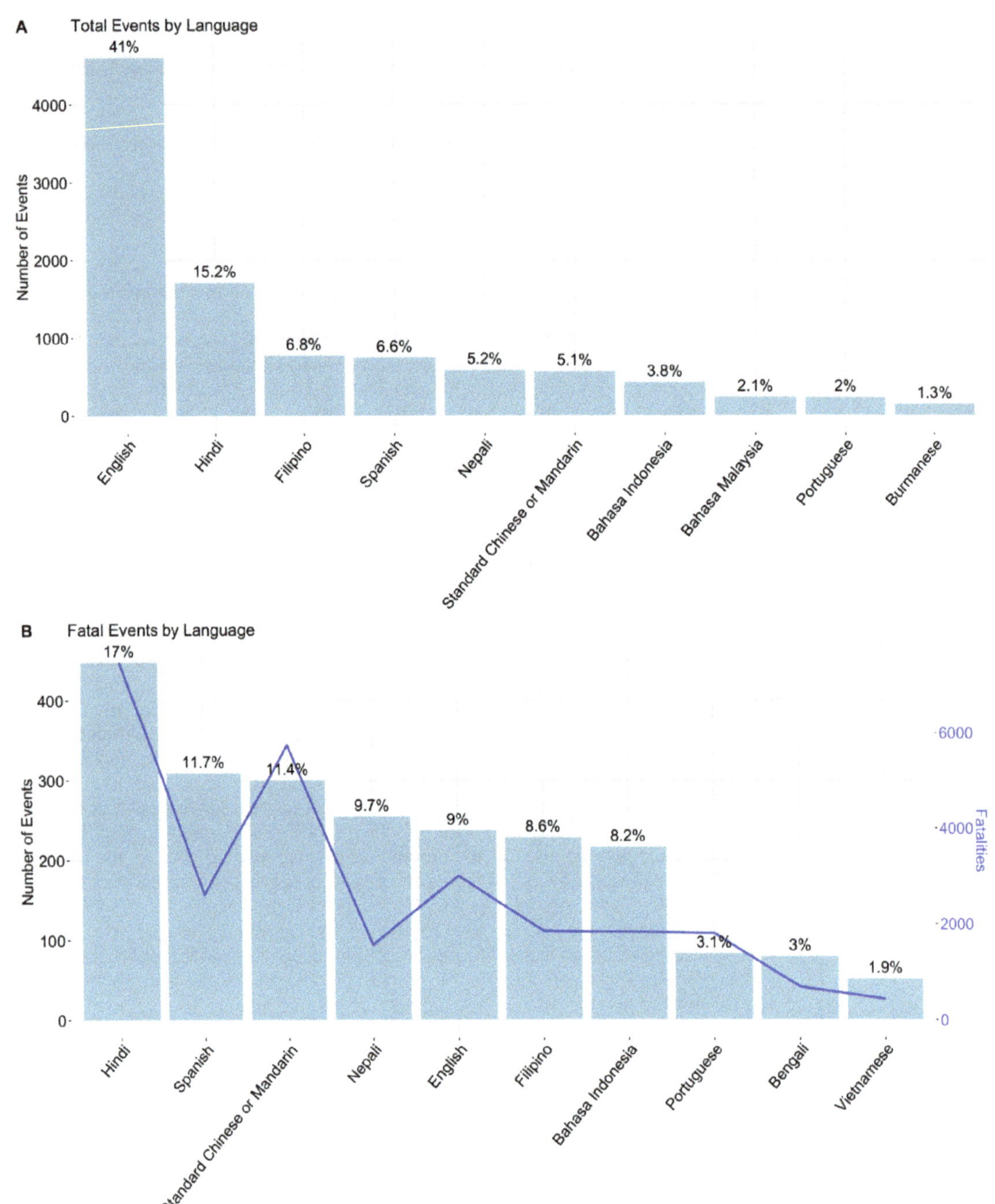

Figure 8. (**A**) Total number and percentage of total events by language; (**B**) Number of fatal events, percentage of total events, and number of fatalities by language.

Table 3. Distribution of reported landslide size for all events.

Size	Number of Events	Percentage of Total Events (%)
Small	3199	28.2
Medium	6880	60.7
Large	900	7.9
Very large	118	1.0
Catastrophic	8	0.1
Unknown	229	2.0

Table 4. Distribution of triggering mechanism for reported events in the GLC.

Triggering Mechanism	Number of Events	Percentage of Total Events (%)
downpour	4836	42.68
rain	2874	25.37
unknown	1174	10.36
continuous rain	839	7.41
tropical cyclone	601	5.30
monsoon	245	2.16
snowfall snowmelt	155	1.37
mining	117	1.03
construction	102	0.90
earthquake	99	0.87
flooding	92	0.81
no apparent trigger	73	0.64
freeze–thaw	41	0.36
other	41	0.36
dam embankment collapse	19	0.17
Leaking pipe	17	0.15
volcano	4	0.04
vibration	1	0.01

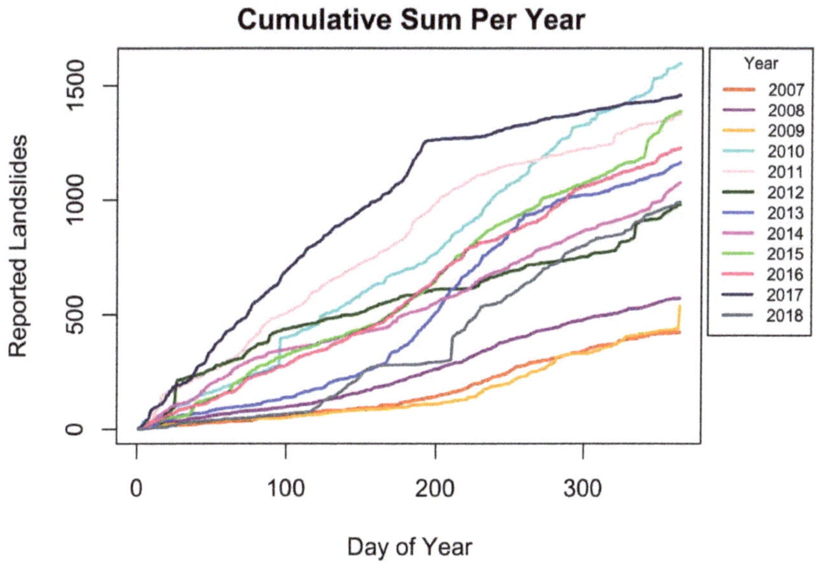

Figure 9. The cumulative sum of reported events in the GLC for each year from 2007 to 2018.

3.4. Inventory Comparison

There are landslide inventories available at global scale as well as focused on individual countries or states. These inventories vary in terms of source, mapping method, and attribute information. To compare the GLC against other landslide inventories, we look at several region-specific inventory databases and their similarities and differences against the GLC methodology. The location, number of points in the inventory, rainfall event date, number of points represented in the GLC, and source of each inventory are shown in Table 5. Only 9 out of the 27 event inventories are represented in the GLC by at least one point. A total of 25 points in the GLC coincide with the extent of the event inventories, which collectively report 17,893 landslides. One small, twelve medium, eight large, and four very large landslides represent the sizes reported by the GLC points and are shown in the reported size column in Table 5. Additionally, the location accuracy reported by the GLC was evaluated using the event inventories and are reported in the location accuracy column of Table 5. Of the 25 points, 22 contained event inventory landslides reported within the location accuracy radius, 3 were near the landslide inventory area but exceeded the reported location accuracy radius, and 1 GLC point that was 5 km from the event inventory reported the location as unknown. These event inventories consist of between 131 and 21,379 points or polygons, and the GLC consists of at most 16 points for the corresponding event in Teresópolis, Brazil, which provides 7268 landslide polygons. Additionally, we compared the GLC events to a local landslide inventory from Vennari et al. (2022) for the Apulian region of southern Italy collected from 2008 to 2016 [39]. We found that no landslides are reported in the GLC in the extent of the Apulian region, while the local inventory reports 107 landslide points. While the GLC does not provide as many points as the event inventories evaluated here, one point in the GLC could be representative of all the landslide activity in that region for the specific event. However, the number of landslides represented in event inventories is up to several orders of magnitude greater than the number of reported landslides in the GLC for the same date and location as the landslide event mapped. This comparison shows that the GLC is less representative of small regions and event-specific landslides.

Table 5. Rainfall-induced landslide event inventories used in this study.

Location	Number of Points	Rainfall Event Date	Points in GLC	Size	Location Accuracy	Source
Khao Phanom, Thailand	225	30 March 2011	2	Large (1) Medium (1)	5 km (2)	[17]
Falam, Myanmar	5086	30–31 July 2015	1	Medium	5 km	[17]
Hakha, Myanmar	1737	30–31 July 2015	1	Small	5 km	[17]
Thaphabath, Laos	242	11 September 2015	0			[17]
Mu Chang Chai, Vietnam	1256	2–3 August 2017	0			[17]
Muong La, Vietnam	758	2–3 August 2017	0			[17]
Bat Xat, Vietnam	99	23–28 August 2017	1	Large	25 km	[17]
Da Bac, Vietnam	1086	10–11 October 2017	1	Large	5 km	[17]
Phu Yen, Vietnam	1368	10–11 October 2017	0			[17]
Tram Tau, Vietnam	1490	10–11 October 2017	0			[17]
Sin Ho, Vietnam	707	23–24 June 2018	0			[17]
Tam Duong, Vietnam	159	23–24 June 2018	0			[17]
Than Uyen, Vietnam	312	23–24 June 2018	0			[17]
Vi Xuyen, Vietnam	157	23–24 June 2018	0			[17]
Hpa-An, Myanmar	992	28–30 July 2018	0			[17]

Table 5. *Cont.*

Location	Number of Points	Rainfall Event Date	Points in GLC	Size	Location Accuracy	Source
Phong Tho, Vietnam	302	3 August 2018	0			[17]
Xieng Ngeun, Laos	1178	30 August 2018	0			[17]
Muong Lat, Vietnam	1718	27 August–1 September 2018	0			[17]
Nha Trang, Vietnam	207	18 November 2018	1	Medium	5 km	[17]
Blumenau, Brazil	597	20–25 November 2008	0			[18]
Teresópolis, Brazil	7268	11–13 January 2011	16	Medium (8) Large (4) Very Large (4)	1 km (1) 5 km (9) 10 km (6)	[18]
Salgar, Colombia	131	17–18 May 2015	0			[18]
Dominica	1756	25–28 August 2015	1	Large	5 km	[38]
Dominica	21,379	18–22 September 2017	0			[38]
Kii Province, Japan	1901	2–5 September 2011	0			[18]
South Taiwan	429	15–18 July 2008	1	Medium	10 km	[18,37]
Taiwan	10,236	6–9 August 2008	0			[18,36,37]

3.5. Economic Status Assessment

Economic variables are compared against landslide activity reported in the GLC to determine the effect of a country's economic status on landslide event reporting. The income status of each country is classified as low, lower middle, upper middle, or high and compared against the number of reported landslide events as well as landslide density. The landslide density is determined by the number of events in the GLC per land area of each country. The comparisons are shown in Figure 10A,B, respectively. The largest percentage of events (40.8%) was reported in lower middle income status countries, which is slightly higher than the percentage of events reported in high income status countries (39.9%), and interestingly, only 18.1% of events were reported in countries associated with upper middle income status (Figure 10A). In both income status comparisons with number of events and landslide density, countries considered to have low income status represent much less landslide activity (less than 2%) compared to higher income statuses. However, countries classified as having high income status experienced greater landslide density than countries with lesser income statuses, and an increase in landslide density with increasing economic status is apparent in Figure 10B. Furthermore, the gross domestic product (GDP) per capita is compared against landslide event density for countries represented in the GLC and is shown in Figure 11, which reveals a positive correlation between the two variables. The countries shown represent the top 25% of countries with the highest GDP per capita to highlight outliers within the comparison. Similarly, the population density of each country represented in the GLC is compared against landslide density in Figure 12. This comparison reveals a positive correlation between population density and landslide reporting, which can be explained by increased landslide exposure near dense populations. However, the trend of increased GLC events in wealthier or more populated countries suggests that more landslides are reported but does not necessarily indicate that more landslide events occur in these countries.

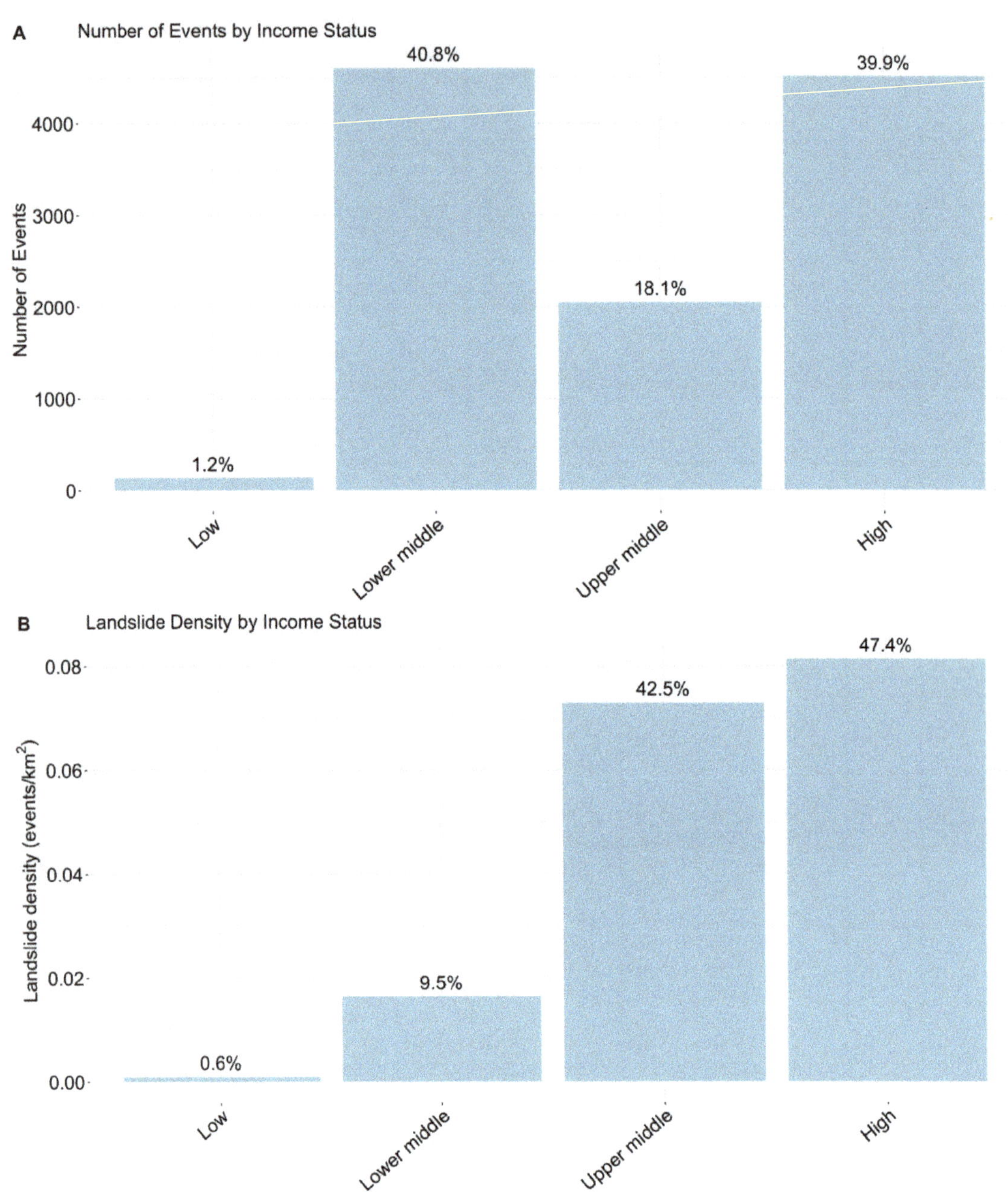

Figure 10. (**A**) Total reported events per income status classification; (**B**) Landslide density (events per km^2) per income status classification. The income status represents the reported income status of each country represented in the GLC.

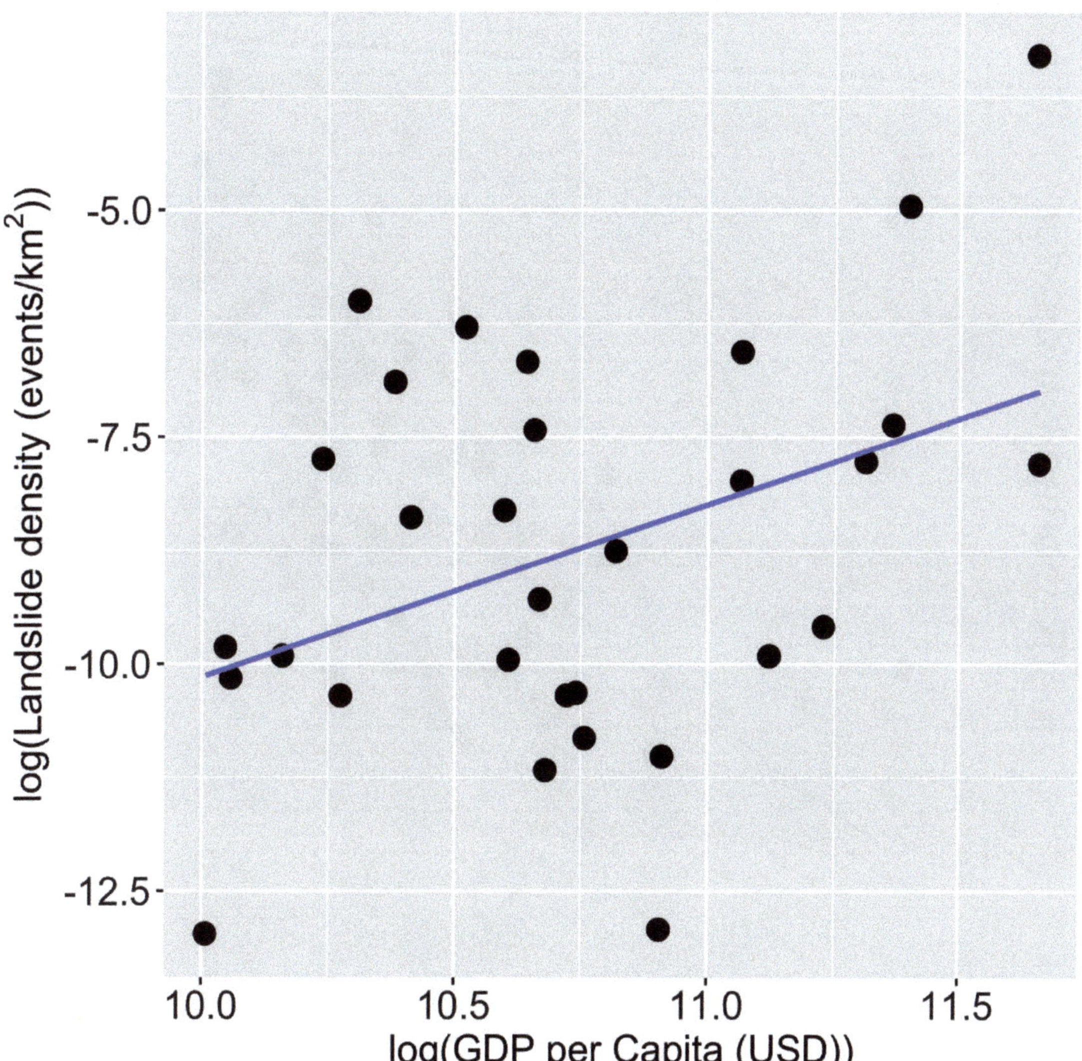

Figure 11. The log of the Gross Domestic Product (GDP) per capita in US dollars versus the log of the landslide density (events per km²) for countries reported in the GLC with the highest reported GDP per capita.

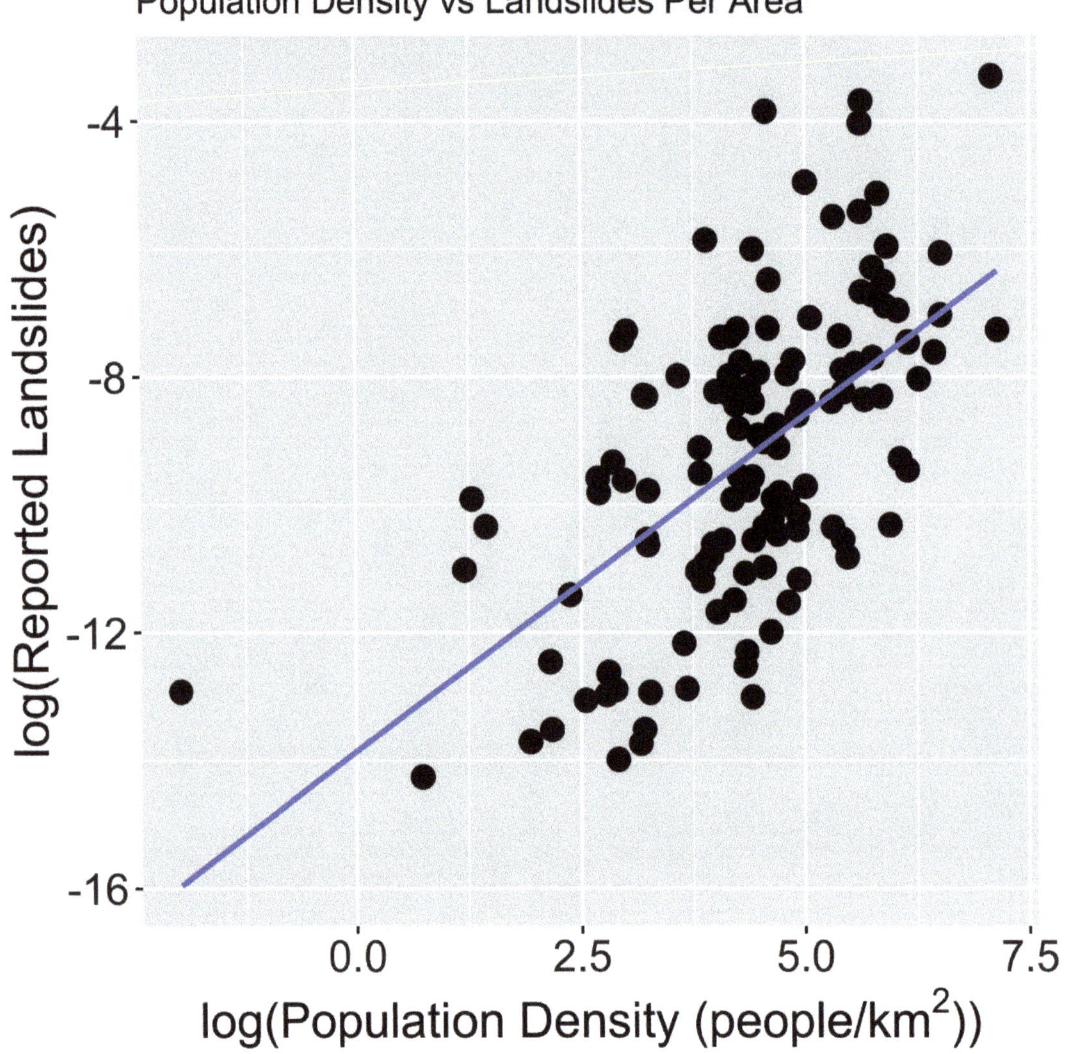

Figure 12. The log of the Population density (people per km^2) versus the log of the landslide density (events per km^2) for each country reported in the GLC.

4. Discussion

This study evaluates the spatial and temporal distribution of global landslide events and global patterns of landslides represented in this dataset including hotspot location and investigation of fatalities associated with landslides. The most notable landslide event hotspots at global scale appear to be in the Pacific Northwest, High Mountain Asia, and the Philippines as reported in the GLC. The GLC contains several outlier events that lead to large spikes in the data trend lines representing fatalities and number of events per month (Figures 1C–7C). For example, the average number of fatalities in August for the continent of Africa is 133, almost twice as high as any other month (Figure 2C). The large number of August fatalities can be attributed to one outlier event that occurred during this month and reported 1141 fatalities. Similar spikes in fatalities can be seen in other geographic locations due to anomaly events with large fatality counts. While the accuracy of these descriptors is not analyzed in depth, important attributes such as landslide category, location accuracy, landslide size, and triggering mechanism reported with each landslide event were assessed and found mostly complete, which further supports the application of the GLC in rainfall-triggered landslide studies. Furthermore, the most landslide events were reported in English-speaking countries, but the most fatal landslide events were reported in India, China, and Nepal. This indicates a clear reporting bias in the GLC towards English-speaking countries.

A positive correlation was discovered between the economic status of countries and the landslide density present in that country's borders, which indicates that richer countries are more prone to reporting landslides than poorer regions with less available resources. Countries with higher economic status have advanced natural disaster mitigation and more detailed landslide recording than low-income countries with less effort to prevent landslide disasters. This does not indicate that low-income countries have fewer landslides within their borders than high-income countries, but landslide events are reported more often and more precisely. Similarly, a positive relationship was found between population density and landslide reporting, which indicates that more landslides are reported in populated areas versus remote locations. Landslides are more likely to be reported in populated locations rather than remote areas due to greater risk of economic damage and casualties. These biases should be considered when applying the GLC. While the GLC shows a level of bias in regards to reporting language and underrepresentation of landslide activity in certain regions, it still can be beneficial for rainfall-triggered landslide research including landslide prediction and hazard awareness.

5. Conclusions

The research presented here evaluates the spatial and temporal distribution of global rainfall-triggered landslide reporting using NASA's Global Landslide Catalog (GLC), which presents a minimum number of events and is recorded from 2007 to 2018. The collection of this dataset is no longer active and will not be updated past 2018. However, the GLC covers one of the largest spatiotemporal ranges of any landslide inventory. Hotspot analysis of global landslide reports and fatalities associated with landslides as well as temporal trends in reporting can be visualized using the GLC and show that most landslides occur in High Mountain Asia and the Pacific Northwest and more landslides occur globally from July to September. The majority of event locations reported in the GLC are known within a small radius of confidence and coincide with NASA's global landslide susceptibility map. The GLC also contains various attributes that describe each event in terms of size, impact, location, and casualties. However, the GLC has several limitations that should be considered when using this dataset. The GLC represents global landslides with less spatial precision than local landslide inventories and inventories mapped via image analysis. Outliers in fatalities associated with landslide events make it difficult to assess temporal trends in global landslide casualties. Additionally, landslide reporting is geographically biased towards English-speaking populations largely due to the method of compiling the GLC. Landslide reporting also occurs more regularly and consistently in areas with high economic status compared to low economic regions. Similarly, this study reveals greater landslide reporting in areas with high population density, which can be attributed to increased landslide exposure near dense populations. Despite its limitations and biases, the GLC is shown to be a useful tool for rainfall-triggered landslide research.

Author Contributions: Conceptualization, C.D., T.A.S., D.B.K. and V.L.; methodology, C.D., T.A.S., D.B.K. and V.L.; validation, C.D., T.A.S., D.B.K. and V.L.; formal analysis, C.D.; data curation, C.D. and T.A.S.; writing—original draft preparation, C.D.; writing—review and editing, T.A.S., D.B.K. and V.L.; visualization, C.D.; supervision, T.A.S., D.B.K. and V.L.; project administration, T.A.S., D.B.K. and V.L.; funding acquisition, V.L. All authors have read and agreed to the published version of the manuscript.

Funding: C.D. was funded by the Graduate Assistance in Areas of National Need (GAANN) and ARCS Washington Metropolitan Chapter Scholarship. D.B.K. and T.A.S. were funded by NASA SERVIR Science Team (NNH18ZDA001N-18-SERVIR18_2-0036) and NASA's Disasters program through the solicitation for Earth Science Applications: Disaster Risk Reduction and Response (NNH18ZDA001N).

Institutional Review Board Statement: Not applicable.

Informed Consent Statement: Not applicable.

Data Availability Statement: Publicly available datasets were analyzed in this study. The landslide inventory data presented in this study are openly available at https://doi.org/10.6084/m9.figshare.14199227; https://doi.org/10.1002/esp.3284; https://doi.org/10.1016/j.geomorph.2013.11.020; https://doi.org/10.5194/esurf-6-903-2018. Data available in a publicly accessible repository that does not issue DOIs can be found here: Gross Domestic Product per Capita: https://data.worldbank.org/indicator/NY.GDP.PCAP.CD (accessed on 15 March 2022); Land Area: https://data.worldbank.org/indicator/AG.SRF.TOTL.K2 (accessed on 15 March 2022); Population Density: https://data.worldbank.org/indicator/EN.POP.DNST (accessed on 15 March 2022); The Global Susceptibility Map: https://gpm.nasa.gov/landslides/projects.html (accessed on 15 March 2022); Hurricane Maria landslide inventory http://www.unitar.org/unosat/node/44/2762 (accessed on 15 March 2022).

Conflicts of Interest: The authors declare no conflict of interest.

References

1. Crozier, M.J. Deciphering the effect of climate change on landslide activity: A review. *Geomorphology* **2010**, *124*, 260–267. [CrossRef]
2. Gariano, S.L.; Guzzetti, F. Landslides in a changing climate. *Earth-Sci. Rev.* **2016**, *162*, 227–252. [CrossRef]
3. Marc, O.; Jucá Oliveira, R.A.; Gosset, M.; Emberson, R.; Malet, J.P. Global assessment of the capability of satellite precipitation products to retrieve landslide-triggering extreme rainfall events. *Earth Interact.* **2022**, *26*, 122–138. [CrossRef]
4. Froude, M.J.; Petley, D.N. Global fatal landslide occurrence from 2004 to 2016. *Nat. Hazards Earth Syst. Sci.* **2018**, *18*, 2161–2181. [CrossRef]
5. Emberson, R.; Kirschbaum, D.B.; Amatya, P.; Tanyas, H.; Marc, O. Insights from the topographic characteristics of a large global catalog of rainfall-induced landslide event inventories. *Nat. Hazards Earth Syst. Sci.* **2022**, *22*, 1129–1149. [CrossRef]
6. Kirschbaum, D.B.; Adler, R.; Hong, Y.; Hill, S.; Lerner-Lam, A. A global landslide catalog for hazard applications: Method, results, and limitations. *Nat. Hazards* **2010**, *52*, 561–575. [CrossRef]
7. Kirschbaum, D.B.; Adler, R.; Adler, D.; Peters-Lidard, C.; Huffman, G. Global distribution of extreme precipitation and high-impact landslides in 2010 relative to previous years. *J. Hydrometeorol.* **2012**, *13*, 1536–1551. [CrossRef]
8. Kirschbaum, D.; Stanley, T.; Zhou, Y. Spatial and temporal analysis of a global landslide catalog. *Geomorphology* **2015**, *249*, 4–15. [CrossRef]
9. Juang, C.S.; Stanley, T.A.; Kirschbaum, D.B. Using citizen science to expand the global map of landslides: Introducing the cooperative open online landslide repository (COOLR). *PLoS ONE* **2019**, *14*, e0218657. [CrossRef]
10. Petley, D. Global patterns of loss of life from landslides. *Geology* **2012**, *40*, 927–930. [CrossRef]
11. Abella, E.A.C.; Van Westen, C.J. Generation of a landslide risk index map for Cuba using spatial multi-criteria evaluation. *Landslides* **2007**, *4*, 311–325. [CrossRef]
12. Chang, K.T.; Chiang, S.H.; Chen, Y.C.; Mondini, A.C. Landslide hazard analysis for Hong Kong using landslide inventory and GIS. *Comput. Geosci.* **2004**, *30*, 429–443. [CrossRef]
13. Guzzetti, F. Landslide fatalities and the evaluation of landslide risk in Italy. *Eng. Geol.* **2000**, *58*, 89–107. [CrossRef]
14. Mirus, B.B.; Jones, E.S.; Baum, R.L.; Godt, J.W.; Slaughter, S.; Crawford, M.M.; Lancaster, J.; Stanley, T.; Kirschbaum, D.B.; Burns, W.J.; et al. Landslides across the USA: Occurrence, susceptibility, and data limitations. *Landslides* **2020**, *17*, 2271–2285. [CrossRef]
15. Hughes, K.S.; Schulz, W.H. *Map Depicting Susceptibility to Landslides Triggered by Intense Rainfall, Puerto Rico*; U.S. Geological Survey Open-File Report 2020–1022; 1 plate, scale 1:150,000; U.S. Geological Survey: Reston, VA, USA, 2020; 91p. [CrossRef]
16. Bhandary, N.P.; Dahal, R.K.; Timilsina, M.; Yatabe, R. Rainfall event-based landslide susceptibility zonation mapping. *Nat. Hazards* **2013**, *69*, 365–388. [CrossRef]
17. Amatya, P.; Kirschbaum, D.; Stanley, T. Rainfall-induced landslide inventories for Lower Mekong based on Planet imagery and a semi-automatic mapping method. *Geosci. Data J.* **2022**, *9*, 315–327. [CrossRef]
18. Marc, O.; Stumpf, A.; Malet, J.P.; Gosset, M.; Uchida, T.; Chiang, S.H. Initial insights from a global database of rainfall-induced landslide inventories: The weak influence of slope and strong influence of total storm rainfall. *Earth Surf. Dyn.* **2018**, *6*, 903–922. [CrossRef]
19. Bessette-Kirton, E.K.; Cerovski-Darriau, C.; Schulz, W.H.; Coe, J.A.; Kean, J.W.; Godt, J.W.; Thomas, M.A.; Stephen Hughes, K. Landslides triggered by Hurricane Maria: Assessment of an extreme event in Puerto Rico. *GSA Today* **2019**, *29*, 4–10. [CrossRef]
20. Guzzetti, F.; Cardinali, M.; Reichenbach, P. The AVI project: A bibliographical and archive inventory of landslides and floods in Italy. *Environ. Manag.* **1994**, *18*, 623–633. [CrossRef]

21. Amatya, P.; Kirschbaum, D.; Stanley, T. Use of very high-resolution optical data for landslide mapping and susceptibility analysis along the Karnali highway, Nepal. *Remote Sens.* **2019**, *11*, 2284. [CrossRef]

22. Lin, Q.; Wang, Y. Spatial and temporal analysis of a fatal landslide inventory in China from 1950 to 2016. *Landslides* **2018**, *15*, 2357–2372. [CrossRef]

23. Chandrasekaran, S.S.; Sayed Owaise, R.; Ashwin, S.; Jain, R.M.; Prasanth, S.; Venugopalan, R.B. Investigation on infrastructural damages by rainfall-induced landslides during November 2009 in Nilgiris, India. *Nat. Hazards* **2013**, *65*, 1535–1557. [CrossRef]

24. Benz, S.A.; Blum, P. Global detection of rainfall-triggered landslide clusters. *Nat. Hazards Earth Syst. Sci.* **2019**, *19*, 1433–1444. [CrossRef]

25. Culler, E.; Livneh, B.; Rajagopalan, B.; Tiampo, K. A data-driven evaluation of post-fire landslide susceptibility. *Nat. Hazards Earth Syst. Sci.* **2021**, 1–24. [CrossRef]

26. Whiteley, J.S.; Chambers, J.E.; Uhlemann, S.; Wilkinson, P.B.; Kendall, J.M. Geophysical Monitoring of Moisture-Induced Landslides: A Review. *Rev. Geophys.* **2019**, *57*, 106–145. [CrossRef]

27. Lin, L.; Lin, Q.; Wang, Y. Landslide susceptibility mapping on a global scale using the method of logistic regression. *Nat. Hazards Earth Syst. Sci.* **2017**, *17*, 1411–1424. [CrossRef]

28. Farahmand, A.; Aghakouchak, A. A satellite-based global landslide model. *Nat. Hazards Earth Syst. Sci.* **2013**, *13*, 1259–1267. [CrossRef]

29. Liao, Z.; Hong, Y.; Wang, J.; Fukuoka, H.; Sassa, K.; Karnawati, D.; Fathani, F. Prototyping an experimental early warning system for rainfall-induced landslides in Indonesia using satellite remote sensing and geospatial datasets. *Landslides* **2010**, *7*, 317–324. [CrossRef]

30. Kirschbaum, D.B.; Stanley, T.; Simmons, J. A dynamic landslide hazard assessment system for Central America and Hispaniola. *Nat. Hazards Earth Syst. Sci.* **2015**, *15*, 2257–2272. [CrossRef]

31. Kirschbaum, D.B.; Adler, R.; Hong, Y.; Kumar, S.; Peters-Lidard, C.; Lerner-Lam, A. Advances in landslide nowcasting: Evaluation of a global and regional modeling approach. *Environ. Earth Sci.* **2012**, *66*, 1683–1696. [CrossRef]

32. Food and Agriculture Organization Surface Area. Available online: https://data.worldbank.org/indicator/AG.SRF.TOTL.K2 (accessed on 15 March 2022).

33. Food and Agriculture Organization and World Bank Population Density. Available online: https://data.worldbank.org/indicator/EN.POP.DNST (accessed on 15 March 2022).

34. World Bank and OECD National Accounts data files GDP Per Capita. Available online: https://data.worldbank.org/indicator/NY.GDP.PCAP.CD (accessed on 15 March 2022).

35. Stanley, T.A.; Kirschbaum, D.B. A heuristic approach to global landslide susceptibility mapping. *Nat. Hazards* **2017**, *87*, 145–164. [CrossRef]

36. Chang, K.; Chiang, S.H.; Chen, Y.C.; Mondini, A.C. Modeling the spatial occurrence of shallow landslides triggered by typhoons. *Geomorphology* **2014**, *208*, 137–148. [CrossRef]

37. Chen, Y.C.; Chang, K.T.; Chiu, Y.J.; Lau, S.M.; Lee, H.Y. Quantifying rainfall controls on catchment-scale landslide erosion in Taiwan. *Earth Surf. Process. Landf.* **2013**, *38*, 372–382. [CrossRef]

38. van Westen, C.J.; Zhang, J. Landslides and Floods Triggered by Hurricane Maria. Unitar-Unosat. 2018. Available online: http://www.unitar.org/unosat/node/44/2762 (accessed on 15 March 2022).

39. Vennari, C.; Salvati, P.; Bianchi, C.; Casarano, D.; Parise, M.; Basso, A.; Marchesini, I. AReGeoDatHa: Apulian Regional GeoDatabase for geo-hydrological Hazards. *J. Environ. Manag.* **2022**, *322*, 116051. [CrossRef]

40. Loo, Y.Y.; Billa, L.; Singh, A. Effect of climate change on seasonal monsoon in Asia and its impact on the variability of monsoon rainfall in Southeast Asia. *Geosci. Front.* **2015**, *6*, 817–823. [CrossRef]

41. Petley, D.N.; Hearn, G.J.; Hart, A.; Rosser, N.J.; Dunning, S.A.; Oven, K.; Mitchell, W.A. Trends in landslide occurrence in Nepal. *Nat. Hazards* **2007**, *43*, 23–44. [CrossRef]

42. Leiba, M. Impact of landslides in Australia to December 2011. *Aust. J. Emerg. Manag.* **2013**, *28*, 30–36.

sustainability

Article

A New Approach to Spatial Landslide Susceptibility Prediction in Karst Mining Areas Based on Explainable Artificial Intelligence

Haoran Fang [1,2,3], Yun Shao [1,2,3], Chou Xie [1,2,3,*], Bangsen Tian [1], Chaoyong Shen [4], Yu Zhu [1], Yihong Guo [1], Ying Yang [1,2], Guanwen Chen [4] and Ming Zhang [1,2]

1 Aerospace Information Research Institute, University of Chinese Academy of Sciences, Beijing 100094, China
2 University of Chinese Academy of Sciences, Beijing 100049, China
3 Laboratory of Target Microwave Properties, Deqing Academy of Satellite Applications, Huzhou 313200, China
4 The Third Surveying and Mapping Institute of Guizhou Province, Guiyang 550004, China
* Correspondence: xiechou@aircas.ac.cn

Abstract: Landslides are a common and costly geological hazard, with regular occurrences leading to significant damage and losses. To effectively manage land use and reduce the risk of landslides, it is crucial to conduct susceptibility assessments. To date, many machine-learning methods have been applied to the landslide susceptibility map (LSM). However, as a risk prediction, landslide susceptibility without good interpretability would be a risky approach to apply these methods to real life. This study aimed to assess the LSM in the region of Nayong in Guizhou, China, and conduct a comprehensive assessment and evaluation of landslide susceptibility maps utilizing an explainable artificial intelligence. This study incorporates remote sensing data, field surveys, geographic information system techniques, and interpretable machine-learning techniques to analyze the sensitivity to landslides and to contrast it with other conventional models. As an interpretable machine-learning method, generalized additive models with structured interactions (GAMI-net) could be used to understand how LSM models make decisions. The results showed that the GAMI-net model was valid and had an area under curve (AUC) value of 0.91 on the receiver operating characteristic (ROC) curve, which is better than the values of 0.85 and 0.81 for the random forest and SVM models, respectively. The coal mining, rock desertification, and rainfall greater than 1300 mm were more susceptible to landslides in the study area. Additionally, the pairwise interaction factors, such as rainfall and mining, lithology and rainfall, and rainfall and elevation, also increased the landslide susceptibility. The results showed that interpretable models could accurately predict landslide susceptibility and reveal the causes of landslide occurrence. The GAMI-net-based model exhibited good predictive capability and significantly increased model interpretability to inform landslide management and decision making, which suggests its great potential for application in LSM.

Keywords: landslides susceptibility map; explainable AI; GIS; Karst landform; coal mining

Citation: Fang, H.; Shao, Y.; Xie, C.; Tian, B.; Shen, C.; Zhu, Y.; Guo, Y.; Yang, Y.; Chen, G.; Zhang, M. A New Approach to Spatial Landslide Susceptibility Prediction in Karst Mining Areas Based on Explainable Artificial Intelligence. *Sustainability* **2023**, *15*, 3094. https://doi.org/10.3390/su15043094

Academic Editors: Chong Xu and Jian Chen

Received: 6 January 2023
Revised: 2 February 2023
Accepted: 6 February 2023
Published: 8 February 2023

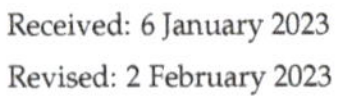

1. Introduction

In Karst regions, landslides are a significant natural hazard distinguished by their widespread occurrence, high frequency, and tremendous damage [1,2]. The frequency, size, and destructiveness of landslides have recently increased due to enhanced human activity and frequent catastrophic weather occurrences [3,4]. Landslides provide a hazard to 74 million people in China alone, with yearly direct economic damages typically running into the hundreds of millions [5]. While landslides may not be entirely avoidable, managers may create effective mitigation measures to lessen mortality and damage by researching landslide condition factors, modeling landslide risk, and mapping and interpreting hazard-prone locations. Landslide susceptibility assessments are valuable for disaster prevention, land resource planning, and risk assessment in landslide-prone areas.

Landslide susceptibility assessment is a method of quantitatively and spatially estimating the likelihood of a landslide occurring in a specific area [6]. It considers various factors contributing to landslides, such as topographic features including slope, aspect, and material properties such as soil type and lithology [7]. In addition, environmental factors such as climate, hydrology, ecology, human activities, and seismic activity could also play a role in the occurrence of landslides [8]. Recent advances in remote sensing technology and spatial datasets have provided valuable data for geological hazard monitoring and landslide susceptibility assessment [9]. The use of remote sensing allows for the collection of a wide range of data, including optical and microwave observations, which can provide information on normalized difference vegetation index (NDVI), soil moisture content, and surface deformation [10,11]. Zhang and Shen et al. discuss the effect of vegetation distribution on LSM implementation and the use of Interferometric Synthetic Aperture Radar (InSAR) surface deformation information to achieve dynamic refinement of LSM [12,13]. However, there are differences in geography, topography, and climatic conditions in different areas, and landslide patterns, mechanisms, and drivers vary in these areas. Therefore, landslide influencing factors need to be selected accordingly to the characteristics of the study area. This study considered the impact of frequent mining activities on landslides in the Nayong area.

Researchers have used landslide influencing factors for landslide susceptibility assessment through physical models, statistical analysis methods, or machine learning techniques [14–21]. Researchers used finite element models and limit equilibrium methods in the numerical simulation methods to assess slope stability [14]. This method simplifies assumptions about some environmental conditions (e.g., soil moisture content and rock characteristics) and landslide structures, and therefore requires costly field exploration and is not conducive to large-scale regional monitoring [15]. Statistical methods are reproducible and objective, and typical methods include logistic regression (LR), frequency ratio methods, and index of entropy [16–18]. Machine learning is the current mainstream method for LSM, which has significantly improved the accuracy and precision of landslide susceptibility assessment models. Common machine learning methods include support vector machines (SVM), random forest (RF), and deep neural networks [19–21].

Previous studies have shown that machine learning (ML) approaches and deep learning (DL) methods can assess landslide susceptibility in various area [22]. However, researchers have tended to neglect the explanatory aspects of models in seeking higher accuracy. Because of the 'black box model', these methods have difficulty explaining how landslide factors contribute to and dominate the occurrence of all or individual landslides in the study area. The lack of interpretation is a significant drawback of the above machine models and a fundamental flaw in their application to high-risk events (e.g., landslides, earthquakes, and lives) [23]. In these applications, decisions can affect lives, property, and generate huge costs. Based on the above analysis, the study provided an explainable artificial intelligence (XAI) method, called generalized additive models with structured interactions (GAMI-net), for generating LSM by integrating landslide influence factors and explained the effects of variables (features) on predictions and their responses in predictors. The GAMI-net model is an interpretable model proposed by Professor Zhang Aijun in 2021, consisting of a generalized additive model and several artificial neural networks (ANN) and sub-networks [24]. The interpretable model results help decision makers better understand the factors contributing to landslide risk and identify areas that may require interventions such as slope stabilization or land management.

The study extracted the corresponding landslide impact factors from satellite remote sensing data and geological information. Specifically, the DEM data derived terrain elevation, slope, aspect, and curvature. The NDVI is a measure of vegetation cover that indicates dense vegetation, soil moisture content, and soil type and directly impacts soil quality movement. Lithology is an essential factor indicating the strength of the soil. Heavy rainfall may lead to soil saturation, resulting in rainfall-induced landslides. In addition, mining, roads, and rivers are also contributing factors to landslides. The GAMI-net model constructed LSM by these factors and provided a global interpretation of landslide susceptibility. Next, the performance of traditional models (LR, SVM, and RF) and GAMI-net models are evaluated and analyzed. Finally, the Shapley adaptive interpretation (SHAP) method was used to analyze the key factors leading to landslides in a typical landslide, taking the town of Zongling in Nayong County as an example.

2. Materials and Methods

2.1. Study Area and Landslide Inventory

The study area is located in Nayong County, southeast of Bijie City, Guizhou Province, at longitude 104°55′–105°38′ East and latitude 26°30′–27°05′ North (Figure 1). The Karst is widely developed in the region, the geological environment is complex, karst action is obvious, and long-term mining activities have caused frequent geological disasters in the area with strong concealment [25]. On 28 August 2017, a large-scale landslide occurred in the Qiaobian group of the Pusha community in Zhangjiawan Town, which caused more than 30 deaths [26]. Since 2010, two hundred ninety-three geological disaster events have been recorded, impacting 1716 people in Nayong [27]. As a result, landslide hazards seriously threaten the safety of people's lives and property in the study area.

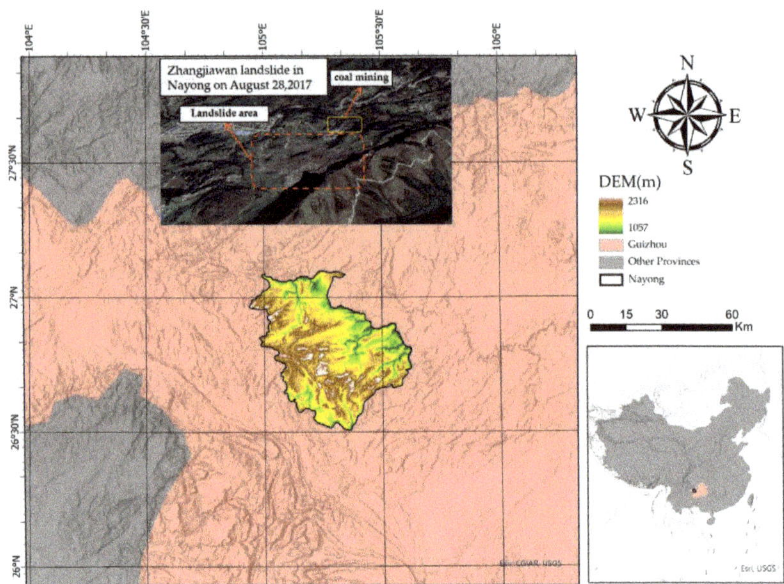

Figure 1. Overview of the study area in Nayong (the illustration of the landslide in Nayong was accessed on Google Earth).

The Third Institute of Surveying and Mapping of Guizhou Province provided the landslides inventory, including a total of 293 historical landslides (2010–2020) in Nayong County. In addition, the same number of non-landslides were used as a reference group. According to the geological hazard survey results, there were 194 small landslides, 88 medium landslides, eight large landslides, and three massive landslides.

2.2. Data Preparation

Landslides are natural hazards influenced by various factors, including geometric conditions such as elevation, slope, and slope orientation, as well as hydrological factors such as the topographic moisture index (TWI) and distance from rivers [28,29]. In addition, geological and environmental factors, such as lithology, land use, and human activity conditions can also impact the likelihood of landslides [30,31]. These factors may vary depending on the region in which the hazard occurs. Accurate detection of landslides requires the consideration of multiple factors, including those related to both natural phenomena and human activities [32,33].

Based on the above considerations and relevant literature, 14 influencing factors were selected for the study, including topography and geology, hydrology, land cover, and human activities [8,9,11,28–33]. The spatial distribution of these influencing factors is shown in Figure 2. Table 1 illustrated the factors selected, resolution and data source. The main landslide factors are the following categories:

Topography: The steepness and orientation of a slope can also play a role in its susceptibility to landslides. Steep slopes are generally more prone to landslides than gentle ones, and slopes facing specific directions may be more susceptible depending on the direction of predominant weather patterns and the location of water sources.

Geology: The type of rock and soil present on a slope can affect its stability. For example, slopes composed of weak, poorly consolidated soils or those with a high clay content may be more prone to landslides.

Hydrology: Water's presence can significantly affect a slope's stability. Water can seep into the ground and weaken the soil, making it more prone to landslides. Additionally, heavy rain or rapid snowmelt can cause landslides by adding weight to the slope or increasing the soil's water content.

Land use: Human activities can also contribute to landslide susceptibility. For example, the construction of buildings or roads on steep slopes can increase the weight on the slope and make it more prone to landslides. Additionally, vegetation removal can destabilize a slope by reducing its ability to absorb water.

Climate: Changes in temperature and precipitation patterns can also influence landslide susceptibility. For example, prolonged drought can cause soil to become dry and brittle, making it more prone to landslides, while heavy rain or rapid snowmelt can increase the water content of the soil and lead to landslides.

To gather data on these factors, the study employed a variety of sources, including DEM, ground surveys, Landsat-8, and Global Precipitation Measurement (GPM). The data processing implemented in this study consisted of the following steps.

(1) Based on Landsat-8 OIL images obtained from the Google Earth Engine, the NDVI of the study area was produced;

(2) Obtaining SRTM DEM data of the study area from USGS to calculate elevation, slope, aspect, plan curvature, and profile curvature (https://urs.earthdata.nasa.gov/ accessed on 12 August 2022);

(3) The GPM data was obtained from the USGS and converted to spatial raster data by Kriging interpolation and meteorological station monitoring data;

(4) Soil type, geological data, and land use and land cover (LULC) were all rasterized, and only resampling was required.

The NDVI time series was calculated from 2016 to 2021, as shown in Figure 3. The 2018 NDVI mean was the lowest. Therefore, the study chose the 2018–2021 NDVI mean as an input factor in this study. Figure 4 depicts a time series graph of NDVI variations from 2016 to 2021, correlating to landslides (purple) and non-landslides (red). The NDVI trends associated with non-landslides showed greater values than the NDVI associated with landslides. This may imply that places with less vegetation are more vulnerable to landslides. For precipitation data, the Global Precipitation Measurement (GPM) mission (IMERG) satellite precipitation products have a spatial resolution of 0.1° (approximately 10 km), which is difficult for regional landslide susceptibility analysis. Therefore, the topographic data (elevation, slope, slope direction) to establish a mapping between precipitation and topographic factors in the study area using the geo-weighted regression approach [34]. Following, descending sampling was used to generate precipitation data for Nayong Town at a resolution of 1 km. Finally, five meteorological station data were used for validation in the study area.

Table 1. List of landslide factors and data scale and source.

Class of Factors	Factors Selected	Scale	Data Source
Topography	Elevation	30 m	Calculated from DEM (NASA)
	Slope angle	30 m	
	Aspect	30 m	
	Topographic wetness index	30 m	
	Profile curvature	30 m	
	Plan curvature	30 m	
Geolithology	Distance to an active fault	30 m	The Third Survey and Mapping Institute of Guizhou Province
	Lithology	30 m	
Land use and land cover	Land use and land cover	30 m	Ministry of Natural Resources Data Center
Morphology	Distance to rivers and road	30 m	Openstreetmap
Climate	Rainfall	Vector/0.1°	Meteorological station monitoring data and GPM
Vegetation	NDVI	30 m	Landsat
Human	Coal mining	Vector	The Third Survey and Mapping Institute of Guizhou Province

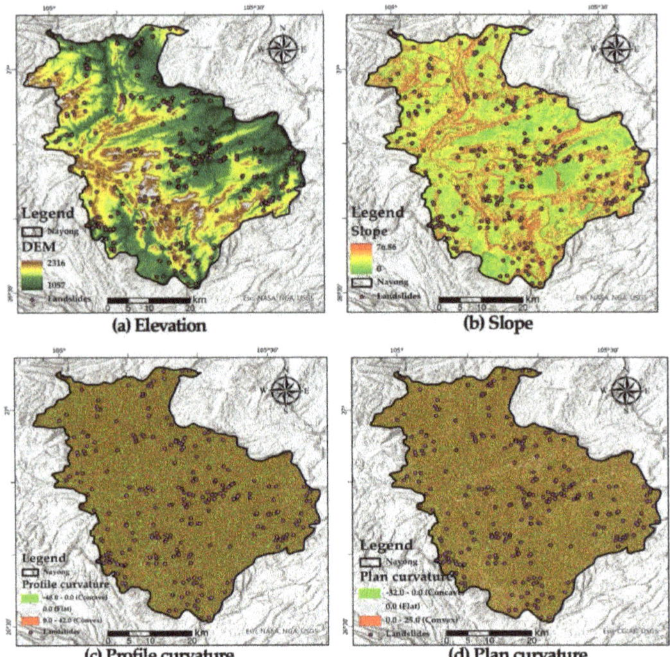

Figure 2. *Cont.*

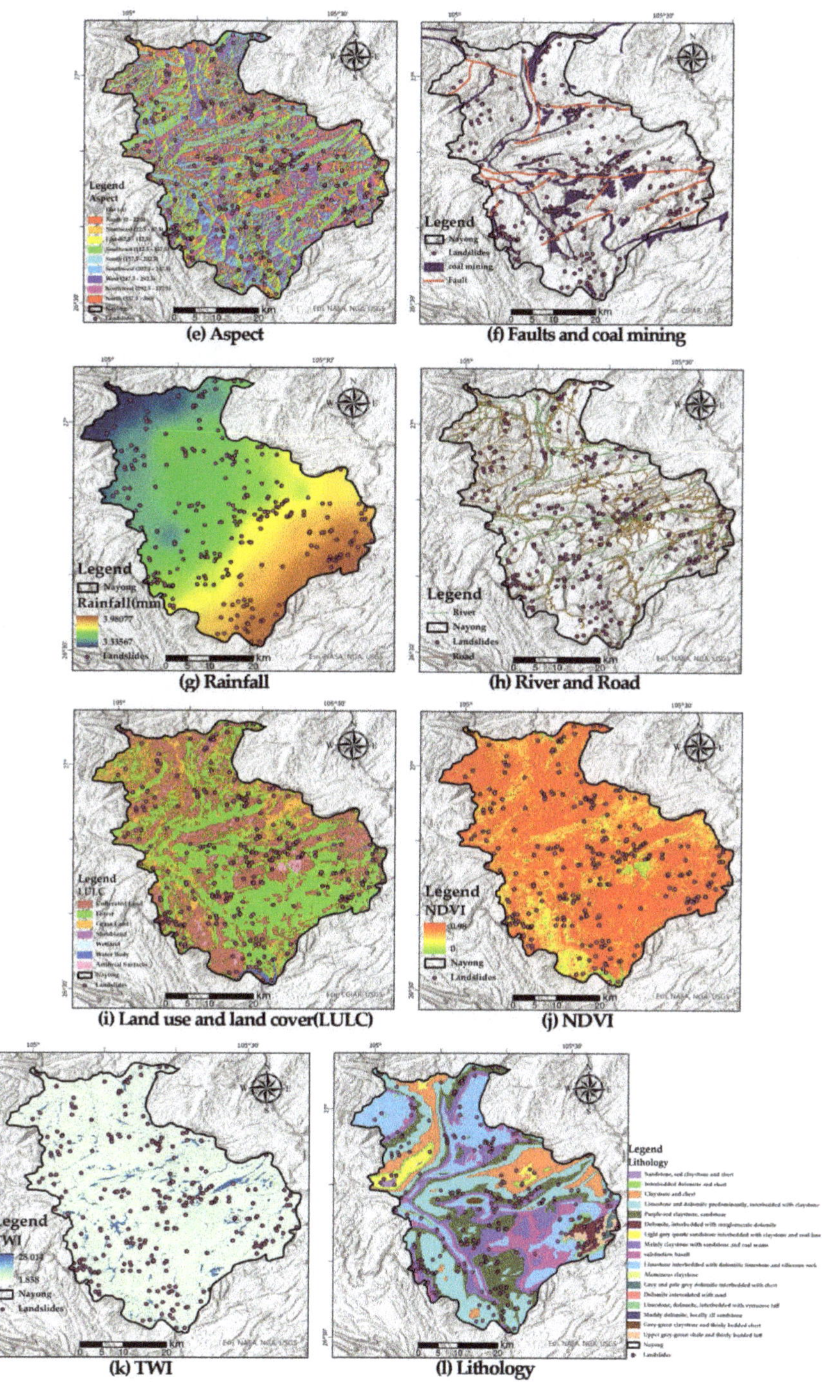

Figure 2. Influencing factors in Nayong. (**a**) Elevation; (**b**) slope; (**c**) profile curvature; (**d**) plan curvature; (**e**) aspect; (**f**) fault and coal mining; (**g**) rainfall; (**h**) road and river; (**i**) land use and land cover; (**j**) NDVI; (**k**) TWI; and (**l**) lithology.

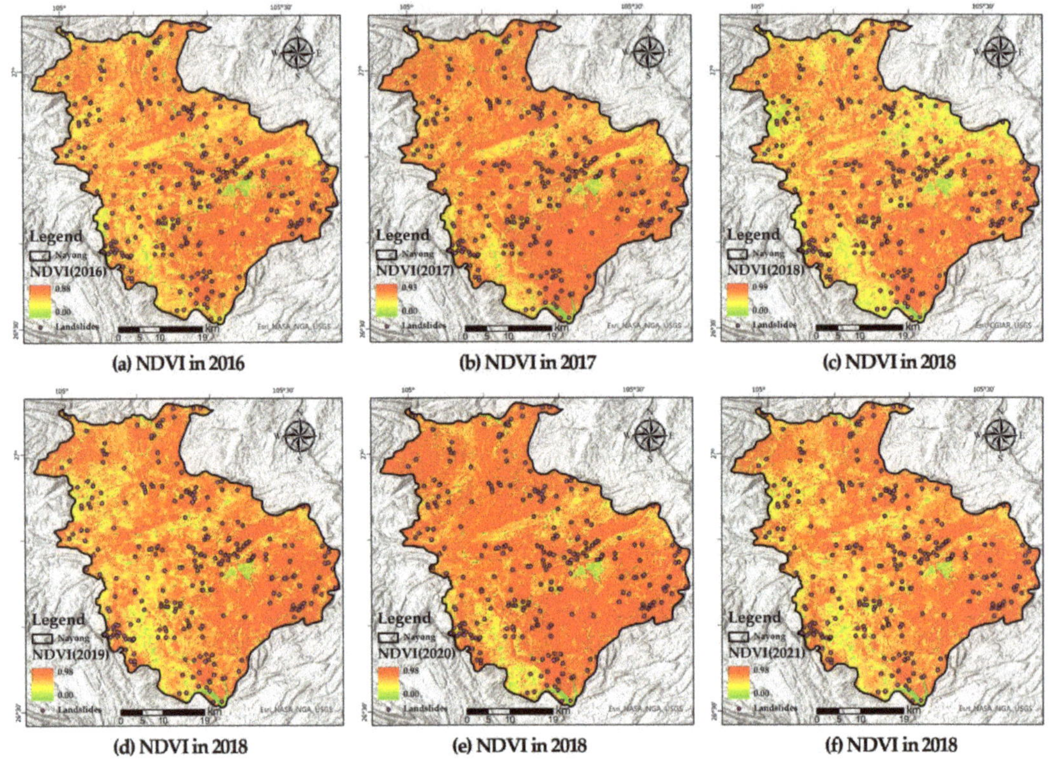

(a) NDVI in 2016   (b) NDVI in 2017   (c) NDVI in 2018

(d) NDVI in 2018   (e) NDVI in 2018   (f) NDVI in 2018

Figure 3. The annual mean NDVI maps in Nayong.

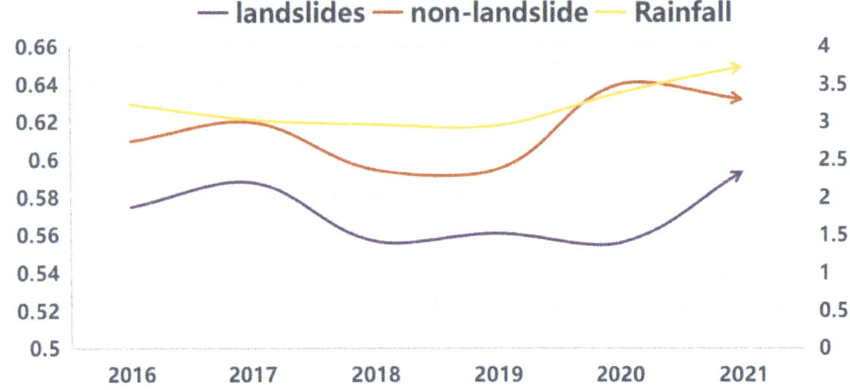

Figure 4. Time series chart for NDVI changes (2016–2021).

3. Methods

3.1. Overall Methodology

As shown in Figure 5, the study began by collecting and extracting various landslide features from various data sources, including topography, lithology, land use and land cover, morphology, climate, vegetation, and human activity. These features were essential for understanding the underlying causes of landslides and identifying areas at high risk. The extracted landslide features were then converted into a stacked dataset using band synthesis and resampling to ensure data consistency and accuracy. The study split the landslides dataset (50% for the landslide sample and 50% for the non-landslide sample) into training (80%) and test (20%) sets upon random permutation. Next, the GAMI-net and other algorithms (LR, SVM, and RF) were applied to the stack dataset for model calculation and parameter optimization. The grid search was utilized to optimize the parameters by adjusting the weights and biases of the model to achieve the best performance. After the model calculation and parameter optimization, the model's performance was evaluated using various metrics, such as accuracy, precision, ROC, and recall. These metrics determine how well the model can predict landslides based on the input data. This evaluation helps to identify any issues or limitations of the model and to improve it if necessary. Finally, the model was interpreted and used to create a landslide susceptibility map (LSM) that can be used to identify areas at high risk of landslides.

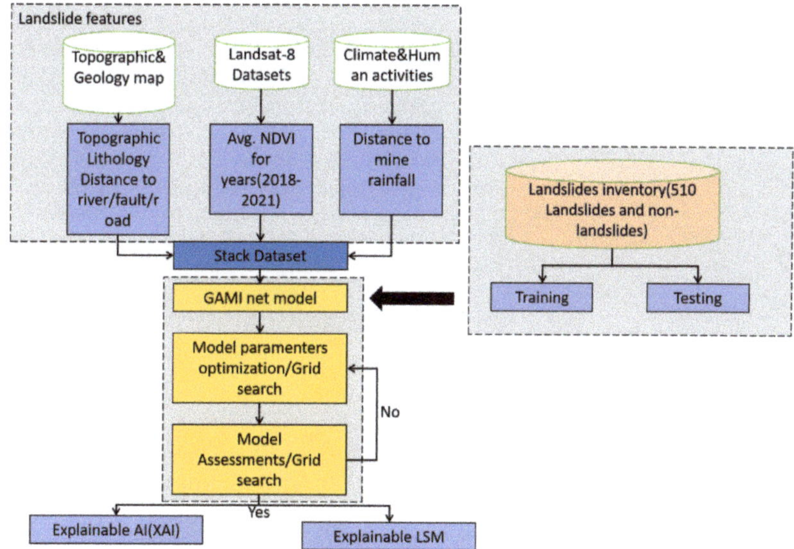

Figure 5. The process for mapping LSM and explainability.

3.2. Machine Learning Methods

3.2.1. Logistic Regression (LR)

Multiple logistic regression was widely used to examine LSM [35]. The basic idea behind logistic regression is to use Equations (1) and (2) to predict the likelihood of landslides based on a set of input factors. By training the model on a dataset of landslides and non-landslides, it is possible to use the trained model to predict the likelihood of a slope failure in a new, unseen location. The probability values of landslide susceptibility were calculated by the model for all landslide factors in the study area. In general, the probability value indicated the landslide susceptibility of the area and the closer it was to 1, the greater the landslide susceptibility. The logistic function is described as:

$$f(x) = a_0 + \sum_{k=1}^{n}(a_k x_k) \tag{1}$$

$$P = \frac{1}{1 + \left(exp^{-f(x)}\right)} \tag{2}$$

where P is the probability of landslide occurrence and $f(x)$ a linear combination of casual x_k factors.

3.2.2. Random Forest (RF)

Random forests are a machine-learning algorithm that can predict the likelihood of a landslide occurring in a particular area [36,37]. The algorithm creates many decision trees, each of which is trained on a random subset of the data. The trees are then combined to form a "forest", and the predictions of the individual trees are averaged to produce a final prediction. To use random forests for landslide susceptibility analysis, a dataset containing information about the characteristics of an area, such as soil type, slope angle, and vegetation cover, is needed. The algorithm is then trained on this dataset to learn the relationship between the predictor variables and the occurrence of landslides in the area. Once the model has been trained, it can predict the likelihood of a landslide occurring in a new area by inputting the relevant predictor variables. The model's output is a probability, indicating the likelihood of a landslide in the given area.

3.2.3. Support Vector Machines (SVM)

Support Vector Machines (SVM) is a supervised learning algorithm that can be adopted for classification and regression tasks [38,39]. In the context of landslide susceptibility, SVM can classify areas into different susceptibility classes (e.g., low, moderate, and high susceptibility) based on a set of input features or predictors.

SVM works by locating the hyperplane in a high-dimensional space that best divides the various classes. The hyperplane is chosen to maximize the distance between the hyperplane and the nearest data points from each class (called the support vectors). This distance is known as the margin. The larger the margin, the lower the generalization error and the better the classifier will perform on unseen data.

The SVM algorithm involves solving a convex optimization problem to find the hyperplane that maximally separates the classes. The optimization problem can be written as:

minimize w, b, ξ,

subject to:

$$y_i\left(W^T x_i + b\right) \geq 1 - \xi_i, i = 1, 2, \ldots, n$$
$$\xi_i \geq 0, i = 1, 2, \ldots, n \tag{3}$$

where w and b are the weight and bias of the hyperplane, respectively, x_i is the input feature vector, y_i is the class label (0 or 1), and ξ_i is the slack variable that allows for misclassification. The optimization problem is subject to the constraints that the distance between the hyperplane and the nearest data points from each class (the support vectors) is at least 1 and the slack variables are non-negative.

In the case of non-linearly separable data, SVM can still find a good hyperplane by introducing additional dimensions through the use of kernels. A kernel is a function that transforms the input data into a higher-dimensional space where the data may be more easily separable.

3.2.4. GAMI-Net

The generalized additive model (GAM) is a semi-parametric smoothing model that provides a valuable tool for investigating factor interaction by Equation (4) [40].

$$g(E(y|x)) = \mu + \sum_{j \in S_1}^{M} h_j(x_j) + \varepsilon \tag{4}$$

where $h_j(x_j)$ is a smooth function, and j denotes various factors. The μ indicates the predictor variable category relationship that is unaffected by nonlinear transformation, while the ε represents the residual.

In a traditional GAM network, a feature was made to have monotonically increasing or decreasing properties on the target value by setting constraints and penalty functions or adjusting the number of spline functions [41]. The GAM model are more explanatory than the other black box ML models; they could not be represented as a single function describing the estimated relationship between independent and dependent variables. The GAM model does not consider the factors' interactions, and the addable spline function does not have sufficient predictive accuracy [42].

To solve the shortcomings of the GAM model, GAMI-net considered factor interactions and replaced the spline function in GAM with an ANN network to improve prediction accuracy. Under the additive model, it was found that it explored and explained a particular variable by being able to control for other variables. As related research has demonstrated, the explanatory nature of ANNs can be achieved under appropriate constraints. This enhancement can explain neural networks with an excellent balance between prediction performance and model interpretation.

$$g(E(y|x)) = \mu + \sum_{j \in S_1}^{M} h_j(x_j) + \sum_{(j,k) \in S_2}^{N} f_{jk}(x_j, x_k) \tag{5}$$

According to Figure 6, the GAMI-net consisted of the individual sub-networks, interaction sub-networks, and three constraints. The all sub-networks estimated the individual and interaction factors. At the same time, the three constraints ensured interpretability. The sparsity constraint ranks the features according to their importance, retaining only the more critical feature ridge functions. The heredity constraint ensures that the interaction feature pair must have at least one feature from a vital feature retained in the individual feature sub-network module. The marginal clarity constraint prevents individual features and their corresponding interaction features from absorbing each other, making the model more stable. According to the GAMI-net structure, the hyperparameters are learning rate, batch size, epoch, the maximum number of features, and crossover features. In this study, the learning rate and batch size were set by default, with learning rates of 0.001 and the batch size of 200. The training process was divided into seven stages as follows (Table 2). First, the individual factors sub-networks were trained iteratively for 1000 epochs, and then the interaction factors subnetwork module was trained for 1000 epochs. In each module, the features were ranked according to their importance. Then, the unimportant tail features were removed according to the magnitude of the loss function value and then trained iteratively for 500 epochs. In this paper, the maximum number of individual features was (4, 8, 12, 16, 20), and the maximum number of interaction features was (0, 6, 9, 12, 15, 20) in the grid search.

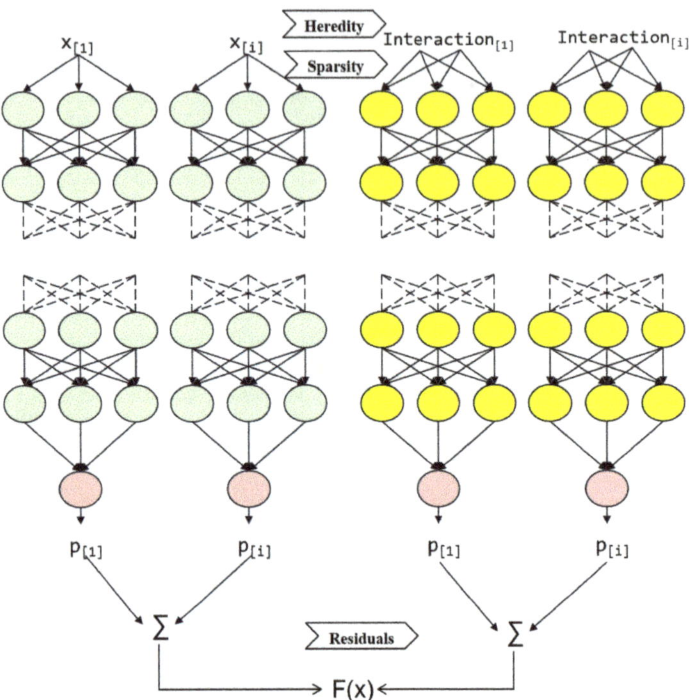

Figure 6. The network of GAMI-net. The main effects are fitted first, and then the top-K ranked pairwise interactions are selected and fitted to the residuals, subject to the heredity constraint. Finally, the trivial subnetworks are pruned.

Table 2. The iterative training process of GAMI-net.

Step	Training Process
1	Training all individual factors sub-networks;
2	Selecting the most important x_i individual factors sub-networks based on factors importance;
3	Fine-tuning the individual factors sub-network modules;
4	Selecting interaction pairwise factors based on the ranking of interaction factors;
5	Training the interaction factors sub-network modules;
6	Selecting the top *interaction_i* significant interaction factors sub-networks based on feature importance;
7	Fine-tuning the interaction factors sub-network modules.

3.3. Explainability of GAMI-Net

3.3.1. Importance Ratio

The importance ratio (IR) values are used to measure each factor contribution to the overall forecast. The definition of IR referenced the Sobol index, the main difference being that the Sobol index was generated under the premise that all variables were independent and uniformly distributed, whereas IR was based on the empirical distribution of the predicted variables [43,44]. Each factor had an importance score based on the model, and the important ratio $IR(j)$ of that factor can be calculated using Equation (6). Correspondingly, the importance of each pairwise factor $IR(j,k)$ could be obtained from Equation (7).

$$IR(j) = D\left(h_j\right)/T \tag{6}$$

$$IR(j,k) = D\left(f_{jk}\right)/T \tag{7}$$

where $D\left(h_j\right)$ and $D\left(f_{jk}\right)$

$$T = \sum_{j \in S_1} D\left(h_j\right) + \sum_{(j,k) \in S_2} D\left(f_{jk}\right) \tag{8}$$

In addition to the IR factors, the ridge function was used to explain the relationship between one or two specific factors and the landslides. The $h_j(x_j)$ value drew one-dimensional linear graphs for continuous factors and the bar charts for categorical factors to show the factor-landslides relationship. Each of the pairwise factors $f_{jk}(x_j, x_k)$ can be visualized using a 2D heat map showing the effect of the factor on the slippage.

3.3.2. Shapley Additive Explanations (SHAP)

For a trained complex machine learning model, SHAP interprets the model by calculating the Shapley values of the features to measure the importance of the features [45,46]. The Shapley value can be calculated for each feature component of each sample, and the Shapley values of the sample feature components are additive. The magnitude of the Shapley value of a sample feature component tells how much the value taken by the sample on each feature affects the prediction result for that sample [47]. The overall feature importance can be obtained by calculating the absolute value of the Shapley values of all samples on each feature to find out the magnitude of each feature's influence on the prediction results of the complex model by averaging the overall importance.

With the SHAP method, GAMI-net may directly extract the value of each additive component, such as the primary factor or the pairwise interaction factor, for every sample x. The sequence of these marginal effects can help to clarify how the input x's choices should be understood. The 1D line plots (or bar charts) and 2D heat maps can be used to statistically analyze how sensitive the forecast is to modest changes in the explanatory factors.

4. Results

4.1. Model Assessments for RF, SVM, and GAMI-Net

The variation of model accuracy (ROC) for the different maximum number of feature parameters (individual and interaction) in the parametric grid search was shown in Figure 7. As the maximum number of features increased, the model accuracy improved. The highest accuracy was obtained when the maximum number of features was 16 and 15 for single and interactive features, respectively. After this point, the model accuracy decreased, probably due to the increase in error in additive models with too many parameters, indicating the need for fine-tuning. Table 3 illustrated the evaluation metrics obtained by predicting 510 samples (20% of the total sample was the test set) within the study area using each algorithm, with LR generally having the weakest algorithm performance and the SVM, RF, and GAMI-net algorithms having better performance. Based on the ROC curves, the algorithms were ranked from smallest to most significant in terms of accuracy: LR, SVM, RF, and GAMI-net.

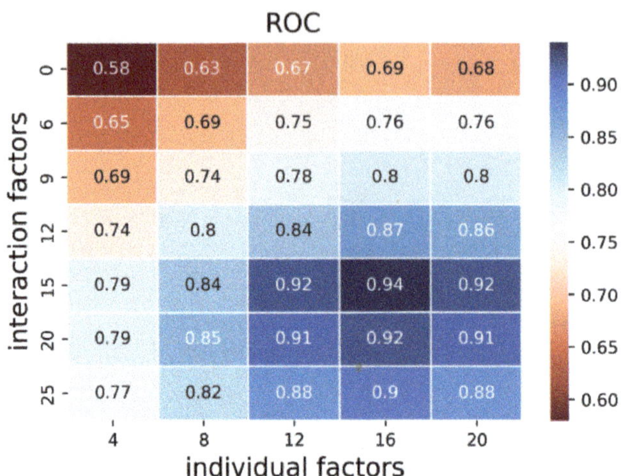

Figure 7. The ROC results of the grid search (blue represents high ROC values, red represents low ROC values).

Table 3. Evaluation of each model.

Algorithm	Accuracy	Precision	F1 Score	Recall	AUC ROC
LR	0.6189	0.6160	0.5754	0.5400	0.6482
SVM	0.6708	0.7013	0.6129	0.5448	0.7530
RF	0.8110	0.8222	0.7897	0.7581	0.9027
GAMI-net	0.8730	0.8648	0.8683	0.8747	0.9442

As shown in Figure 8, the GAMI-net method provided comparable performance to the RF methods, with prediction accuracies all greater than 0.8, F1 scores of 0.7897 and 0.8683, respectively, and ROC values all greater than 0.9. The GAMI-net model outperformed LR and SVM and slightly outperformed RF in predicting landslide susceptibility in the study area, demonstrating that the algorithm could achieve good results.

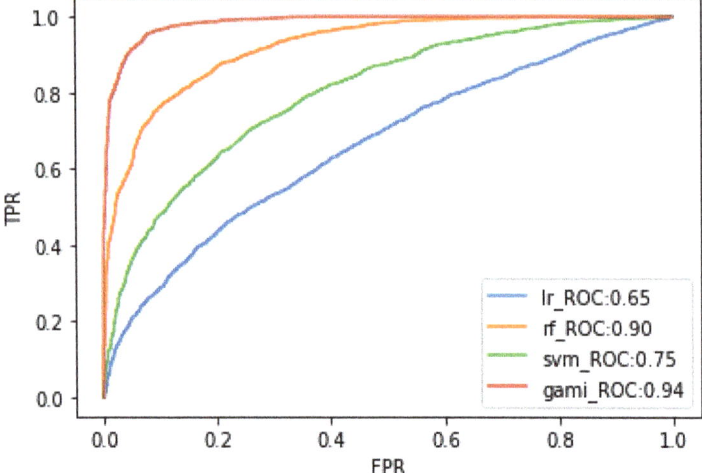

Figure 8. The ROC of LR, RF, SVM and GAMI-net.

Figure 9 shows the training and validation losses of the GAMI-net with 25 interaction effects considered. In the first stage, the loss rate does not change significantly when only the main effects are considered, indicating that landslide susceptibility cannot be effectively predicted using only the main effects. According to Table 3, as paired interactions were added to the network, the loss rate reduced dramatically, demonstrating the necessity to add pairwise interactions to the GAMI-net. There was a significant surge in training and validation loss at the start of the fine-tuning phase, which coincided with a fine-tuning operation on the paired factors. Figure 10 also depicts the validation loss for establishing the ideal number of main effects and pairwise interactions. The number of main effects and pairwise interactions included is indicated on the left and right image x-axes, respectively. The red star symbols represent the number of main features or pairwise interactions that should be used. According to the findings, the GAMI-net had 15 main features and 13 pairwise interactions. The rise in the primary and pair effects beyond this period had no significant influence on the loss rate.

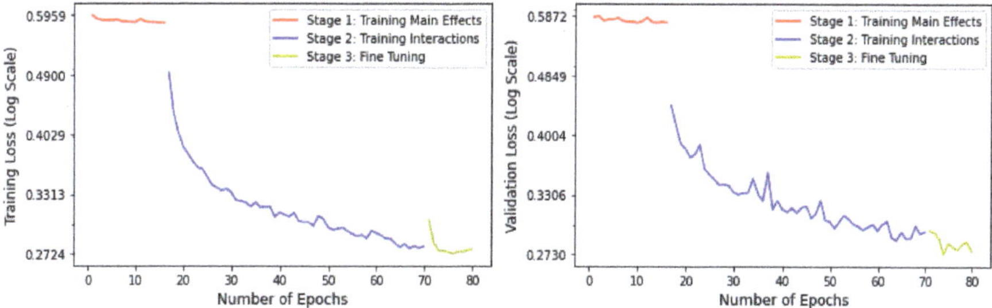

Figure 9. The GAMI-net loss function variation on the training and validation datasets (the red part is training main effects, the blue part is training interactions, and the yellow part is fine-tuning).

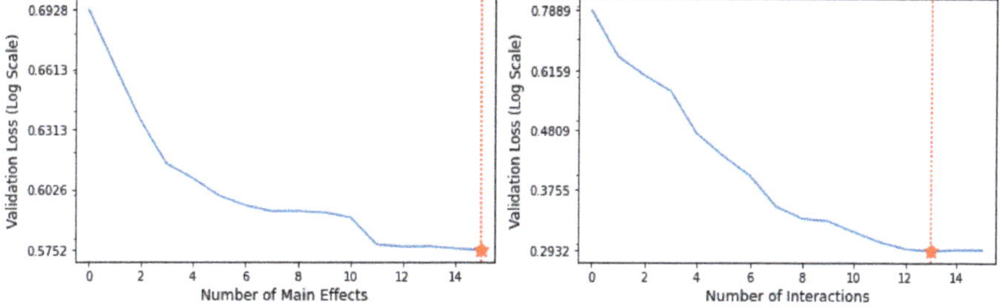

Figure 10. The validation loss for landslide factors. The red dots indicate the number of features that minimize the loss function and the red stars indicate the number of features selected for retention (the dots and stars have been overlapped in the figure).

4.2. Comparing Landslides Susceptibility Map

Landslide susceptibility maps can be generated using a variety of methods, including support vector machines (SVM), random forests, and GAMI-net. Each of these methods has its own advantages and disadvantages, and the most appropriate method will depend on the specific needs of the application.

The map divides Nayong into several categories of landslide susceptibility, including high, moderate, low, and very low. High-susceptibility areas are those with a high probability of landslides occurring, while low-susceptibility areas have a low probability of landslides.

In Figure 11, the map showed that the highest susceptibility areas are found in the coal mining area, particularly in the central and southern part of the study area. As coal is extracted from underground mines, the void left behind can cause the overlying rock and soil to collapse, leading to landslides. This is especially true in Karst areas, where the rock is naturally porous and prone to landslides. In addition, the southern region is characterized by a highly weathered layer of loose dolomite, which is subject to intense erosion due to the impact of precipitation. This erosion process significantly disrupts the mechanical equilibrium of the otherwise fragile grass and irrigation-dominated system in a short period, leading to an increase in the regional susceptibility to landslides. This underscores the need for careful management of erosion and landslides in this region. Moderate susceptibility areas are found in the northern and central parts of the state, including the Mazong Hill. These areas are characterized by a mix of steep and gentle slopes, as well naturally porous rock and prone-to-collapse slope, along with some coal mining operations. They are also prone to mining and heavy rainfall, which can increase the risk of landslides. Low-susceptibility areas are found in the western part of the state. These areas are characterized by gentle slopes and sandy soils. They are generally less prone to landslides, although they can still be affected by earthquakes and other natural disasters.

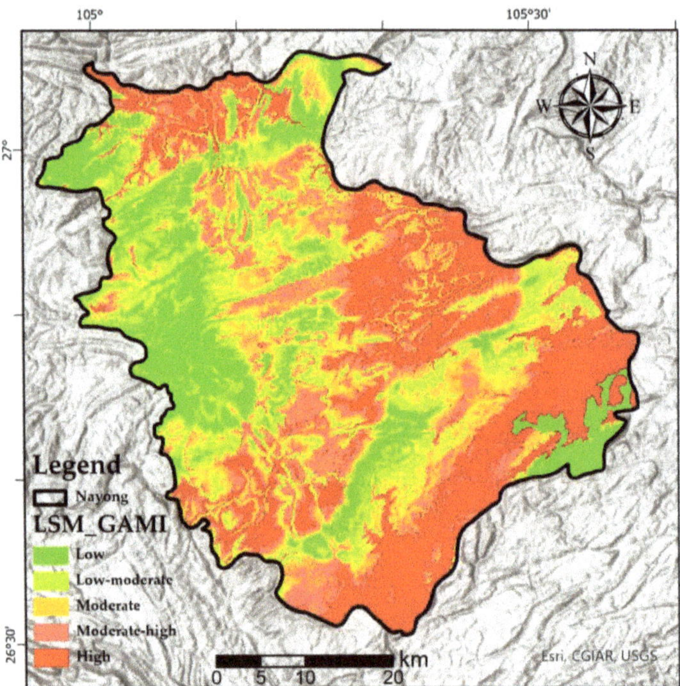

Figure 11. The XAI-LSM produced by GAMI-net.

4.3. Interpretative Results for GAMI-Net and Shapley Additive Explanations

Figure 12 showed the importance estimates for all impact factor variables for the GAMI-net model. The results for the study area indicated that rainfall, mining areas, lithology, elevation, rivers, and faults are the main factors. In brief, human activity and climate had a strong influence on landslides.

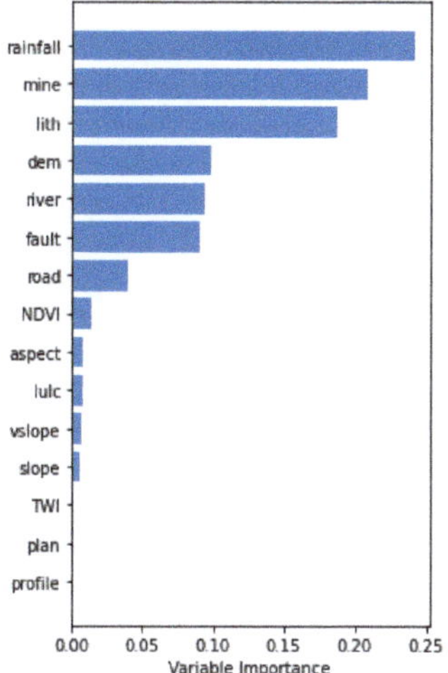

Figure 12. Global feature importance of GAMI-net LSM.

The global interpretation of the GAMI-net model factors was shown in Figure 13. With the introduction of 13 main effects and 25 interaction effects, it is found that rainfall, mine, river, failure, and lithology are the factors that account for 6.8%, 5.8%, 2.8%, 1.8%, and 1.7% of the landslide susceptibility factors, respectively. In terms of interaction effects, lithology rainfall, mining, rainfall, and elevation-rainfall account for 32.8% and 20.7% of the landslide susceptibility factors. The other susceptibility factors and their interaction effects were below 1%. In relative terms, the importance of the main effect is 18.9%, and the importance of the pairwise effect is 81.1%.

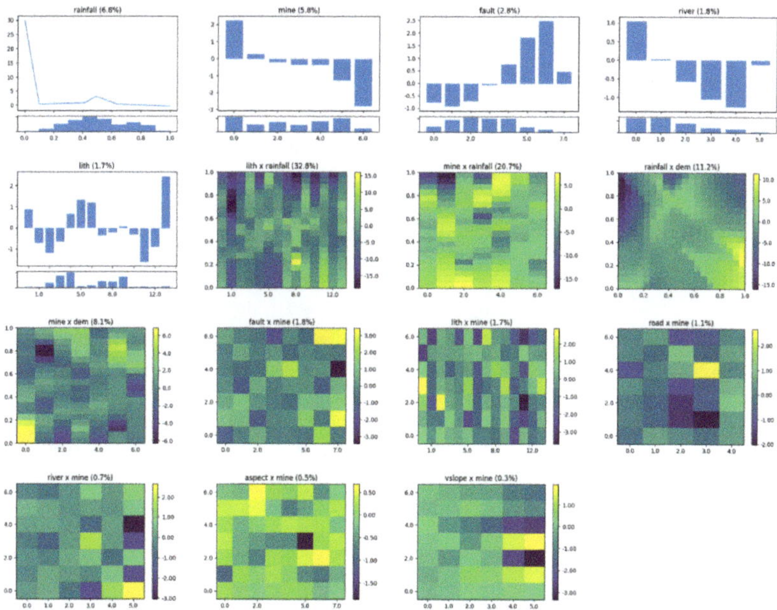

Figure 13. The factor ridge function images output by GAMI-Net global interpretation (reflects changes in IR values due to changes in landslide factors).

Firstly, rainfall was the most dominant influence factor. Rainfall causes soil destabilization and is highly susceptible to landslides. Extreme rainfall (class 1) significantly impacts landslides, with subsequent impacts on landslides decreasing as rainfall values decrease. Rainfall can cause landslides in several ways. For the distance factor of the mining area, the closer the mining area (the smaller the classification value in the graph), the higher the model score, indicating a higher landslide susceptibility factor, which suggests that mining activities are the main factor for landslide susceptibility in the area. Certain types of rock, such as shale (class 13) and clay (class 7), have high water retention capabilities due to their high porosity and low permeability. When these rocks become saturated with water, they can become unstable and prone to landslides. In contrast, rocks with low water retention capabilities, such as sandstone (class 3) and granite (class 11), are less prone to landslides because they are able to drain excess water more effectively. Earthquake-induced landslides are less common in the region due to few earthquakes and low seismic intensities. In addition, faults can create cracks and other types of pressure in the ground, which can increase the likelihood of landslides, and this probably will increase under the influence of mining and rainfall.

Secondly, in terms of interaction effects, the interaction between rainfall and lithology, mining, and elevation are the main factors, and the vulnerability of fragile rocks to landslides under rainfall conditions is evident. For the Nayong area, long-term mining activities have increased the fragile surface of the mountain. Landslide susceptibility was then significantly increased under external conditions such as rainfall.

Overall, the research measured the influence of various factors on landslide susceptibility using 1D/2D heat maps. For the study area, rainfall, mining, and lithology are the most significant influences.

4.4. Factorial Interpretability Analysis of the Zongling Landslides

The location of the Karst landslide in Zongling Town, Guizhou Province, China, was shown in Figure 14. It is a typical karst erosion landscape in the Guizhou plateau. Figure 14 is an optical image and InSAR deformation of Zongling, with the black polygon showing the landslide area. The InSAR surface deformation monitoring results were from our previous study [13,48]. Since the 1980s, Zongling Town has gradually become a central coal mining area, and a large amount of coal mining has created many landslide hazards. In particular, there is a cluster of landslides on the southern side of the Mazongling Mountains, running from west to east for a length of 5.6 km, and these landslides seriously threaten the people's safety. The rear edge of the slope is an almost vertical escarpment over 30 m on top and below. There are numerous ground fissures on the summit, ranging from several hundred meters to several kilometers in length, with fissures ranging from nearly 2 to 6 m in width. The study used Zongling landslides as an example for interpretable analysis based on the landslide susceptibility results above. As shown in Figure 14, points A and B are located in the more deformed areas of the Zongling landslide. The results of the interpretable model analysis in Figure 15 reveal that rainfall, mining, lithology, and their coupling were the key factors in the landslides at these two places, which is congruent with the ground research results. As shown in Figure 16, mining causes stress damage to the Karst hills in the area; lack of vegetation, extreme climatic background, and rock desertification further exacerbate landslide susceptibility and threaten the safety of the roads and people below. It suggests that our interpretable results are credible and can provide some basis for disaster mitigation and resilience.

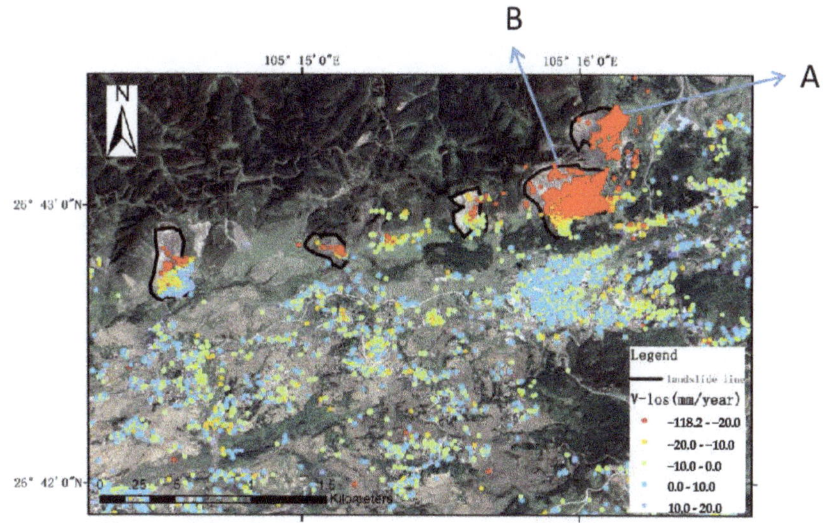

Figure 14. Optical and radar interpretation of the Zongling landslide (image and InSAR). Points A and B were the two most prominent landslides in the Zongling landslides, located on the side (A) and the front (B) of Mount Zongling, respectively.

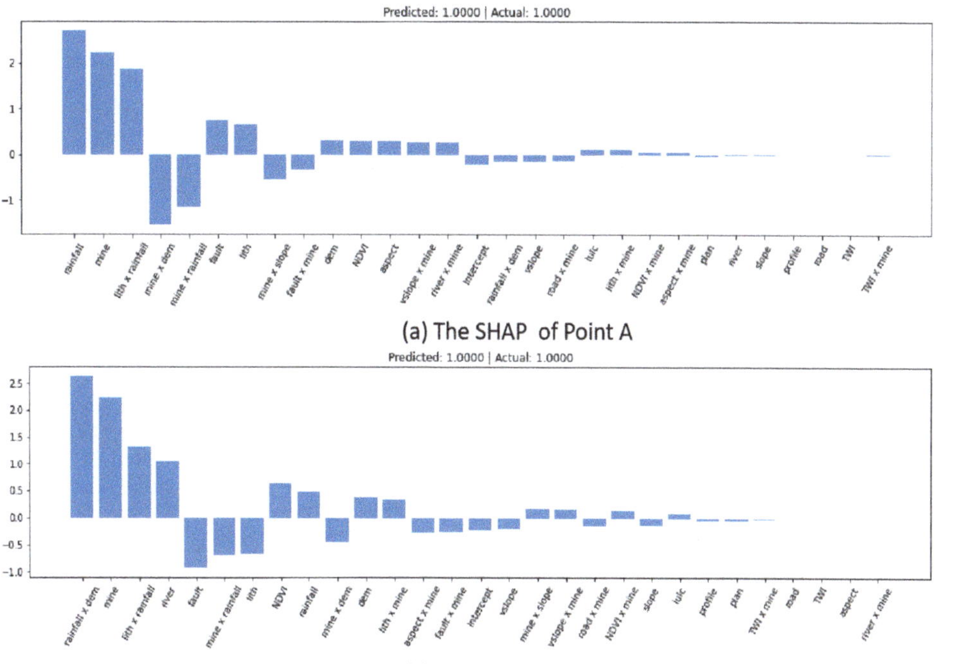

Figure 15. The SHAP results of Point A and B. The horizontal axis is the landslide impact factor and the vertical axis is the IR value (the 1 in the forecast represents the landslide).

(a) Front view of the Zongling landslide

(b) Side view of the Zongling landslide

Figure 16. Ground survey pictures of the Zongling landslide.

5. Discussion

In this study, we used a combination of GIS, remote sensing, and machine learning to analyze landslide susceptibility in Nayong County, Guizhou Province. It analyzed complex functional relationships through ANN networks of main effect factors and pairwise interaction factors and considered the coupling between the landslide factors, increasing the model's interpretability. In addition, the relationship between the influence of landslide factors on landslide susceptibility can be easily assessed using 1D line/bar diagrams and 2D heat maps. Our results showed that landslide susceptibility in the region is significantly correlated with several landslide influence factors, including rainfall, slope, and lithology. By introducing interaction factors, we also found that coupling several factors had significant effects on landslide susceptibility, such as rainfall and mining, lithology and rainfall, rainfall and elevation, and elevation and mining.

Validation of the four models showed that GAMI-net was the most reliable method, with the best model performance on the landslide dataset, followed by RF and SVM, while LR had the worst. The result suggests that GAMI-net can provide a strong interpretation while guaranteeing better accuracy. Compared with the four machine-learning models, LR has fast classification speed and easy interpretation of classification results but has a specific error rate in classification decisions; RF has high training efficiency and high practicality but ignores the correlation between data. SVM can avoid overfitting but depends on the choice of kernel function; GAMI-net has the highest accuracy, considers factors' coupling effect, has good interpretability, and quantifies the influence of landslide features.

In addition, our study uses interpretable algorithms to account for the changing influence of these factors on landslide susceptibility in the region. This method adds credibility to the landslide susceptibility model and provides a method for quantitatively assessing the influence of natural and human factors on landslides. Ranking rainfall and mining as more influential features in landslide prediction was in line with a series of regional studies [25,49,50]. Tao and Shi et al. used the Jinhaihu landslide event in Bijie as an example of a physical experiment and field investigation to show that lithology, construction, and rainfall (snow) were the main factors contributing to the landslide [49]. Wang et al. used time-series InSAR techniques to illustrate the persistent influence of mining activities on the Zongling landslide [50]. Unusually, the slope factor was classified as a less influential feature in Nayong, contrary to the findings of Zhao and Masoumi et al. in Iran [51], which may reflect the characteristics and regional effects of landslide events, landslide causal factors, and the dependence of the algorithm from one study area to another. In the present study, although the slope was positively correlated, its effect was small (Figure 13). The Nayong region was Karst landform with steep slopes at high elevations. The slopes in most parts of the region were eligible for landslide occurrence, making slope a relatively low-impact feature in the internal modeling process.

Our results showed that targeted measures, such as rational mining and rock desertification reduction, can effectively reduce landslide risks. In addition, our study could be used to inform land resource managers and policymakers. There have several limitations in the study that should be acknowledged. First, our analysis is based on remotely sensed data, which may only capture some relevant variables affecting landslides, such as cohesion and angle of internal friction. In addition, our study area is limited to mining areas and karst landscapes, so extension to other areas with different climatic and land use patterns will require corresponding changes.

6. Conclusions

LSM is an essential tool for landslide hazard reduction. Existing LSM research has mainly focused on machine-learning algorithms, with less interpretable research on LSM. Therefore, the GAMI-net method was applied as an interpretable algorithm and three traditional machine-learning models (LR, SVM, RF) to generate landslide sensitivity maps for Nayong region of Guizhou Province.

Firstly, the performance of the GAMI-net model and the three traditional ML models were examined and compared against several evaluation metrics. The results showed that GAMI-net outperformed the other three models (accuracy = 87.3%, ROC = 0.944). GAMI-net shows excellent potential for susceptibility assessment and interpretation, and it would be advantageous to explore its predictive capabilities in landslide susceptibility assessment.

Second, the LSM shows that most of the high-susceptibility study areas are located in mining and heavy rainfall areas, which are affected by erosion processes and rock desertification. The low susceptibility areas are the main areas of gentle slopes and lithological stability.

Thirdly, GAMI-net allows for an assessment of the significance of the characteristics of the sample (global and local interpretation). The findings suggest rainfall, mining, and lithology are the main causal factors for local landslides. In addition, some interaction factors such as rainfall and mining, lithology and rainfall, rainfall and elevation, and elevation and mining are also worth noting.

Overall, GAMI-net could be used to develop LSM and provide clear explanations for why the model made that prediction. This could help decision-makers better understand the risks of landslides in their area and take appropriate actions to prevent or mitigate them. The results of this work could be useful for decision-makers and planners in planning land use in areas prone to landslides.

Author Contributions: Conceptualization, Y.S. and C.X.; data curation, C.S. and H.F.; funding acquisition, Y.S. and C.X.; investigation, G.C., Y.Z. and Y.Y.; methodology, H.F. and C.X.; project administration, Y.G. and M.Z.; software, H.F. and B.T.; supervision, Y.S. and B.T.; writing—original draft, H.F.; writing—review and editing, C.X. All authors have read and agreed to the published version of the manuscript.

Funding: This research was funded by National Key R&D Program of China, grant number 2022YFC3005601, Outstanding Youth Science and Technology program of Guizhou Province of China ([2021]5615), and Multi-Source Remote Sensing Regional Landslide Hazard Risk Mapping and Key Landslide Fine Survey, Science and Technology Bureau of Fuzhou City, Fujian Province, grant number E1D7110100.

Institutional Review Board Statement: Not applicable.

Informed Consent Statement: Not applicable.

Data Availability Statement: Not applicable.

Acknowledgments: The generalized additive models with structured interactions—PyTorch version (https://github.com/SelfExplainML/GamiNet-PyTorch, accessed on 12 August 2022) and the Third Surveying and Mapping Institute of Guizhou Province.

Conflicts of Interest: The authors declare no conflict of interest.

References

1. Wistuba, M.; Malik, I.; Gärtner, H.; Kojs, P.; Owczarek, P. Application of eccentric growth of trees as a tool for landslide analyses: The example of Picea abies Karst. in the Carpathian and Sudeten Mountains (Central Europe). *Catena* **2013**, *111*, 41–55. [CrossRef]
2. Chen, L.; Zhao, C.; Li, B.; He, K.; Ren, C.; Liu, X.; Liu, D. Deformation monitoring and failure mode research of mining-induced Jianshanying landslide in karst mountain area, China with ALOS/PALSAR-2 images. *Landslides* **2021**, *18*, 2739–2750. [CrossRef]
3. Guzzetti, F. Landslide fatalities and the evaluation of landslide risk in Italy. *Eng. Geol.* **2000**, *58*, 89–107.
4. Gariano, S.L.; Guzzetti, F. Landslides in a changing climate. *Earth-Sci. Rev.* **2016**, *162*, 227–252.
5. Zheng, H.; Liu, B.; Han, S.; Fan, X.; Zou, T.; Zhou, Z.; Gong, H. Research on landslide hazard spatial prediction models based on deep neural networks: A case study of northwest Sichuan, China. *Environ. Earth Sci.* **2022**, *81*, 1–15. [CrossRef]
6. Yong, C.; Jinlong, D.; Fei, G.; Bin, T.; Tao, Z.; Hao, F.; Li, W.; Qinghua, Z. Review of landslide susceptibility assessment based on knowledge mapping. *Stoch. Environ. Res. Risk Assess.* **2022**, *36*, 2399–2417.
7. Dikshit, A.; Sarkar, R.; Pradhan, B.; Acharya, S.; Alamri, A.M. Spatial landslide risk assessment at Phuentsholing, Bhutan. *Geosciences* **2020**, *10*, 131. [CrossRef]
8. Chen, X.; Chen, W. GIS-based landslide susceptibility assessment using optimized hybrid machine learning methods. *Catena* **2021**, *196*, 104833. [CrossRef]
9. Kalantar, B.; Ueda, N.; Saeidi, V.; Ahmadi, K.; Halin, A.A.; Shabani, F. Landslide susceptibility mapping: Machine and ensemble learning based on remote sensing big data. *Remote Sens.* **2020**, *12*, 1737.
10. Wentao, Y.; Ming, W.; Peijun, S. Using MODIS NDVI Time Series to Identify Geographic Patterns of Landslides in Vegetated Regions. *IEEE Geosci. Remote Sens. Lett.* **2013**, *10*, 707–710. [CrossRef]
11. Zhao, F.; Meng, X.; Zhang, Y.; Chen, G.; Su, X.; Yue, D. Landslide Susceptibility Mapping of Karakorum Highway Combined with the Application of SBAS-InSAR Technology. *Sensors* **2019**, *19*, 2685. [CrossRef] [PubMed]
12. Zhang, Y.; Shen, C.; Zhou, S.; Luo, X. Analysis of the Influence of Forests on Landslides in the Bijie Area of Guizhou. *Forests* **2022**, *13*, 1136.
13. Shen, C.; Feng, Z.; Xie, C.; Fang, H.; Zhao, B.; Ou, W.; Zhu, Y.; Wang, K.; Li, H.; Bai, H. Refinement of Landslide Susceptibility Map Using Persistent Scatterer Interferometry in Areas of Intense Mining Activities in the Karst Region of Southwest China. *Remote Sens.* **2019**, *11*, 2821. [CrossRef]
14. Bai, D.; Lu, G.; Zhu, Z.; Zhu, X.; Tao, C.; Fang, J. Using Electrical Resistivity Tomography to Monitor the Evolution of Landslides' Safety Factors under Rainfall: A Feasibility Study Based on Numerical Simulation. *Remote Sens.* **2022**, *14*, 3592. [CrossRef]
15. Sengani, F.; Mashao, F.M.; Allopi, D. An integrated approach to develop a slope susceptibility map based on a GIS-based approach, soft computing technique and finite element formulation of the bound theorems. *Transp. Geotech.* **2022**, *36*, 100818. [CrossRef]
16. Yan, F.; Zhang, Q.; Ye, S.; Ren, B. A novel hybrid approach for landslide susceptibility mapping integrating analytical hierarchy process and normalized frequency ratio methods with the cloud model. *Geomorphology* **2019**, *327*, 170–187.
17. Mondal, S.; Mandal, S. Landslide susceptibility mapping of Darjeeling Himalaya, India using index of entropy (IOE) model. *Appl. Geomat.* **2019**, *11*, 129–146. [CrossRef]
18. Das, G.; Lepcha, K. Application of logistic regression (LR) and frequency ratio (FR) models for landslide susceptibility mapping in Relli Khola river basin of Darjeeling Himalaya, India. *SN Appl. Sci.* **2019**, *1*, 1–22.
19. Azarafza, M.; Azarafza, M.; Akgün, H.; Atkinson, P.M.; Derakhshani, R. Deep learning-based landslide susceptibility mapping. *Sci. Rep.* **2021**, *11*, 1–16. [CrossRef]
20. Sun, D.; Wen, H.; Wang, D.; Xu, J. A random forest model of landslide susceptibility mapping based on hyperparameter optimization using Bayes algorithm. *Geomorphology* **2020**, *362*, 107201.
21. Luo, X.; Lin, F.; Zhu, S.; Yu, M.; Zhang, Z.; Meng, L.; Peng, J. Mine landslide susceptibility assessment using IVM, ANN and SVM models considering the contribution of affecting factors. *PLoS ONE* **2019**, *14*, e0215134. [CrossRef]
22. Mohan, A.; Singh, A.K.; Kumar, B.; Dwivedi, R. Review on remote sensing methods for landslide detection using machine and deep learning. *Trans. Emerg. Telecommun. Technol.* **2021**, *32*, e3998. [CrossRef]
23. Yin, Z.; Xu, M. Explainable Neural Network Cancer Diagnosis of Unstained Tissue Samples Measured by Chemometric Microscope. In *Novel Techniques in Microscopy*; Optica Publishing Group: Washington, DC, USA, 2021; p. NM1C. 5.
24. Yang, Z.; Zhang, A.; Sudjianto, A. GAMI-Net: An explainable neural network based on generalized additive models with structured interactions. *Pattern Recognit.* **2021**, *120*, 108192. [CrossRef]
25. Yao, K.; Yang, S.; Wu, S.; Tong, B. Landslide Susceptibility Assessment Considering Spatial Agglomeration and Dispersion Characteristics: A Case Study of Bijie City in Guizhou Province, China. *ISPRS Int. J. Geo-Inf.* **2022**, *11*, 269. [CrossRef]
26. Petley, D. Zhangjiawan Landslide: A Massive Rockslope Collapse with 35 Fatalities Caught on a Remarkable Video. Available online: https://blogs.agu.org/landslideblog/2017/08/29/zhangjiawan-landslide-1/ (accessed on 24 January 2022).

27. Shen, C.; Xie, C.; Ji, S.; Zhou, D. Chapter 1. In *New Technology for Geological Hazard Potential Identification and Vulnerability Assessment in Karst Mountains*; Wang, Y., Ed.; Science Press: Beijing, China, 2022; pp. 2–3, ISBN 978-7-03-071326-1.
28. Liao, M.; Wen, H.; Yang, L. Identifying the essential conditioning factors of landslide susceptibility models under different grid resolutions using hybrid machine learning: A case of Wushan and Wuxi counties, China. *Catena* **2022**, *217*, 106428.
29. Tang, Y.; Feng, F.; Guo, Z.; Feng, W.; Li, Z.; Wang, J.; Sun, Q.; Ma, H.; Li, Y. Integrating principal component analysis with statistically-based models for analysis of causal factors and landslide susceptibility mapping: A comparative study from the loess plateau area in Shanxi (China). *J. Clean. Prod.* **2020**, *277*, 124159.
30. Chowdhuri, I.; Pal, S.C.; Chakrabortty, R.; Malik, S.; Das, B.; Roy, P.; Sen, K. Spatial prediction of landslide susceptibility using projected storm rainfall and land use in Himalayan region. *Bull. Eng. Geol. Environ.* **2021**, *80*, 5237–5258. [CrossRef]
31. Huang, F.; Ye, Z.; Jiang, S.-H.; Huang, J.; Chang, Z.; Chen, J. Uncertainty study of landslide susceptibility prediction considering the different attribute interval numbers of environmental factors and different data-based models. *Catena* **2021**, *202*, 105250. [CrossRef]
32. Li, Y.; Wang, X.; Mao, H. Influence of human activity on landslide susceptibility development in the Three Gorges area. *Nat. Hazards* **2020**, *104*, 2115–2151.
33. Chen, L.; Guo, Z.; Yin, K.; Shrestha, D.P.; Jin, S. The influence of land use and land cover change on landslide susceptibility: A case study in Zhushan Town, Xuan'en County (Hubei, China). *Nat. Hazards Earth Syst. Sci.* **2019**, *19*, 2207–2228.
34. Sharifi, E.; Saghafian, B.; Steinacker, R. Downscaling satellite precipitation estimates with multiple linear regression, artificial neural networks, and spline interpolation techniques. *J. Geophys. Res. Atmos.* **2019**, *124*, 789–805. [CrossRef]
35. Yordanov, V.; Brovelli, M. Comparing model performance metrics for landslide susceptibility mapping. *Int. Arch. Photogramm. Remote Sens. Spat. Inf. Sci.* **2020**, *43*, 1277–1284. [CrossRef]
36. Dou, J.; Yunus, A.P.; Bui, D.T.; Merghadi, A.; Sahana, M.; Zhu, Z.; Chen, C.-W.; Khosravi, K.; Yang, Y.; Pham, B.T. Assessment of advanced random forest and decision tree algorithms for modeling rainfall-induced landslide susceptibility in the Izu-Oshima Volcanic Island, Japan. *Sci. Total Environ.* **2019**, *662*, 332–346.
37. Chen, W.; Xie, X.; Wang, J.; Pradhan, B.; Hong, H.; Bui, D.T.; Duan, Z.; Ma, J. A comparative study of logistic model tree, random forest, and classification and regression tree models for spatial prediction of landslide susceptibility. *Catena* **2017**, *151*, 147–160. [CrossRef]
38. Yao, X.; Tham, L.G.; Dai, F.C. Landslide susceptibility mapping based on Support Vector Machine: A case study on natural slopes of Hong Kong, China. *Geomorphology* **2008**, *101*, 572–582. [CrossRef]
39. Pham, B.T.; Prakash, I.; Khosravi, K.; Chapi, K.; Trinh, P.T.; Ngo, T.Q.; Hosseini, S.V.; Bui, D.T. A comparison of Support Vector Machines and Bayesian algorithms for landslide susceptibility modelling. *Geocarto Int.* **2019**, *34*, 1385–1407. [CrossRef]
40. Ravindra, K.; Rattan, P.; Mor, S.; Aggarwal, A.N. Generalized additive models: Building evidence of air pollution, climate change and human health. *Environ. Int.* **2019**, *132*, 104987.
41. Timoori Yansari, Z.; Hosseinzadeh, S.R.; Kavian, A.; Pourghasemi, H.R. Comparison of Landslide Susceptibility Maps using Logistic Regression (LR) and Generalized Additive Model (GAM). *J. Watershed Manag. Res.* **2019**, *9*, 208–219.
42. Shi, Y.; Han, Y.; Zhang, Q.; Kuang, X. Adaptive iterative attack towards explainable adversarial robustness. *Pattern Recognit.* **2020**, *105*, 107309.
43. Archer, G.; Saltelli, A.; Sobol, I.M. Sensitivity measures, ANOVA-like techniques and the use of bootstrap. *J. Stat. Comput. Simul.* **1997**, *58*, 99–120.
44. Pawluszek-Filipiak, K.; Borkowski, A. On the Importance of Train–Test Split Ratio of Datasets in Automatic Landslide Detection by Supervised Classification. *Remote Sens.* **2020**, *12*, 3054. [CrossRef]
45. Zhou, X.; Wen, H.; Li, Z.; Zhang, H.; Zhang, W. An interpretable model for the susceptibility of rainfall-induced shallow landslides based on SHAP and XGBoost. *Geocarto Int.* **2022**, 1–32. [CrossRef]
46. Mangalathu, S.; Hwang, S.-H.; Jeon, J.-S. Failure mode and effects analysis of RC members based on machine-learning-based SHapley Additive exPlanations (SHAP) approach. *Eng. Struct.* **2020**, *219*, 110927. [CrossRef]
47. Van den Broeck, G.; Lykov, A.; Schleich, M.; Suciu, D. On the tractability of SHAP explanations. *J. Artif. Intell. Res.* **2022**, *74*, 851–886.
48. Zhu, Y.; Tian, B.; Xie, C.; Guo, Y.; Fang, H.; Yang, Y.; Wang, Q.; Zhang, M.; Shen, C.; Wei, R. Multi-Temporal InSAR Deformation Monitoring Zongling Landslide Group in Guizhou Province Based on the Adaptive Network Method. *Sustainability* **2023**, *15*, 894. [CrossRef]
49. Tao, T.; Shi, W.; Liang, F.; Wang, X. Failure mechanism and evolution of the Jinhaihu landslide in Bijie City, China, on 3 January 2022. *Landslides* **2022**, *19*, 2727–2736. [CrossRef]
50. Wang, J.; Wang, C.; Xie, C.; Zhang, H.; Tang, Y.; Zhang, Z.; Shen, C. Monitoring of large-scale landslides in Zongling, Guizhou, China, with improved distributed scatterer interferometric SAR time series methods. *Landslides* **2020**, *17*, 1777–1795.
51. Zhao, P.; Masoumi, Z.; Kalantari, M.; Aflaki, M.; Mansourian, A. A GIS-Based Landslide Susceptibility Mapping and Variable Importance Analysis Using Artificial Intelligent Training-Based Methods. *Remote Sens.* **2022**, *14*, 211. [CrossRef]

Article

Integrated Approach for Landslide Risk Assessment Using Geoinformation Tools and Field Data in Hindukush Mountain Ranges, Northern Pakistan

Nisar Ali Shah [1,2,*], **Muhammad Shafique** [1,2], **Muhammad Ishfaq** [1,2], **Kamil Faisal** [3] **and Mark Van der Meijde** [4]

1   GIS and Space Application in Geosciences (GSAG) Lab, National Center of GIS and Space Application (NCGSA), Islamabad 44000, Pakistan
2   National Centre of Excellence in Geology, University of Peshawar, Peshawar 25130, Pakistan
3   Department of Geomatics, Faculty of Architecture and Planning, King Abdulaziz University, Jeddah 21589, Saudi Arabia
4   Department of Earth System Analysis, University of Twente, 7500 AE Enschede, The Netherlands
*   Correspondence: geologist.nisar07@gmail.com

Abstract: Landslides are one of the most recurring and damaging natural hazards worldwide, with rising impacts on communities, infrastructure, and the environment. Landslide hazard, vulnerability, and risk assessments are critical for landslide mitigation, land use and developmental planning. They are, however, often lacking in complex and data-poor regions. This study proposes an integrated approach to evaluate landslide hazard, vulnerability, and risk using a range of freely available geospatial data and semi-quantitative techniques for one of the most landslide-prone areas in the Hindukush mountain ranges of northern Pakistan. Very high-resolution satellite images and their spectral characteristics are utilized to develop a comprehensive landslide inventory and predisposing factors using bi-variate models to develop a landslide susceptibility map. This is subsequently integrated with landslide-triggering factors to derive a Landslide Hazard Index map. A geospatial database of the element-at-risk data is developed from the acquired remote sensing data and extensive field surveys. It contains the building's footprints, accompanied by typological data, road network, population, and land cover. Subsequently, it is analyzed using a spatial multi-criteria evaluation technique for vulnerability assessment and further applied in a semi-quantitative technique for risk assessment in relative risk classes. The landslide risk assessment map is classified into five classes, i.e., very low (18%), low (39.4%), moderate (26.3%), high (13.3%), and very high (3%). The developed landslide risk index map shall assist in highlighting the landslide risk hotspots and their subsequent mitigation and risk reduction.

Keywords: landslides; susceptibility; hazard; vulnerability; risk; northern Pakistan

Citation: Shah, N.A.; Shafique, M.; Ishfaq, M.; Faisal, K.; Van der Meijde, M. Integrated Approach for Landslide Risk Assessment Using Geoinformation Tools and Field Data in Hindukush Mountain Ranges, Northern Pakistan. *Sustainability* **2023**, *15*, 3102. https://doi.org/10.3390/su15043102

Academic Editor: William Frodella

Received: 12 December 2022
Revised: 27 January 2023
Accepted: 3 February 2023
Published: 8 February 2023

1. Introduction

Around the globe, landslides are becoming more frequent and devastating, with 200,000 casualties in the twentieth century and an estimated exposure of around 3.7 million km^2 area and 300 million population [1–4]. Given changes in the global climate and microclimate, increases in population, urbanization, ever-expanding communication networks, and other infrastructural development on fragile slopes, the landslide-induced risk to society and the economy is growing [5–8]. The assessment of landslide risk is critical to demarcate landslide-prone areas, infrastructure, and communities and accordingly plan and implement risk reduction strategies. Space and airborne remote sensing offer a variety of geospatial data for Geographic Information Systems (GIS) to perform landslide hazard, vulnerability, and risk assessments from a local to regional scale [9–11].

Landslide susceptibility is the probability of the spatial distribution of landslides driven by the topography, geology, tectonics, climate, land cover, and human activities [12,13]. A landslide susceptibility analysis begins by developing a comprehensive landslide inventory, which is often developed from space and airborne remote sensing data using a range of image classification techniques [14,15]. Techniques for landslide susceptibility assessment can be categorized in the evaluation of landslide inventories and include heuristic techniques, geomorphological mapping, and physical and statistical modeling approaches [16–18]. Usually, the landslide susceptibility assessment is further analyzed with the temporal distribution, landslide frequency-magnitude relationship and major triggers, including rainfall and earthquakes, for the landslide hazard assessment [19–21]. The population, buildings, economic activities, public utilities, infrastructure, and environmental features are often impacted by landslides and thus considered as the elements at risk [22–27]. The degree of loss for elements at risk is determined by the landslide vulnerability assessment, which is influenced by the landslide type and magnitude and the nature of the exposed elements [22,28,29]. The typological information of the buildings, including their foundation type, building material, design etc., strongly affects the vulnerability of the buildings to landslides and other hazards [26]. Landslide risk refers to the expected damage to the people, property, and economy in a region and a given period [30].

Landslide risk can be assessed through qualitative and quantitative techniques, where the qualitative methods mostly rely on subjective evaluation, and the quantitative methods aim to quantify the landslide-induced losses [31]. Contrary to the qualitative risk assessment that leads to relative rankings (high, moderate, low), quantitative risk assessment (QRA) leads to quantifying the landslide-induced risk in terms of integrated costs of the exposed element at risk [31]. QRA is effective in comparing the landslide-prone sites and administrative units, and therefore, it is important for community awareness, resource prioritization, landslide mitigation planning, and policy development for landslide risk reduction [26,32,33]. However, a comprehensive QRA requires a range of data on the spatial and temporal distribution of landslides with associated characteristics, historical records of the potential landslide triggers, calculated costs of the elements at risk, and scenarios of human activities with time, which are rarely available in the developing countries [34–36]. Alternatively, semi-quantitative risk assessment techniques have been developed that apply weighting and rating of the input parameters, which are eventually integrated to compute the risk [35,37]. A semi-quantitative risk assessment approach is encouraging in landslide studies [38,39]. Crawford et al., (2022) [40] proposed that a practical level exists between qualitative and quantitative risk assessments for a significant landslide risk assessment at a regional scale with limitations regarding hazard behavior and asset data.

Given the rough terrain, active tectonic and denudation, glaciation, climatic conditions, and human factors, landslides are one of the most damaging natural disasters in the Hindukush mountains in northern Pakistan [41–43]. Landslides in the region are investigated by various researchers using field surveys, geological and tectonic records, and geological settings to derive the landslide susceptibility map [42,44–48]. However, despite the increasing landslide frequency and associated damages that are exacerbated by climate change, an integrated risk assessment is lacking for the region. The aim of the present study is to utilize the freely available geospatial and extensive field data and integrate these data in a semi-quantitative approach for landslide risk assessment. The proposed study is initiated with quantitative data for hazard and vulnerability assessments, and the derived risk is classified into relative classes from very high to very low risk levels. The developed methodology can be replicated in other landslide-prone regions to assist in risk reduction.

2. Study Area

The study area is located in the Hindukush mountain ranges in the Ghizer district of northern Pakistan, with a total area of 7040 km^2 and elevation ranging from 1672 to 6239 m above sea level (ASL) (Figure 1).

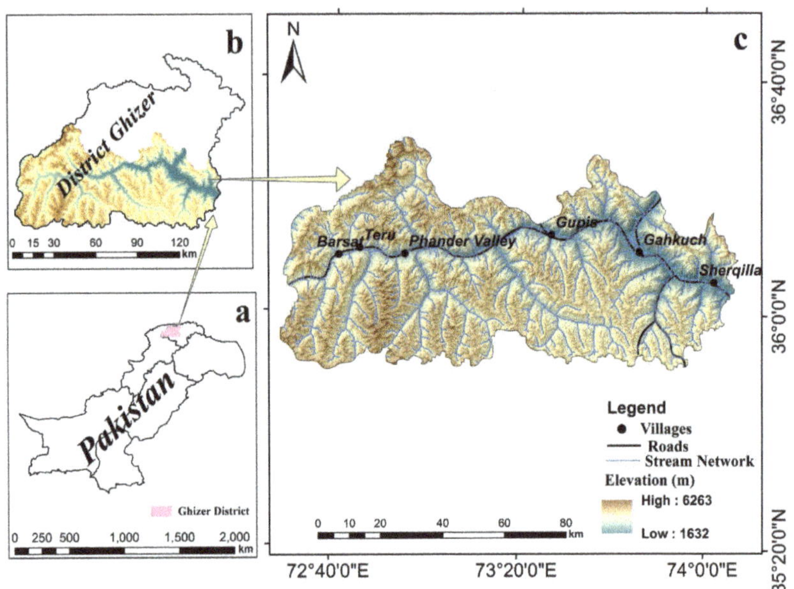

Figure 1. Location map of the study area; (**a**) Pakistan boundary, (**b**) study area boundary, and (**c**) topography, locations, roads, and drainage network.

The area is located in an active tectonics zone attributed to the collision between the Eurasian plate and the Indian tectonic plate. The area lies at the Kohistan Island arc and Karakorum block, with the Main Karakorum Thrust between them. The lithology of the area is comprised of the Kohistan Batholiths, Chalt Volcanic group, Greenstone complex, Quaternary deposits, Darkot group, and Yasin group [49]. The topography of the area is characterized by steep to gentle slopes favoring widespread landslides, rock falls, scree slopes and debris flows [50–52] (Figure 2).

Figure 2. Field photographs of landslides in steep slopes. (**a**) Catchment area of debris flow in Khalti village; (**b**) debris slide; (**c**) scree captured from the toe; (**d**) rockfall along the main road in Teru village.

3. Materials and Methods

3.1. Materials

The landslide boundaries and road network were visually digitized from Google Earth Pro and subsequently verified in the field. The resolution of the data has a significant impact on the parameters for deriving the causative factors of landslides, landslide hazards, presentation of the element-at-risk database, and terrain representation [53,54] The ALOS PALSAR Digital Elevation Model (DEM) with a 12.5 m spatial resolution was exploited to compute the topographic features, including the terrain elevation, aspect, slope, curvature, and stream network (Table 1). The geology and fault lines were digitized from the geological map after [49]. The landcover map was developed from the Sentinel-2 data (https://sentinel.esa.int/web/sentinel/missions/sentinel-2, accessed on 25 July 2022). For landslide hazard assessment, the precipitation record was derived from the global precipitation measurements (GPM), and the peak ground acceleration (PGA) values were adopted from [55]. For the landslide exposure and vulnerability analysis, the population data were acquired from the 2017 census records. The building footprints were visually demarcated from the base map in the ArcGIS software, and the typological information, including building use (houses, schools, and hospitals), roof and wall materials, physical condition, and age of the buildings, was collected for each of the buildings through an extensive field survey and subsequently utilized for the landslide vulnerability and risk assessments.

Table 1. Data and source used in this study.

Data	Factors Determined	Resolution	Data Source
DEM	Elevation, aspect, slope, curvature and stream network	12.5 m	ALOS PALSAR
Google Earth Pro Images	Landslide inventory and road network	2 m	Digital globe
Sentinel-2	Land cover	10 m	ESA (European Space Agency)
Average annual rainfall Precipitation	Precipitation map	10 km	(NASA) National Aeronautics and Space Administration
GIS base Map	Building footprints	2 m	Digital globe

3.2. Methods

3.2.1. Landslide Susceptibility and Hazard Assessments

Landslide Inventory

A landslide inventory depicts the type, volume, magnitude, date, and place of occurrence of landslides in an area. A landslide inventory of the study area was developed through the visual interpretation of Google Earth images following [56]. The landslides were demarcated considering their morphology, changes in vegetation, and the presence of eroded deposits and subsequently updated in the field and classified after [57]. Among the mapped landslides, 70% of the landslide data were used for susceptibility modeling, and 30% were used for validation.

Landslide Causative Factors and Susceptibility Assessment

Considering the influence on the spatial distribution of landslides in the area, nine landslide causative parameters were selected for the susceptibility mapping. Elevation has an influence on the weather, drainage, and other factors influencing the spatial distribution of landslides [58,59] (Figure 3a). Terrain slope determines the shear forces acting on hill slopes and thus impacting the occurrence of landslides [60–62] (Figure 3b). The terrain aspect (Figure 3c) is the measurement of the direction of the slope, impacting the vegetation cover, moisture retention, and strength of the soil [63,64]. Terrain curvature (Figure 3d) is the change in slope along a small arc of the curve and is often applied as a controlling factor for landslides [65]. Water runoff contributes to slope stability through infiltration, saturation and slope toe-cutting [66,67] (Figure 3e). Excavation for road construction and repair in mountainous areas often destabilizes the slope and triggers landslides [68] (Figure 3f). The

Euclidean distance from the selected parameters is classified into five classes following literature and field observations. Variation in land cover affects the hydrological conditions, shear strength and slope instability [41]. A land cover map of the area was developed from the Sentinel-2 data using the maximum livelihood classification algorithm and classified into six categories, i.e., water bodies, settlements, barren land, snow, forest, and agriculture (Figure 3g). Proximity to a fault (Figure 3h) was reclassified following [61]. Lithology and faults influence the rock's strength and largely drive the spatial concentration of landslides [69]. The area of study is comprised of eight lithological units (Figure 3i).

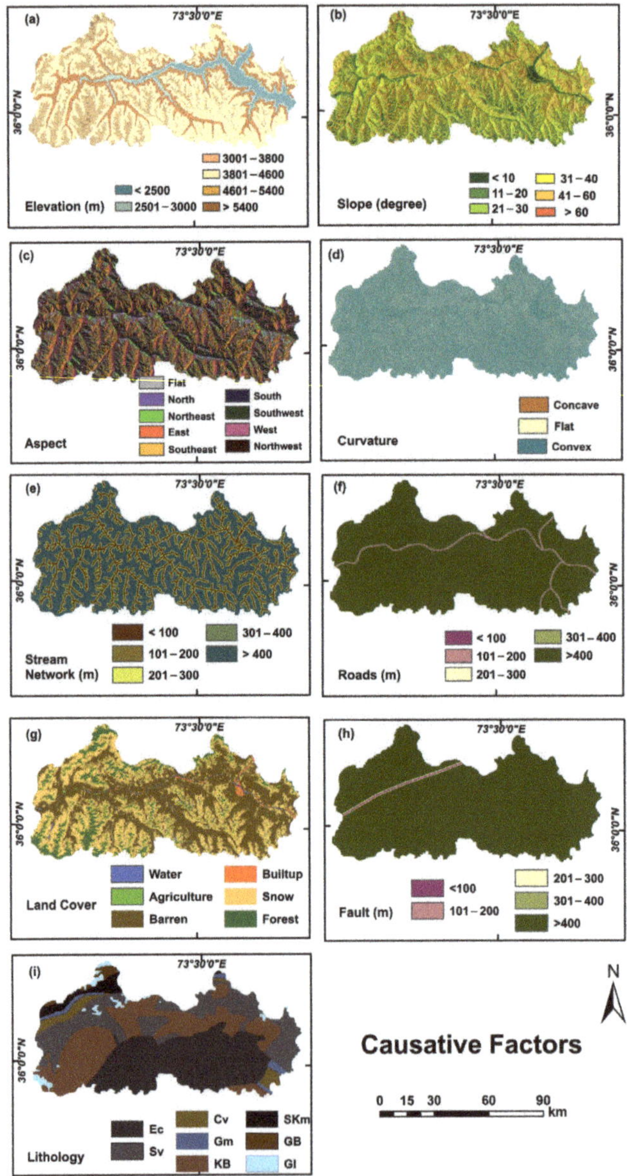

Figure 3. Showing landslide causative factors selected for the landslide susceptibility model: (**a**) elevation in m, (**b**) slope in degrees, (**c**) aspect, (**d**) curvature, (**e**) stream networks, (**f**) roads, (**g**) land use,

(**h**) faults, and (**i**) lithology (Ec represents Basal Eclogites, Sv represents the Shyok Suture Zone, Cv stands for the Chalt group, Gm stands for Gilgit complex metasedimentary rocks, KB represents Kohistan Batholiths, SKm represents the Southern Karakorum metamorphic complex, GB represents the Karakorum Batholiths, and Gl represents glaciers).

For the landslide susceptibility assessment, the frequency ratio (FR) approach evaluating the relationship between the landslide inventory and the selected causative factors is frequently and effectively applied [70–74]. The frequency ratio (FR) is the ratio of the landslide's occurrence area to the total area and leads to higher accuracy compared to other models [46,75–77].

Landslide Triggers and Hazard Mapping

The intensity and duration of rainfall are among the major triggers for widespread landslides and are often used for landslide hazard assessments [41]. The area-specific threshold of rainfall intensity and duration are often used to predict landslides [78]. Global precipitation measurement (GPM) data for seven years (1st January 2015 to 31st December 2021) were utilized with mean rainfall ranges from 1002 to 2873 mm/year [79]. The earthquake-induced shaking depicted through the peak ground acceleration (PGA) leads to widespread co-seismic landslides.

For the landslide hazard assessment, the temporal distribution of landslides and their frequency-magnitude records are important; however, they are rarely available in data-poor regions, such as the study area [35]. Due to the lack of temporal and intensity records of landslides, a static hazard map was developed in this research work. Following [35,80–83], weights of 0.3 to PGA and 0.7 to rainfall are assigned and integrated into the landslide susceptibility to derive the landslide hazard map.

$$TI = (PGA \times 0.3) + (RF \times 0.7) \tag{1}$$

where TI is the triggering indicator, PGA is the peak ground acceleration, and RF is the rainfall precipitation. The exposure of buildings and roads to landslides was assessed through an overlay analysis [31].

3.2.2. Landslide Vulnerability and Risk Assessment

Vulnerability indicates the susceptibility of the exposed elements to the impacts of landslides [22,82,84–86]. Physical vulnerability refers to damage to buildings, utilities, and infrastructure, social vulnerability is the proneness of people and relates to fatalities or injuries [26], and environmental vulnerability refers to the impact on land cover, i.e., stream environments, forest cover, agriculture, etc. In this study, physical, social, and environmental vulnerabilities were derived and, using the multi-criteria analysis approach, integrated into a vulnerability index map (Figure 4).

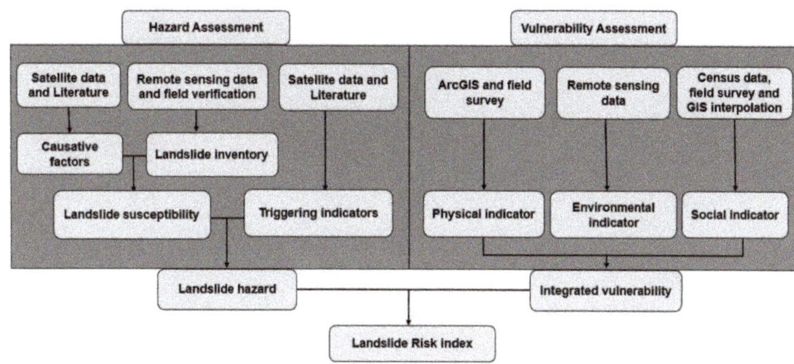

Figure 4. Methodology for landslide hazard, vulnerability, and risk assessment.

Physical Vulnerability

An extensive database of 23,456 building footprints was developed through onscreen digitization on the ArcGIS base map and subsequently verified in the field. For each building, the building use and associated typological data comprising the structural system, geometry, material properties, state of maintenance, levels of design codes, foundation and superstructure details, number of floors, and other factors, were recorded during field visits [26]. Applying a multi-criteria analysis, weights were assigned to each parameter using expert opinion and AHP after [87] (Table 2). Recently constructed buildings with no structural damage are generally more resistant to hazards compared to older and damaged buildings [88]. Therefore, buildings with no structural damage were assigned a weight of 0.15, and buildings with minor structural damage were assigned a weight of 0.35; those with major structural damage were assigned a weight of 0.5. Dry rubble stone without mortar was given a weight of 0.4, dressed stone masonry with mortar was given a weight of 0.2, cement block masonry was given a weight of 0.3, and reinforced confined masonry was assigned a weight of 0.1. Horizontal metallic sheets supported from underneath were given a weight of 0.4, metallic sheets with pitched sloping were given a weight of 0.3, wood planks covered by mud were given a weight of 0.2, and reinforced concrete cement was assigned a weight of 0.1. Considering the use of the building, educational institutes, i.e., schools/colleges were assigned a weight of 0.35, hospitals/dispensaries/clinics were assigned a weight of 0.25, houses were assigned a weight of 0.2, commercial buildings/hotels/offices were assigned a weight of 0.13, and worship places were assigned a weight of 0.07. Mapped roads are classified as main roads traversing through the main valley and link roads in the side valleys. The main roads were assigned a weight of 0.7, and the linking roads were assigned a weight of 0.3. The vulnerability of roads was integrated into the buildings to generate a physical vulnerability map.

Table 2. Weights assigned to buildings' typology parameters and roads for physical vulnerability.

Typology Parameters	Assigned Weights
Physical Condition	
No Structural Damage	0.15
Minor Cracks in the Walls	0.35
Major Cracks in the Walls/Roofs	0.5
Building Age (Years)	
= <10	0.1
11–20	0.15
21–30	0.2
31–40	0.25
>40	0.3
Wall Material	
Dry Stone Masonry	0.4
Dressed Stone Masonry	0.2
Cement Block Masonry	0.3
Reinforced Confined Masonry	0.1
Roof Material	
Horizontal Metal Sheet	0.4
Pitched Sloping Sheet	0.3
Wood Planks Covered by Mud	0.2
Reinforced Concrete Slab	0.1

Table 2. *Cont.*

Typology Parameters	Assigned Weights
Building Use	
Educational Institutes	0.35
Hospitals/Dispensaries/Clinics	0.25
Residential Houses	0.2
Shops/Hotels/ offices	0.13
Houses of Worship	0.07
Roads	
Main Roads	0.7
Linking Roads	0.3

Social and Environmental Vulnerability

Social vulnerability refers to the likelihood of human losses determined by the nature, magnitude, and intensity of the landslide and the coping capacity of the community [89]. Using a dissymmetric mapping approach, the average population per building was computed. Following [83], the population data were divided into four classes ranging from low to very high population, and accordingly, weights were assigned.

Using the developed land cover map, the highest weightage was assigned to agriculture and forest, i.e., 0.5 and 0.4, respectively, due to their environmental and economic significance. Water bodies were given a weight of 0.1. The physical, social, and environmental vulnerabilities were integrated to develop a vulnerability index map.

Landslide Risk Assessment

A landslide risk map was developed by integrating the developed landslide hazard and vulnerability index maps [33,90]. For landslide risk assessment, the landslide hazard and vulnerability indexes were classified into five classes ranging from very low to very high. The landslide risk index was developed by integrating the normalized landslide vulnerability map with the landslide hazard map through multi-criteria analysis following Equation (2) and after [83].

$$R = (H \times 0.6) + (V \times 0.4) \tag{2}$$

where R is the landslide risk index, H is the landslide hazard, and V is the landslide vulnerability index. The derived risk values are normalized and classified into 5 classes, including very high, high, moderate, low, and very low. The classes are the relative ranking of the risk classes. The very high risk is the riskiest area with very high hazard and very high vulnerability. The elements at risk overlay these very high risks and are very prone to damage due to landslides.

4. Results

4.1. Landslide Inventory

In the studied area, 1748 landslides were mapped with a total area of 344 km^2 (Figure 5). The types of landslides distributed in the area included rock falls (102), rockslides (62), scree (1382), debris flows (78), and debris cones (124). Landslides are largely concentrated along streams. A few large rock falls were identified in the study area along the Gilgit river, leading to the blockage of the river and natural dams, which were later drained due to overflow. The rock avalanches are composed of boulders and scree deposits of the Cretaceous to Jurassic age volcanic and volcanoclastic sediments, containing greenschist and slates.

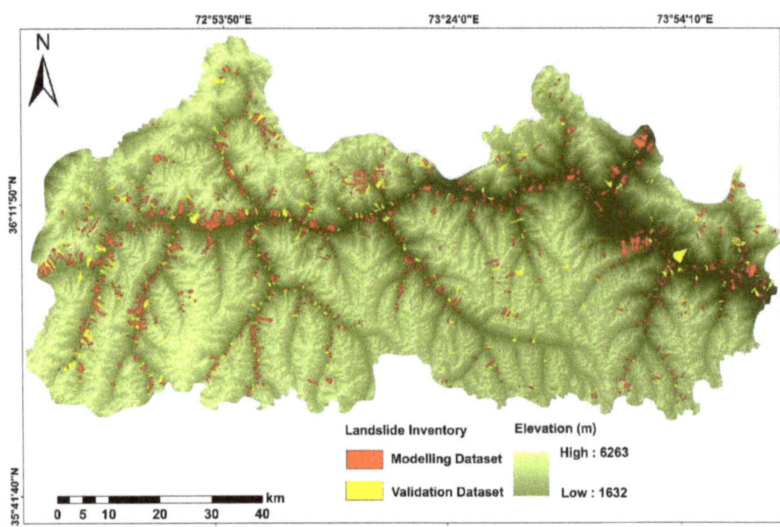

Figure 5. Spatial distribution of landslides in the area.

4.2. Landslide Causative Factors

Landslides in the area are largely concentrated along the slope ranging from 20–40°. The frequency ratio (FR) for the landslides reaches up to 1.27 for the slope class of 20–30° and 1.15 for the 30–40° class. Landslides in class < 10° and >60° have the lowest FR. The slope aspects toward the north, northeast, west, south, and southwest show the highest FR and lowest for flat areas. The FR is at its maximum for the elevation classes of <2500, 2500–3000, and 3000–3800, and it decreases in higher elevations, which are usually covered with glaciers. The FR for concave and flat terrain is 1.03, and the FR for convex terrain is 0.68. Landslides have an inverse relation with distance from streams. In summers, glaciers melt, increasing water in the streams, and monsoon rainfalls also add more water, triggering slope-toe erosion that often leads to landslides. The stream class of 100–200 m has the highest frequency ratio of 1.82, and the ratio decreases to 0.80 as the distance from the stream decreases (Figure 6). The road distance buffer zones along the roads have the highest FR values, ranging from 0.96–2.81 due to slope cut-offs, uncontrolled blasting during construction, and vibration due to heavy traffic. The distance from fault lines shows that the highest FR is computed for <100 m, i.e., 1.49, reflecting increased landslides along the faults. The geological units of the Shyok Suture Zone (Sv) and Kohistan Batholiths (KB) are highly susceptible to landslides, with the highest FR ratio, i.e., 1.64 and 1.13, respectively. The barren land landcover class has an FR value of 1.8 (Figure 6).

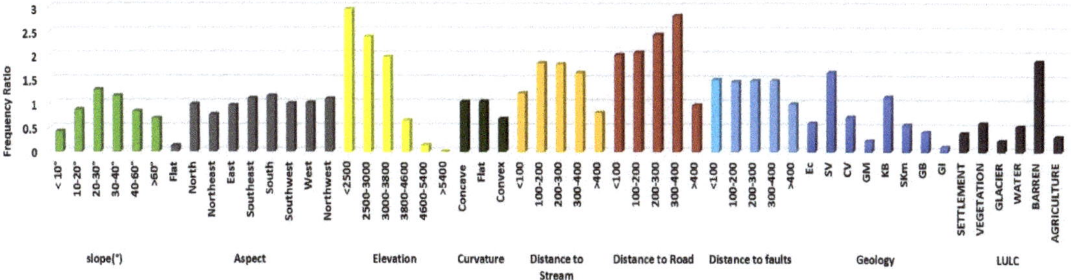

Figure 6. Graph showing the influence of classes of causative factors on the landslides using the frequency ratio modal. The blue lines show the values of frequency ratio in each class of causative factors.

4.3. Landslide Susceptibility and Hazard Mapping

The FR weights were integrated to develop a Landslide Susceptibility Index map (LSI) and classified into five classes (Figure 7a). The assessment of the accuracy of the susceptibility map was performed using the 30% dataset of landslides. The area under the curve (AUC) is a useful indicator to validate the prediction performance of the model [91]. The susceptibility map was classified into 100 classes with a 1% cumulative interval. Subsequently, the validation landslide set was correlated with the classified susceptibility map to identify the landslide area in each of the 100 susceptible classes. Qualitative analysis of the area under the curve of the success rate curve shows 88% accuracy for the susceptibility map (Figure 7b). The landslide susceptibility is further integrated with the weights of the landslide triggers (precipitation and PGA) to compute the landslide hazard map (Figure 8). The landslide hazard map indicates that 14% of the area lies in the very low hazard class, while 29.7% lies in the low hazard class, 30.3% lies in the medium hazard class, 19.2% lies in the high hazard class, and 6.6% is in the very high hazard class.

The spatial comparison of the buildings and roads with the hazard map shows that 118 buildings (0.38%) are located in the very low hazard zone, 2838 buildings (12.1%) are located in the low hazard zone, 7037 buildings (30%) are located in the moderate hazard zone, 8233 (35.1%) are located in the high hazard zone, and 5230 buildings (22.3%) are located in the high hazard class. The exposure assessment of roads indicates that 33% of roads are in the very high hazard zone, 27% of roads are in the high hazard zone, 20% are in the moderate hazard zone, 13% are in the low hazard zone, and 7% are in a very low hazard zone.

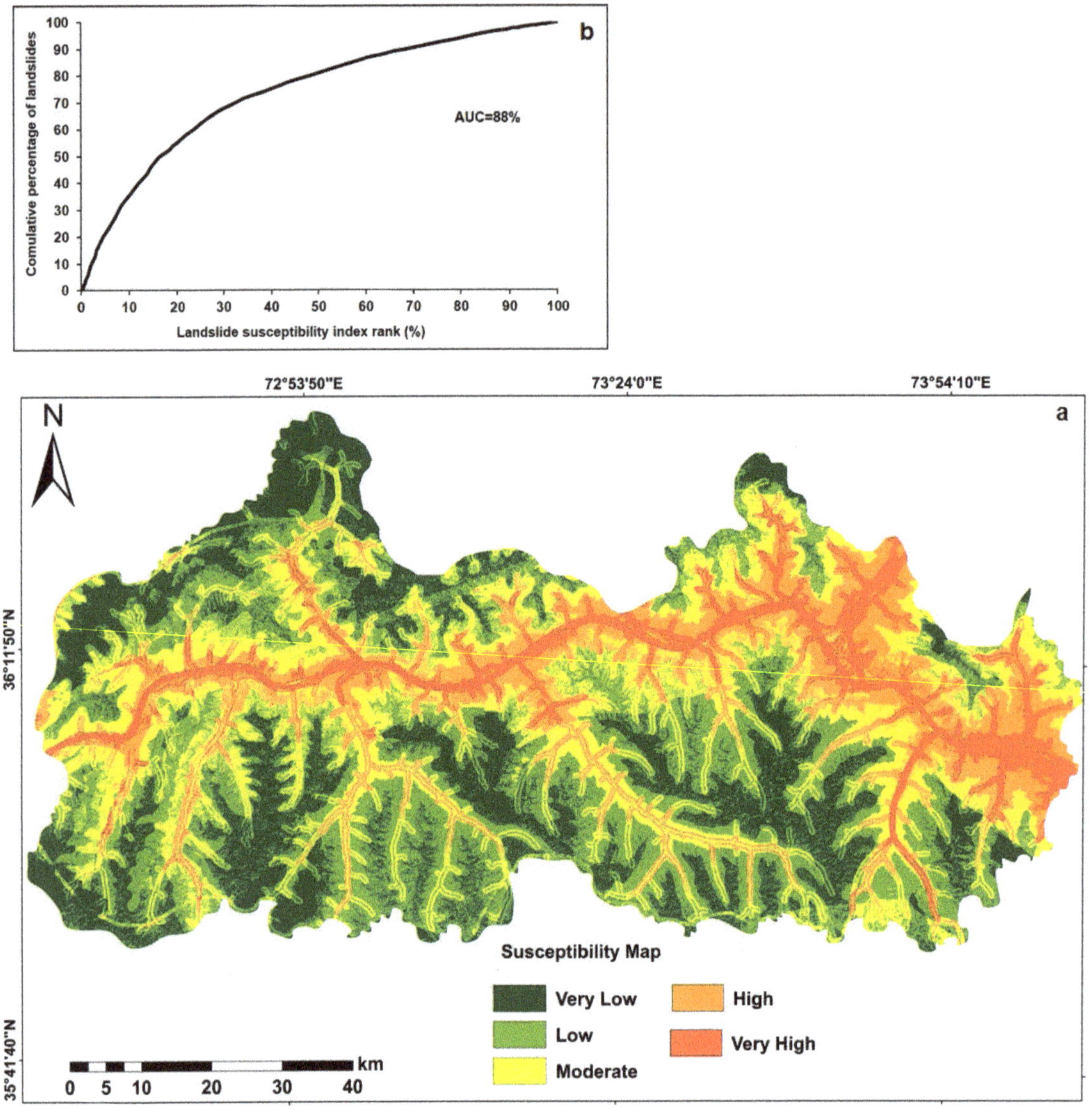

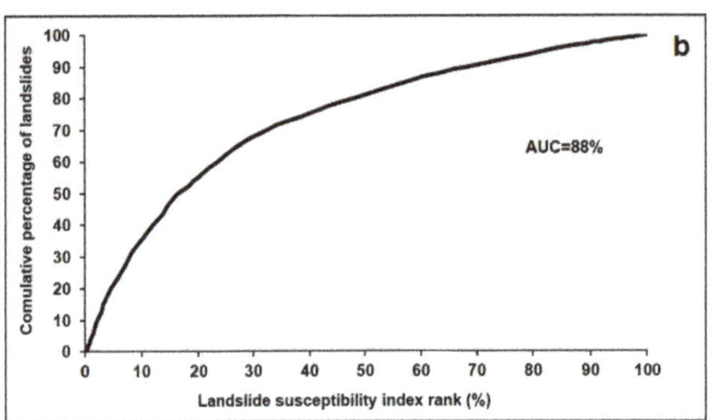

Figure 7. Landslide susceptibility index map and its validation. (**a**) Susceptibility map showing the zones prone to landslides, developed using frequency ratio model and (**b**) accuracy of the model by success rate curve, shown on the x-axis.

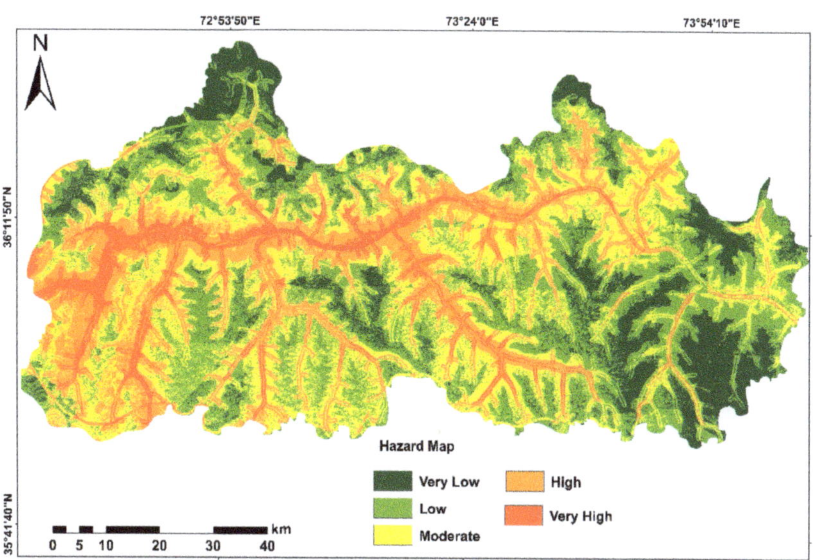

Figure 8. Landslide hazard index map of the study area, showing the landslide hazard zones from very low to very high hazard.

4.4. Landside Vulnerability Assessment

4.4.1. Physical Vulnerability

The typological information of 23456 buildings mapped in the area (Table 3) indicates that the wall materials were dry rubble stone without mortar in 9645 buildings (41%), dressed stone masonry with mortar in 5741 buildings (24%), cement block masonry in 7867 buildings (34%), and reinforced concrete masonry in 203 buildings (1%). The roof materials were horizontal metallic sheets supported from underneath by wood/steel studs in 6310 buildings (27%), metallic sheets with pitched sloping in 9437 buildings (40%), wood planks covered by mud in 5709 buildings (24%) and reinforced concrete cement (RCC) in 2055 buildings (9%). There were 16996 buildings with different physical conditions showing no structural damage (72%), 4383 with minor cracks in the walls/plaster and broken seepage (19%), and 2077 with cracks in the beams, reinforcement exposed, and major cracks in the walls and roofs (9%). The ages of the buildings show that 5537 (24%) buildings are less than 10 years, 3861 (16%) buildings are 11–20 years of age, 5931 (25%) buildings are 21–30 years old, 4929 (21%) buildings are 31–40 years of age, and 3891 (14%) buildings are older than 40 years. Although a large part of the study area lies in the low and intermediate hazard classes, a large portion of the population is situated in high and very high hazard zones. Buildings with weak physical conditions, older age, and walls and roofs with building materials having less resistance occur in the high vulnerability class. All the educational institutes (schools and colleges), hospitals, and dispensaries are located in very highly vulnerable classes. Following the multi-criteria analysis technique and the weights derived from Table 2, a physical vulnerability map was developed and subsequently classified into five classes ranging from very low to very highly vulnerable (Figure 9). The vulnerability of roads indicates that the main road is highly vulnerable to landslides because of the highly hazardous zone along the main road and the high traffic passing through it. The linking roads are less vulnerable due to less traffic and are largely located in the low to medium vulnerability class.

Table 3. Information on the typology data of buildings.

Building Typology Data	Number	Percentage
Educational Institutes	281	1.2
Hospitals/Dispensaries/Clinics	68	0.3
Residential Houses	20,942	89.2
Shops/Hotels/Offices	1956	8.3
Houses of Worships	209	1
Wall Materials		
Dry Stone Masonry	9645	41
Dressed Stone Masonry	5741	24
Cement Block Masonry	7867	34
Reinforced Confined Masonry	203	1
Roof Materials		
Horizontal Metal Sheet	6310	27
Pitched Sloping Sheet	9437	40
Wood Planks Covered by Mud	5709	24
Reinforced Concrete Slab	2055	9
Building Age (Years)		
<10	5537	24
11–20	3861	16
21–30	5931	25
31–40	4929	21
>40	3198	14
Physical Condition		
No Structural Damage	16,996	72
Minor Cracks in the Walls	4383	29
Major Cracks in the Walls/Roofs	2077	9
Total	**23,456**	**100**

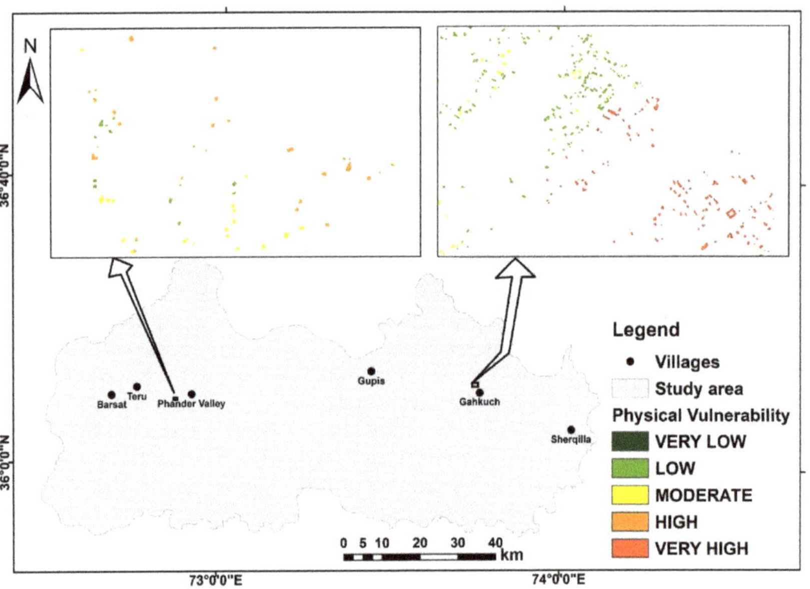

Figure 9. Physical vulnerability map of the study area. The two insets show the zoom view of some buildings.

4.4.2. Social Vulnerability

Population density is directly related to residential buildings. In the study area, locations such as the Gahkuch, Sherkilla, Gupis, and Teru are more populated, leading to high social vulnerability (Figure 10a). The social vulnerability map was classified into five classes (very low, low, medium, high, and very high). Of the total population of 117,280, 14,425 (12.3 %) are in the very highly vulnerable class, 29,785 (25.4 %) people are in the highly vulnerable class, 34,011 (29 %) people are in the medium class, and the remaining 39,288 (33.5 %) are in very low and low vulnerable classes.

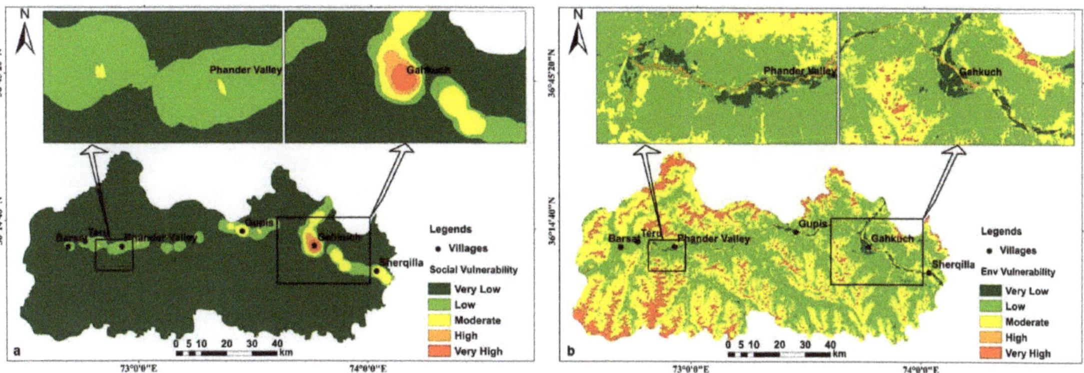

Figure 10. Social and environmental vulnerability maps of the area: (**a**) social vulnerability and (**b**) environmental vulnerability. The black rectangles show a zoomed-in view of some villages.

4.4.3. Environmental Vulnerability

Forest and agriculture have high environmental and economic impacts, and therefore, the highest weights were assigned to these areas. Barren land, snow, and built-up land are in the medium to low vulnerable classes in the environmental vulnerability map (Figure 10b).

The physical, social, and environmental vulnerability maps were combined into an integrated vulnerability map (Figure 11).

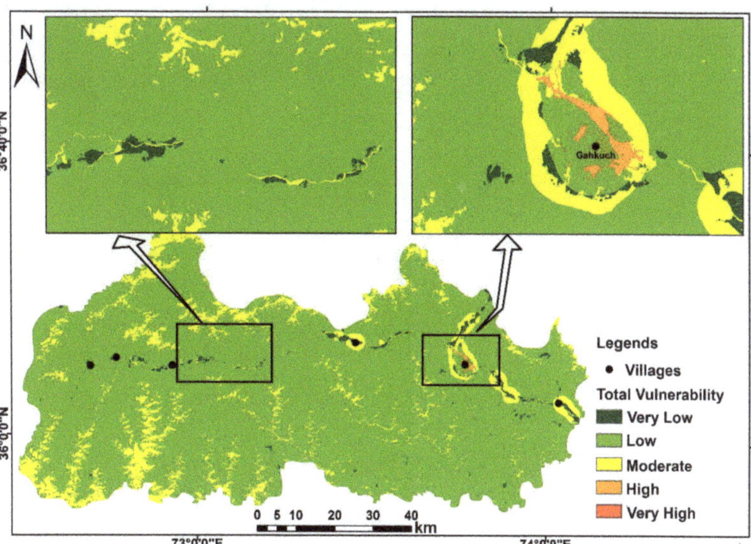

Figure 11. Integrated vulnerability map.

4.4.4. Landslide Risk Map

The landslide vulnerability map and landslide hazard map were integrated by assigning 0.6 weightage to the hazard map and 0.4 weightage to the vulnerability map to compute the landslide risk assessment map (Figure 12). The landslide risk map was reclassified into five classes according to the natural break method, namely, very low risk, low risk, moderate risk, high risk, and very high risk.

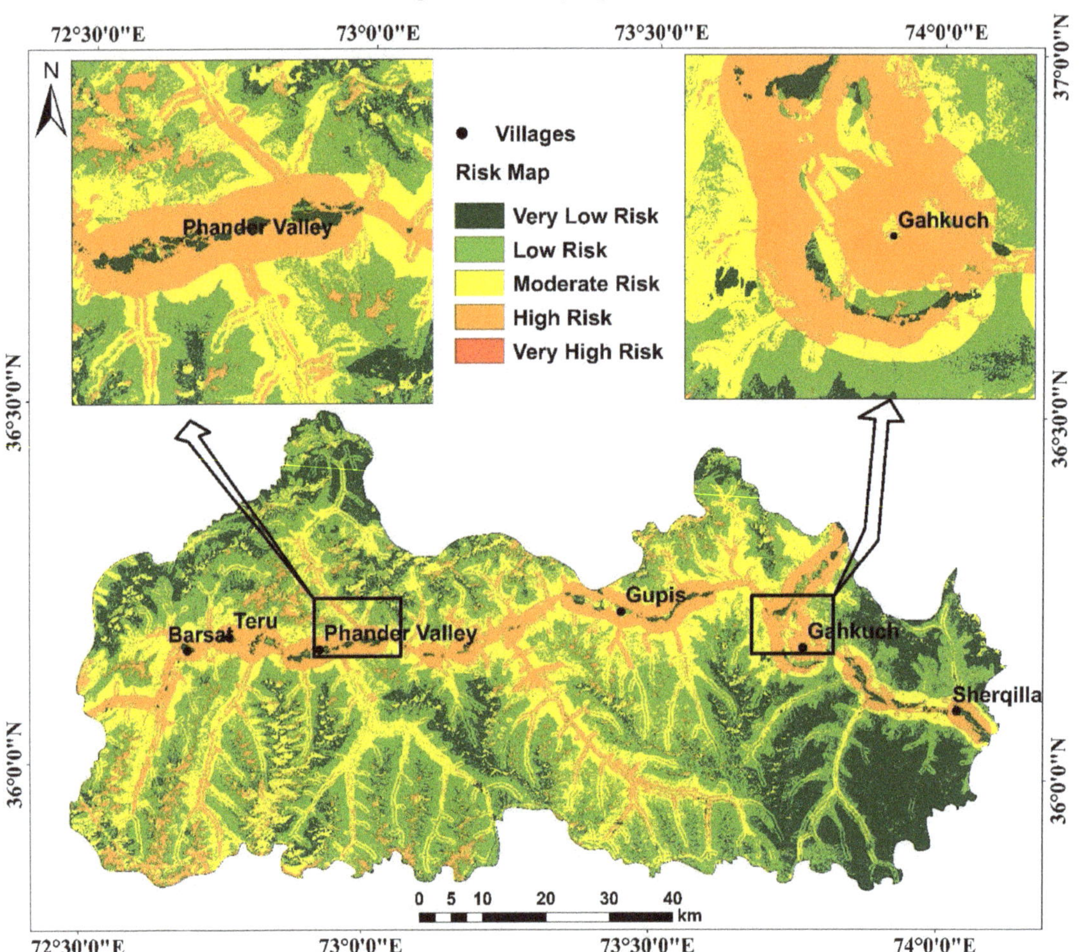

Figure 12. Landslide risk map of the study area. The black rectangles show a zoomed-in view of some villages.

In the study area, 18% of the area is in the very low-risk class, 39.4% is in the low-risk class, 26.3% is in the moderate-risk class, 13.3% is in the high-risk class, and 3% of the area is in the very high-risk class.

5. Discussion

This is one of the first studies on landslide risk assessment in the complex topographic, climatic, and tectonic environment of northern Pakistan. Landslides are one of the most frequent and damaging hazards in the region; however, the available literature on the area is limited to landslide susceptibility assessments [47,92–97]. From the global perspective, most of the landslide risk papers employing semi-quantitative risk assessments utilize the elements of at-risk data from secondary sources and often map the settlements collectively as large polygons without the demarcation of individual houses. In this study, 23456 buildings were mapped separately, and subsequently, their typological information was collected from extensive fieldwork to assess the landslide vulnerability and risk to each of the buildings considering their spatial location in the landslide hazard class and typological information determining its vulnerability to landslides. In this paper, specific buildings are identified that are high risk, as well as the underlying reasons for these classifications.

The number of landslides in this study varies from other studies in the area, including [46,98,99]. A comprehensive landslide inventory was developed from the VHR images, which were subsequently verified and updated in the field. The developed landslide inventory is largely concentrated at the lower elevations. In this area, landslides rarely exist at higher elevations due to the presence of snow cover and glaciers. The majority of the landslides in the study area are scree slopes, rock falls, and debris flows. Debris flows are very fast to extremely fast-flowing slides of saturated debris that occur in steep channels, with a significant impact on the downstream settlements and road networks [57,100,101]. Considering the fine resolution of images from Google Earth Pro, a detailed landslide inventory can record smaller landslides, which are often difficult to map from the relatively coarser resolution satellite images provided by Landsat, ASTER and MODIS (Moderate Imaging Spectroradiometer) [56,99]. Moreover, it is often convenient to visualize landslides in 3D through Google Earth Pro to precisely demarcate their boundaries.

The frequency ratio model used for the susceptibility assessment in this study lead to higher accuracy than the other frequently applied techniques, including the weight of evidence, logistic regression, and artificial neural networks [94]. The critical issue for the landslide hazard assessment is largely the lack of temporal records of landslides in developing countries, and therefore, the standard procedure for the landslide hazard assessment using the spatial, temporal, and magnitude/size probabilities is often difficult to apply in the study area [102]. Eventually, following [90,103], we integrated the landslide triggers, including the rainfall and PGA, to acquire the landslide hazard map. However, the inclusion of the temporal probability of landslides shall improve the reliability of the landslide hazard map.

A vulnerability assessment methodology relies heavily on the typology of exposed elements. The structural system, geometry, material properties, state of maintenance, levels of design codes, foundation and superstructure details, number of floors, and other factors are among the typological parameters that determine a building's ability to withstand landslide actions [26]. Researchers have used different approaches for vulnerability assessment. Ref. [104] proposed the back analysis of real event damage data to acquire the correlations between the intensity of landslides and building vulnerability through regression analysis. Ref. [105] calculated vulnerability curves as a function of the differential settlements of a reinforced concrete frame building. Ref. [106] developed a debris flow intensity index that considers the flow height and velocity to calculate the probability of damage. Ref. [88] generated the vulnerability of buildings according to the type of building structure by the visual interpretation of high-resolution imagery, while current research work interprets individual buildings by performing building typology.

For landslide risk assessments, many researchers have focused on the quantitative assessment of vulnerability [107–109]. Refs. [32,34,110] performed qualitative landslide risk assessments at a regional scale. In this study, a semi-quantitative approach was adopted for risk assessment through the analysis of the collected extensive field-based data of the element-at-risk and applied methods. The landslide hazard and vulnerability maps were derived using quantitative approaches. However, to integrate the hazard and vulnerability maps into a risk map, both maps were normalized and then integrated into the risk map. Field visits verified that Gahkuch, the most populated zone in the study area with a high hazard probability of landslide, mostly lies in the moderate- to high-risk classes. Similarly, Phander and Teru (Figure 12) are the other populated areas that lie in the high-risk class. The quantitative validation of a risk index is often ignored due to the lack of high-quality data [111] and, therefore, could not be included in this study; however, it shall be accomplished in future quantitative landslide risk assessment studies in the region. This study is an intermediate methodology between purely qualitative and quantitative approaches, including quanitative data, and providing relative risk levels. As the performance of risk mapping is largely dependent on data availability, the lack of temporal probability and more detailed data for elements at risk lead to some limitations in the research work.

6. Conclusions

In this research work, we conducted a semi-quantitative risk assessment for landslides at a regional scale in the Ghizer district in northern Pakistan based on the integration of statistical analysis, field surveys, and spatial (GIS) analysis. The study concentrated on the potential damage to buildings, loss of life for the populations in the buildings, roads, and environmental elements at risk. Due to the lack of a temporal record of landslides, the landslide susceptibility map was integrated with landslide triggers (rainfall and PGA) to compute the landslide hazard map. The results show that the major causative factors for landslides in the study area are roads and stream networks; other parameters, such as the slope, elevation, and faults, also dominantly influence the occurrence of landslides, and the main triggering indicator is rainfall. The extensive database of the mapped building footprints associated with the field-based typology data assists in the realistic vulnerability and risk assessment. The derived vulnerability and risk maps identify and quantify the number of buildings, roads, schools, and hospitals prone to different levels of landslide hazards. The derived landslide risk map shall assist the concerned organizations in developing and implementing landslide mitigation strategies and prioritizing resource allocations and land use planning.

Author Contributions: Conceptualization, N.A.S.; Methodology, M.S. and M.I.; Formal analysis, K.F.; Resources, M.V.d.M. All authors have read and agreed to the published version of the manuscript.

Funding: This research received no external funding.

Institutional Review Board Statement: Not relevant in this study.

Informed Consent Statement: Not relevant in this study.

Data Availability Statement: The links to the data sources used in the study are given in the text and references.

Conflicts of Interest: The authors declare no conflict of interest.

References

1. Carrara, A.; Crosta, G.; Frattini, P. Geomorphological and historical data in assessing landslide hazard. *Earth Surf. Process. Landf. J. Br. Geomorphol. Res. Group* **2003**, *28*, 1125–1142. [CrossRef]
2. Lee, S.-Y. *Structural Equation Modeling: A Bayesian Approach*; John Wiley & Sons: Hoboken, NJ, USA, 2007.
3. Schlögel, R.; Torgoev, I.; de Marneffe, C.; Havenith, H.B. Evidence of a changing size–frequency distribution of landslides in the Kyrgyz Tien Shan, Central Asia. *Earth Surf. Process. Landf.* **2011**, *36*, 1658–1669. [CrossRef]
4. Dilley, M. *Natural Disaster Hotspots: A Global Risk Analysis*; World Bank Publications: Washington, DC, USA, 2005; Volume 5.
5. Igwe, O.; Una, C.O. Landslide impacts and management in Nanka area, Southeast Nigeria. *Geoenviron. Disasters* **2019**, *6*, 5. [CrossRef]
6. Zhong, C.; Liu, Y.; Gao, P.; Chen, W.; Li, H.; Hou, Y.; Nuremanguli, T.; Ma, H. Landslide mapping with remote sensing: Challenges and opportunities. *Int. J. Remote Sens.* **2020**, *41*, 1555–1581. [CrossRef]
7. Liu, C. Genetic mechanism of landslide tragedy happened in Hong'ao dumping place in Shenzhen. *China* **2016**, *27*, 1–5.
8. Psomiadis, E.; Charizopoulos, N.; Efthimiou, N.; Soulis, K.X.; Charalampopoulos, I. Earth observation and GIS-based analysis for landslide susceptibility and risk assessment. *ISPRS Int. J. Geo-Inf.* **2020**, *9*, 552. [CrossRef]
9. Yalcin, A.; Reis, S.; Aydinoglu, A.; Yomralioglu, T. A GIS-based comparative study of frequency ratio, analytical hierarchy process, bivariate statistics and logistics regression methods for landslide susceptibility mapping in Trabzon, NE Turkey. *Catena* **2011**, *85*, 274–287. [CrossRef]

10. Plank, S.; Twele, A.; Martinis, S. Landslide mapping in vegetated areas using change detection based on optical and polarimetric SAR data. *Remote Sens.* **2016**, *8*, 307. [CrossRef]

11. Kayastha, P.; Dhital, M.R.; de Smedt, F. Application of the analytical hierarchy process (AHP) for landslide susceptibility mapping: A case study from the Tinau watershed, west Nepal. *Comput. Geosci.* **2013**, *52*, 398–408. [CrossRef]

12. Brabb, E.E. The world landslide problem. *Epis. J. Int. Geosci.* **1991**, *14*, 52–61. [CrossRef]

13. VanDine, D.; Moore, G.; Wise, M. A comparison of landslide risk terminology. In *Landslide Risk Management*; CRC Press: Boca Raton, FL, USA, 2005; pp. 567–572.

14. Shafique, M.; van der Meijde, M.; Khan, M.A. A review of the 2005 Kashmir earthquake-induced landslides; from a remote sensing prospective. *J. Asian Earth Sci.* **2016**, *118*, 68–80. [CrossRef]

15. Guzzetti, F.; Mondini, A.C.; Cardinali, M.; Fiorucci, F.; Santangelo, M.; Chang, K.-T. Landslide inventory maps: New tools for an old problem. *Earth-Sci. Rev.* **2012**, *112*, 42–66. [CrossRef]

16. Reichenbach, P.; Rossi, M.; Malamud, B.D.; Mihir, M.; Guzzetti, F. A review of statistically-based landslide susceptibility models. *Earth-Sci. Rev.* **2018**, *180*, 60–91. [CrossRef]

17. Guzzetti, F.; Carrara, A.; Cardinali, M.; Reichenbach, P. Landslide hazard evaluation: A review of current techniques and their application in a multi-scale study, Central Italy. *Geomorphology* **1999**, *31*, 181–216. [CrossRef]

18. Aleotti, P.; Chowdhury, R. Landslide hazard assessment: Summary review and new perspectives. *Bull. Eng. Geol. Environ.* **1999**, *58*, 21–44. [CrossRef]

19. Arnone, M.P.; Small, R.V.; Chauncey, S.A.; McKenna, H.P. Curiosity, interest and engagement in technology-pervasive learning environments: A new research agenda. *Educ. Technol. Res. Dev.* **2011**, *59*, 181–198. [CrossRef]

20. Longo, M.R.; Azañón, E.; Haggard, P. More than skin deep: Body representation beyond primary somatosensory cortex. *Neuropsychologia* **2010**, *48*, 655–668. [CrossRef]

21. Glade, T.; Crozier, M.J. Hazard Impact. In *Landslide Hazard and Risk*; John Wiley & Sons: Hoboken, NJ, USA, 2005; p. 43.

22. Glade, T. Vulnerability assessment in landslide risk analysis. *Erde* **2003**, *134*, 123–146.

23. Cassidy, M.J.; Uzielli, M.; Lacasse, S. Probability risk assessment of landslides: A case study at Finneidfjord. *Can. Geotech. J.* **2008**, *45*, 1250–1267. [CrossRef]

24. Bründl, M.; Romang, H.E.; Bischof, N.; Rheinberger, C.M. The risk concept and its application in natural hazard risk management in Switzerland. *Nat. Hazards Earth Syst. Sci.* **2009**, *9*, 801–813. [CrossRef]

25. Fell, R.; Corominas, J.; Bonnard, C.; Cascini, L.; Leroi, E.; Savage, W.Z. Guidelines for landslide susceptibility, hazard and risk zoning for land-use planning. *Eng. Geol.* **2008**, *102*, 99–111. [CrossRef]

26. Corominas, J.; van Westen, C.; Frattini, P.; Cascini, L.; Malet, J.-P.; Fotopoulou, S.; Catani, F.; van den Eeckhaut, M.; Mavrouli, O.; Agliardi, F. Recommendations for the quantitative analysis of landslide risk. *Bull. Eng. Geol. Environ.* **2014**, *73*, 209–263. [CrossRef]

27. Papathoma-Köhle, M.; Neuhäuser, B.; Ratzinger, K.; Wenzel, H.; Dominey-Howes, D. Elements at risk as a framework for assessing the vulnerability of communities to landslides. *Nat. Hazards Earth Syst. Sci.* **2007**, *7*, 765–779. [CrossRef]

28. Fell, R. Landslide risk assessment and acceptable risk. *Can. Geotech. J.* **1994**, *31*, 261–272. [CrossRef]

29. Pazzi, V.; Morelli, S.; Fidolini, F.; Krymi, E.; Casagli, N.; Fanti, R. Testing cost-effective methodologies for flood and seismic vulnerability assessment in communities of developing countries (Dajç, northern Albania). *Geomat. Nat. Hazards Risk* **2016**, *7*, 971–999. [CrossRef]

30. Opolot, E. Application of remote sensing and geographical information systems in flood management: A review. *Res. J. Appl. Sci. Eng. Technol.* **2013**, *6*, 1884–1894. [CrossRef]

31. Erener, A.; Düzgün, H.B.S. A regional scale quantitative risk assessment for landslides: Case of Kumluca watershed in Bartin, Turkey. *Landslides* **2013**, *10*, 55–73. [CrossRef]

32. Caleca, F.; Tofani, V.; Segoni, S.; Raspini, F.; Rosi, A.; Natali, M.; Catani, F.; Casagli, N. A methodological approach of QRA for slow-moving landslides at a regional scale. *Landslides* **2022**, *19*, 1539–1561. [CrossRef]

33. Quesada-Román, A. Landslide risk index map at the municipal scale for Costa Rica. *Int. J. Disaster Risk Reduct.* **2021**, *56*, 102144. [CrossRef]

34. Chang, M.; Cui, P.; Dou, X.; Su, F. Quantitative risk assessment of landslides over the China-Pakistan economic corridor. *Int. J. Disaster Risk Reduct.* **2021**, *63*, 102441. [CrossRef]

35. Biçer, Ç.T.; Ercanoglu, M. A semi-quantitative landslide risk assessment of central Kahramanmaraş City in the Eastern Mediterranean region of Turkey. *Arab. J. Geosci.* **2020**, *13*, 732. [CrossRef]

36. Wang, H.B.; Wu, S.R.; Shi, J.S.; Li, B. Qualitative hazard and risk assessment of landslides: A practical framework for a case study in China. *Nat. Hazards* **2011**, *69*, 1281–1294. [CrossRef]

37. Achour, Y.; Boumezbeur, A.; Hadji, R.; Chouabbi, A.; Cavaleiro, V.; Bendaoud, E.A. Landslide susceptibility mapping using analytic hierarchy process and information value methods along a highway road section in Constantine, Algeria. *Arab. J. Geosci.* **2017**, *10*, 194. [CrossRef]

38. Pereira, S.; Santos, P.; Zêzere, J.; Tavares, A.; Garcia, R.; Oliveira, S. A landslide risk index for municipal land use planning in Portugal. *Sci. Total Environ.* **2020**, *735*, 139463. [CrossRef]

39. Segoni, S.; Caleca, F. Definition of environmental indicators for a fast estimation of landslide risk at national scale. *Land* **2021**, *10*, 621. [CrossRef]

40. Crawford, M.M.; Dortch, J.M.; Koch, H.J.; Zhu, Y.; Haneberg, W.C.; Wang, Z.; Bryson, L.S. Landslide Risk Assessment in Eastern Kentucky, USA: Developing a Regional Scale, Limited Resource Approach. *Remote Sens.* **2022**, *14*, 6246. [CrossRef]

41. Ali, S.; Biermanns, P.; Haider, R.; Reicherter, K. Landslide susceptibility mapping by using a geographic information system (GIS) along the China–Pakistan Economic Corridor (Karakoram Highway), Pakistan. *Nat. Hazards Earth Syst. Sci.* **2019**, *19*, 999–1022. [CrossRef]

42. Rahman, G.; Rahman, A.U.; Bacha, A.S.; Mahmood, S.; Moazzam, M.F.U.; Lee, B.G. Assessment of landslide susceptibility using weight of evidence and frequency ratio model in Shahpur Valley, Eastern Hindu Kush. *Nat. Hazards Earth Syst. Sci. Discuss.* **2020**, *2020*, 1–19.

43. Rahman, G.; Rahman, A.-U.; Ullah, S.; Miandad, M.; Collins, A.E. Spatial analysis of landslide susceptibility using failure rate approach in the Hindu Kush region, Pakistan. *J. Earth Syst. Sci.* **2019**, *128*, 59. [CrossRef]

44. Ahmed, M.; Rogers, J.; Abu Bakar, M.Z. *Hunza River Watershed Landslide and Related Features Inventory Mapping*; Springer: Berlin/Heidelberg, Germany, 2016; Volume 75.

45. Ahmed, M.F.; Rogers, J.D.; Ismail, E.H. A regional level preliminary landslide susceptibility study of the upper Indus river basin. *Eur. J. Remote Sens.* **2014**, *47*, 343–373. [CrossRef]

46. Hussain, S.; Hongxing, S.; Ali, M.; Ali, M. PS-InSAR based validated landslide susceptibility modelling: A case study of Ghizer valley, Northern Pakistan. *Geocarto Int.* **2021**, *37*, 3941–3962. [CrossRef]

47. Hussain, S.; Hongxing, S.; Ali, M.; Sajjad, M.M.; Ali, M.; Afzal, Z.; Ali, S. Optimized landslide susceptibility mapping and modelling using PS-InSAR technique: A case study of Chitral valley, Northern Pakistan. *Geocarto Int.* **2022**, *37*, 5227–5248. [CrossRef]

48. Maqsoom, A.; Aslam, B.; Khalil, U.; Kazmi, Z.A.; Azam, S.; Mehmood, T.; Nawaz, A. Landslide susceptibility mapping along the China Pakistan Economic Corridor (CPEC) route using multi-criteria decision-making method. *Model. Earth Syst. Environ.* **2022**, *8*, 1519–1533. [CrossRef]

49. Searle, M.; Khan, M. *Geological Map of North Pakistan and Adjacent Areas of Northern Landmark and Western Tibet (Western Himalaya, Salt Ranges, Kohistan, Karakoram, Hindu Kush), 1: 650 000*; Shell International Exploration and Production: The Hague, The Netherlands, 1996.

50. Hewitt, K.; Gosse, J.; Clague, J.J. Rock avalanches and the pace of late Quaternary development of river valleys in the Karakoram Himalaya. *Bulletin* **2011**, *123*, 1836–1850. [CrossRef]

51. Korup, O.; Densmore, A.L.; Schlunegger, F. The role of landslides in mountain range evolution. *Geomorphology* **2010**, *120*, 77–90. [CrossRef]

52. Hewitt, K. Styles of rock-avalanche depositional complexes conditioned by very rugged terrain, Karakoram Himalaya, Pakistan. *Rev. Eng. Geol.* **2002**, *15*, 345–377.

53. Shafique, M.; van der Meijde, M.; Kerle, N.; van der Meer, F. Impact of DEM source and resolution on topographic seismic amplification. *Int. J. Appl. Earth Obs. Geoinf.* **2011**, *13*, 420–427. [CrossRef]

54. Tian, Y.; Xiao, C.; Liu, Y.; Wu, L. Effects of raster resolution on landslide susceptibility mapping: A case study of Shenzhen. *Sci. China Ser. E Technol. Sci.* **2008**, *51*, 188–198. [CrossRef]

55. Rehman, K.; Burton, P.W.; Weatherill, G.A. Application of Gumbel I and Monte Carlo methods to assess seismic hazard in and around Pakistan. *J. Seismol.* **2018**, *22*, 575–588. [CrossRef]

56. Pham, B.T.; Pradhan, B.; Bui, D.T.; Prakash, I.; Dholakia, M. A comparative study of different machine learning methods for landslide susceptibility assessment: A case study of Uttarakhand area (India). *Environ. Model. Softw.* **2016**, *84*, 240–250. [CrossRef]

57. Hungr, O.; Leroueil, S.; Picarelli, L. The Varnes classification of landslide types, an update. *Landslides* **2014**, *11*, 167–194. [CrossRef]

58. Shirzadi, A.; Soliamani, K.; Habibnejhad, M.; Kavian, A.; Chapi, K.; Shahabi, H.; Chen, W.; Khosravi, K.; Thai Pham, B.; Pradhan, B. Novel GIS based machine learning algorithms for shallow landslide susceptibility mapping. *Sensors* **2018**, *18*, 3777. [CrossRef]

59. Pham, B.T.; Prakash, I.; Dou, J.; Singh, S.K.; Trinh, P.T.; Tran, H.T.; Le, T.M.; van Phong, T.; Khoi, D.K.; Shirzadi, A. A novel hybrid approach of landslide susceptibility modelling using rotation forest ensemble and different base classifiers. *Geocarto Int.* **2020**, *35*, 1267–1292. [CrossRef]

60. Bui, D.T.; Lofman, O.; Revhaug, I.; Dick, O. Landslide susceptibility analysis in the Hoa Binh province of Vietnam using statistical index and logistic regression. *Nat. Hazards* **2011**, *59*, 1413–1444. [CrossRef]

61. Dou, J.; Paudel, U.; Oguchi, T.; Uchiyama, S.; Hayakavva, Y.S. Shallow and Deep-Seated Landslide Differentiation Using Support Vector Machines: A Case Study of the Chuetsu Area, Japan. *Terr. Atmos. Ocean. Sci.* **2015**, *26*, 227. [CrossRef]

62. Javier, D.; Kumar, L. Frequency Ratio Landslide Susceptibility Estimation in a Tropical Mountain Region. *Int. Arch. Photogramm. Remote Sens. Spat. Inf. Sci.* **2019**, *XLII-3/W8*, 173–179. [CrossRef]

63. Magliulo, P.; di Lisio, A.; Russo, F.; Zelano, A. Geomorphology and landslide susceptibility assessment using GIS and bivariate statistics: A case study in southern Italy. *Nat. Hazards* **2008**, *47*, 411–435. [CrossRef]

64. García-Rodríguez, M.J.; Malpica, J.; Benito, B.; Díaz, M. Susceptibility assessment of earthquake-triggered landslides in El Salvador using logistic regression. *Geomorphology* **2008**, *95*, 172–191. [CrossRef]

65. Dou, J.; Oguchi, T.; Hayakawa, Y.S.; Uchiyama, S.; Saito, H.; Paudel, U. GIS-based landslide susceptibility mapping using a certainty factor model and its validation in the Chuetsu Area, Central Japan. In *Landslide Science for a Safer Geoenvironment*; Springer: Berlin/Heidelberg, Germany, 2014; pp. 419–424.

66. Tiwari, B.; Pradel, D.; Ajmera, B.; Yamashiro, B.; Khadka, D. Landslide movement at Lokanthali during the 2015 earthquake in Gorkha, Nepal. *J. Geotech. Geoenvironmental Eng.* **2018**, *144*, 05018001. [CrossRef]

67. Gorum, T.; Fan, X.; van Westen, C.J.; Huang, R.Q.; Xu, Q.; Tang, C.; Wang, G. Distribution pattern of earthquake-induced landslides triggered by the 12 May 2008 Wenchuan earthquake. *Geomorphology* **2011**, *133*, 152–167. [CrossRef]

68. Pradhan, S.; Toll, D.G.; Rosser, N.J.; Brain, M.J. An investigation of the combined effect of rainfall and road cut on landsliding. *Eng. Geol.* **2022**, *307*, 106787. [CrossRef]

69. Kanwal, S.; Atif, S.; Shafiq, M. GIS based landslide susceptibility mapping of northern areas of Pakistan, a case study of Shigar and Shyok Basins. *Geomat. Nat. Hazards Risk* **2017**, *8*, 348–366. [CrossRef]

70. Chen, W.; Chai, H.; Sun, X.; Wang, Q.; Ding, X.; Hong, H. A GIS-based comparative study of frequency ratio, statistical index and weights-of-evidence models in landslide susceptibility mapping. *Arab. J. Geosci.* **2016**, *9*, 204. [CrossRef]

71. Yan, F.; Zhang, Q.; Ye, S.; Ren, B. A novel hybrid approach for landslide susceptibility mapping integrating analytical hierarchy process and normalized frequency ratio methods with the cloud model. *Geomorphology* **2019**, *327*, 170–187. [CrossRef]

72. Solaimani, K.; Mousavi, S.Z.; Kavian, A. Landslide susceptibility mapping based on frequency ratio and logistic regression models. *Arab. J. Geosci.* **2013**, *6*, 2557–2569. [CrossRef]

73. Wu, Y.; Li, W.; Wang, Q.; Liu, Q.; Yang, D.; Xing, M.; Pei, Y.; Yan, S. Landslide susceptibility assessment using frequency ratio, statistical index and certainty factor models for the Gangu County, China. *Arab. J. Geosci.* **2016**, *9*, 84. [CrossRef]

74. Wang, Q.; Li, W. A GIS-based comparative evaluation of analytical hierarchy process and frequency ratio models for landslide susceptibility mapping. *Phys. Geogr.* **2017**, *38*, 318–337. [CrossRef]

75. Akgun, A.; Dag, S.; Bulut, F. Landslide susceptibility mapping for a landslide-prone area (Findikli, NE of Turkey) by likelihood-frequency ratio and weighted linear combination models. *Environ. Geol.* **2007**, *54*, 1127–1143. [CrossRef]

76. Lee, S.; Sambath, T. Landslide susceptibility mapping in the Damrei Romel area, Cambodia using frequency ratio and logistic regression models. *Environ. Geol.* **2006**, *50*, 847–855. [CrossRef]

77. Panchal, S.; Shrivastava, A.K. A comparative study of frequency ratio, Shannon's entropy and analytic hierarchy process (AHP) models for landslide susceptibility assessment. *ISPRS Int. J. Geo-Inf.* **2021**, *10*, 603. [CrossRef]

78. Lissak, C.; Maquaire, O.; Davidson, R.; Malet, J.-P. Piezometric thresholds for triggering landslides along the Normandy coast, France. *Géomorphologie Relief Process. Environ.* **2014**, *20*, 145–158. [CrossRef]

79. Kempler, S. Goddard Earth Sciences Data and Information Services Center. Disc. Sci. Gsfc. NASA Gov. 2013. Available online: https://disc.gsfc.nasa.gov/ (accessed on 27 August 2022).

80. Akgun, A.; Kıncal, C.; Pradhan, B. Application of remote sensing data and GIS for landslide risk assessment as an environmental threat to Izmir city (west Turkey). *Environ. Monit. Assess.* **2012**, *184*, 5453–5470. [CrossRef]

81. Nefeslioglu, H.A.; Gokceoglu, C.; Sonmez, H.; Gorum, T. Medium-scale hazard mapping for shallow landslide initiation: The Buyukkoy catchment area (Cayeli, Rize, Turkey). *Landslides* **2011**, *8*, 459–483. [CrossRef]

82. Dai, F.; Lee, C.F.; Ngai, Y.Y. Landslide risk assessment and management: An overview. *Eng. Geol.* **2002**, *64*, 65–87. [CrossRef]

83. Abella, E.; van Westen, C. Generation of a landslide risk index map for Cuba using spatial multi-criteria evaluation. *Landslides* **2007**, *4*, 311–325. [CrossRef]

84. Leone, F.; Asté, J.; Leroi, E. Vulnerability assessment of elements exposed to mass-movement: Working toward a better risk perception. In *Landslides-Glissements de Terrain*; Balkema: Rotterdam, The Netherlands, 1996; pp. 263–270.

85. Bednarik, M.; Yilmaz, I.; Marschalko, M. Landslide hazard and risk assessment: A case study from the Hlohovec–Sered'landslide area in south-west Slovakia. *Nat. Hazards* **2012**, *64*, 547–575. [CrossRef]

86. Zêzere, J.; Garcia, R.; Oliveira, S.; Reis, E. Probabilistic landslide risk analysis considering direct costs in the area north of Lisbon (Portugal). *Geomorphology* **2008**, *94*, 467–495. [CrossRef]

87. Saaty, T. *The Analytic Hierarchy Process: Planning, Priority Setting, Resources Allocation*; McGraw-Hill: New York, NY, USA, 1980.

88. Priyono, K.D.; Saputra, A.; Fikriyah, V.N. Risk analysis of landslide impacts on settlements in Karanganyar, Central Java, Indonesia. *Geomate J.* **2020**, *19*, 100–107. [CrossRef]

89. Uzielli, M.; Nadim, F.; Lacasse, S.; Kaynia, A.M. A conceptual framework for quantitative estimation of physical vulnerability to landslides. *Eng. Geol.* **2008**, *102*, 251–256. [CrossRef]

90. Abdulwahid, W.M.; Pradhan, B. Landslide vulnerability and risk assessment for multi-hazard scenarios using airborne laser scanning data (LiDAR). *Landslides* **2017**, *14*, 1057–1076. [CrossRef]

91. Dou, J.; Tien Bui, D.P.; Yunus, A.; Jia, K.; Song, X.; Revhaug, I.; Xia, H.; Zhu, Z. Optimization of causative factors for landslide susceptibility evaluation using remote sensing and GIS data in parts of Niigata, Japan. *PLoS ONE* **2015**, *10*, e0133262. [CrossRef] [PubMed]

92. Bacha, A.S.; Shafique, M.; van der Werff, H.; van der Meijde, M.; Hussain, M.L.; Wahid, S. Spatio-temporal landslide inventory and susceptibility assessment using Sentinel-2 in the Himalayan mountainous region of Pakistan. *Environ. Monit. Assess.* **2022**, *194*, 845. [CrossRef] [PubMed]

93. Khan, H.; Shafique, M.; Khan, M.A.; Bacha, M.A.; Shah, S.U.; Calligaris, C. Landslide susceptibility assessment using Frequency Ratio, a case study of northern Pakistan. *Egypt. J. Remote Sens. Space Sci.* **2018**, *22*, 11–24. [CrossRef]

94. Hussain, M.L.; Shafique, M.; Bacha, A.S.; Chen, X.-Q.; Chen, H.-Y. Landslide inventory and susceptibility assessment using multiple statistical approaches along the Karakoram highway, northern Pakistan. *J. Mt. Sci.* **2021**, *18*, 583–598. [CrossRef]

95. Ahmed, M.F.; Rogers, J. First-Approximation Landslide Inventory Maps for Northern Pakistan, Using ASTER DEM Data and Geomorphic Indicators. *Environ. Eng. Geosci.* **2014**, *20*, 67–83. [CrossRef]
96. Bacha, A.S.; Shafique, M.; van der Werff, H. Landslide inventory and susceptibility modelling using geospatial tools, in Hunza-Nagar valley, northern Pakistan. *J. Mt. Sci.* **2018**, *15*, 1354–1370. [CrossRef]
97. Farooq Ahmed, M.; David Rogers, J. Regional level landslide inventory maps of the Shyok River watershed, Northern Pakistan. *Bull. Eng. Geol. Environ.* **2016**, *75*, 563–574. [CrossRef]
98. Rahim, I.; Ali, S.M.; Aslam, M. GIS Based landslide susceptibility mapping with application of analytical hierarchy process in District Ghizer, Gilgit Baltistan Pakistan. *J. Geosci. Environ. Prot.* **2018**, *6*, 34–49. [CrossRef]
99. Ahmed, M.F.; Awan, U.; Rogers, J.D. Use of anomalous topographic features for landslide inventory mapping of Gilgit area, Gilgit-Baltistan, Pakistan. *Arab. J. Geosci.* **2021**, *14*, 2416. [CrossRef]
100. Mikoš, M.; Četina, M.; Brilly, M. Hydrologic conditions responsible for triggering the Stože landslide, Slovenia. *Eng. Geol.* **2004**, *73*, 193–213. [CrossRef]
101. Mikoš, M.; Majes, B. Delineation of risk area in Log pod Mangartom due to debris flows from the Stože landslide. *Acta Geogr. Slov.* **2007**, *47*, 171–198. [CrossRef]
102. Van Westen, C.; van Asch, T.W.; Soeters, R. Landslide hazard and risk zonation—Why is it still so difficult? *Bull. Eng. Geol. Environ.* **2006**, *65*, 167–184. [CrossRef]
103. Althuwaynee, O.F.; Pradhan, B.; Ahmad, N. Estimation of rainfall threshold and its use in landslide hazard mapping of Kuala Lumpur metropolitan and surrounding areas. *Landslides* **2015**, *12*, 861–875. [CrossRef]
104. Agliardi, F.; Crosta, G.; Frattini, P. Integrating rockfall risk assessment and countermeasure design by 3D modelling techniques. *Nat. Hazards Earth Syst. Sci.* **2009**, *9*, 1059–1073. [CrossRef]
105. Negulescu, C.; Foerster, E. Parametric studies and quantitative assessment of the vulnerability of a RC frame building exposed to differential settlements. *Nat. Hazards Earth Syst. Sci.* **2010**, *10*, 1781–1792. [CrossRef]
106. Jakob, M.; Stein, D.; Ulmi, M. Vulnerability of buildings to debris flow impact. *Nat. Hazards* **2012**, *60*, 241–261. [CrossRef]
107. Ko Ko, C.; Flentje, P.; Chowdhury, R. Landslides qualitative hazard and risk assessment method and its reliability. *Bull. Eng. Geol. Environ.* **2004**, *63*, 149–165. [CrossRef]
108. Catani, F.; Casagli, N.; Ermini, L.; Righini, G.; Menduni, G. Landslide hazard and risk mapping at catchment scale in the Arno River basin. *Landslides* **2005**, *2*, 329–342. [CrossRef]
109. Galli, M.; Guzzetti, F. Landslide vulnerability criteria: A case study from Umbria, Central Italy. *Environ. Manag.* **2007**, *40*, 649–665. [CrossRef]
110. Wong, C.; Lee, C. Applications of quantitative landslide risk assessment in Hong Kong. In *Slope Stability Engineering*; Routledge: London, UK, 2021; pp. 1303–1307.
111. Caleca, F.; Tofani, V.; Segoni, S.; Raspini, F.; Franceschini, R.; Rosi, A. How can landslide risk maps be validated? Potential solutions with open-source databases. *Front. Earth Sci.* **2022**, *10*, 998885. [CrossRef]

 *sustainability*

Article

Study on Disaster Mechanism of Oil and Gas Pipeline Oblique Crossing Landslide

Fa-You A [1,2,*], Teng-Hui Chen [3], Cheng Yang [1], Yu-Feng Wu [1] and Shi-Qun Yan [1]

[1] Faculty of Land Resource Engineering, Kunming University of Science and Technology, Kunming 650093, China
[2] Key Laboratory of Geohazard Forecast and Geoecological Restoration in Plateau Mountainous Area, Ministry of Natural Resources of the People's Republic of China, Kunming 650000, China
[3] Shenzhen Water Planning & Design Institute Co., Ltd., Shenzhen 518000, China
*   Correspondence: afayou@163.com

Abstract: Landslides are one of the most serious geological disasters in oil and gas pipelines. According to investigations, the cross-cutting relationship between landslides and pipelines can be divided into three types: pipeline longitudinal crossing landslide, pipeline transversely crossing landslide, and pipeline oblique crossing landslide. This different cross-cutting relationship is one of the important factors affecting pipeline landslide disasters. As a result, it is necessary to study the stress and deformation characteristics of oil and gas pipelines under different cross-cutting relationships, which is of great significance for the prevention and control of oil and gas pipeline landslides. In this paper, an ideal pipe-soil coupling interaction model of oil and gas pipeline oblique crossing landslide was established using FLAC3D. The influence of the buried depth of the pipeline, the displacement of the sliding body, and the different intersection angles of landslide and pipeline on the deformation and stress of the pipeline under the action of a landslide is analyzed, and a typical case of pipeline oblique crossing landslide is used for analysis. The results demonstrated that the stress of pipeline oblique crossing landslide is complex, and the stress concentration is obvious at the shear outlet and the trailing edge of the landslide. The stress at the shear outlet is the largest, which should be regarded as the key location. The displacement and stress of pipeline oblique crossing landslide are obviously affected by the displacement of the sliding body and the buried depth of the pipeline. The displacement and stress of the pipeline increase significantly with the increase of the displacement of the sliding body. With the increase of pipeline buried depth, the displacement of the pipeline shows an overall decrease, and when the buried depth of the pipeline is 3–3.5 m, the displacement and stress are close to the peak, indicating that the buried depth is of great risk. The intersection angle between the pipeline and landslide has a significant effect on the stress of the pipeline. The smaller the intersection angle, the safer the pipeline is.

Keywords: landslide; oil and gas pipeline; pipeline oblique crossing landslide; disaster mechanism; numerical simulation

Citation: A, F.-Y.; Chen, T.-H.; Yang, C.; Wu, Y.-F.; Yan, S.-Q. Study on Disaster Mechanism of Oil and Gas Pipeline Oblique Crossing Landslide. *Sustainability* **2023**, *15*, 3012. https://doi.org/10.3390/su15043012

Academic Editors: Chong Xu and Jian Chen

Received: 5 January 2023
Revised: 27 January 2023
Accepted: 28 January 2023
Published: 7 February 2023

1. Introduction

As one of the important transportation modes of onshore oil and gas, pipeline transportation has the advantages of safety, environmental friendliness, and high efficiency. It has become one of the most important oil and gas transportation modes in the world. As a result, pipeline safety affects the stable and healthy development of the global economy [1–3]. However, as a long-distance linear project, oil and gas pipelines traverse different geological environments. Especially in mountainous areas, geological disasters have the characteristics of diversity, complexity, and universality, which bring great challenges to the safety prevention and control of oil and gas pipelines. By 2021, the length of China's oil and gas pipeline network reached 150,000 km, and it is expected that by 2025, the length of China's oil and gas pipeline will reach 240,000 km [4]. Due to the increasing mileage of pipeline construction and the susceptibility of geological disasters in mountainous areas, geological disasters have become an important threat to the safe operation of pipelines. According to incomplete statistics, up to 2014, there were 282 geological disasters along the Lanzhou-Chengdu-Chongqing oil pipeline, Zhonggui gas pipeline, and West-East gas pipeline in China, among which 107 were landslides, accounting for 38% [5]. Thus, landslides are one of the main geological disasters threatening the safe operation of pipelines [6–8]. Under the action of a landslide, the pipeline is pushed or pulled by soil, resulting in pipeline deformation and rupture, which brings great challenges to pipeline operation and management (Table 1) [9–11].

Table 1. Pipeline disaster cases caused by landslides.

Pipeline	Time	Accident Cause	Effect of Pipeline Failure
Ecuadorian Pipeline	1983.03	Earthquake	Pipeline fracture
Brazilian refined oil pipeline	2001.02	Heavy rainfall	Pipeline fracture, oil leak
Zichang section of West-East gas pipeline	2007.07	Heavy rainfall	The pipes were pushed
Baiyangping section of Sichuan gas transmission	2007.08	Slope creeping	The pipes were pushed
Qingdao oil pipeline	2013.11	Pipeline corrosion caused by tidal movement	The oil leak exploded, killing 62 people
Chuijiaba section of Eastern Sichuan gas transmission	2016.07	Heavy rainfall	The pipes were pushed
Qinglong sand section, Guizhou Province	2017.07	Continuous rainfall	The gas leak exploded, killing 8 people

The way of pipeline crossing landslide is very important to study the deformation and failure characteristics of pipelines. Therefore, Brodavkin [12] first divided pipelines into oblique and parallel types according to the relationship between pipelines and slide direction. The classification method is simple, but it does not show the characteristics of landslides. On the basis of Brodavkin, Lin et al. [13] established the classification system of pipeline landslide by scientific induction in 2009. According to the cross-cutting relationship between pipelines and landslides, the pipeline crossing landslide was divided into three types: pipeline transversely crossing landslide, pipeline longitudinal crossing landslide, and pipeline oblique crossing landslide (Figure 1). At present, this naming method has been widely used.

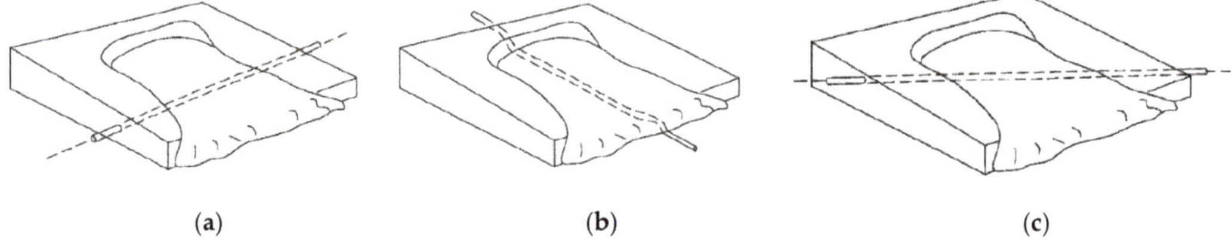

(a)   (b)   (c)

Figure 1. Cross-cutting relationship between pipelines and landslides: (**a**) pipeline transversely crossing landslide; (**b**) pipeline longitudinal crossing landslide; and (**c**) pipeline oblique crossing landslide [13].

The response state of pipelines under landslides and the pipe–soil interaction are the basis of studying pipeline-landslide disaster mechanism. Since the 1970s, scholars at home and abroad have done many studies on the mechanical response state of pipeline landslide and the law of pipe–soil interaction by using three methods: analytical method, numerical simulation, and model test. Despite the rapid development in recent years, no obvious consensus has been achieved. In 1995, Rajani et al. [14] analyzed the force and deformation characteristics of the pipeline for the first time by the analytic method; they pointed out that with the aggravation of the landslide deformation, the pipeline deviation would increase continuously. On the basis of Rajani, O'Rourke [15] established an elastic-plastic model of the pipe–soil interaction and concluded that the maximum stress of the pipeline increased with the increase of the buried depth of the pipeline, soil cohesion, soil density, and internal friction angle. Cocchetti et al. [16,17] assumed the pipe as a beam element, on the basis of the elastoplastic model of pipe–soil interaction established by O'Rourke, and applied nonlinear theory to study the pipe–soil contact interaction. Long et al. [18] adopted the elastic foundation beam method based on the Winkler model when analyzing the characteristics of pipeline deformation and failure. Audibert [19] studied the influence of different pipe diameters on the side earth pressure of the pipeline through the test method, and believed that the smaller the pipe diameter, the greater the influence of the side earth pressure on the pipeline. Calvetti et al. [20] used the pulley wire rope to simulate the axial force system of the pipeline, and then used the sinking soil box to simulate the landslide. Georgiadis et al. [21] carried out an analysis experiment on the response state of the pipeline and believed that the decisive factor of the force on the pipeline was the sliding velocity of the sliding body. Yoshizaki and Karimain et al. [22,23] studied the force law of the pipeline under the action of backfill soil and sliding body through a series of experiments; they believed that backfill material with light density could significantly reduce the influence of the sliding body on the damage of the pipeline. Yatabe [24] used Abaqus to establish a pipeline landslide model and analyzed the displacement variation rule of elbow under large ground deformation. Evans et al. [25] established a pipeline crossing landslide model. Assuming that the pipe is a linear elastic material and the contact between the pipe and soil adopts the common node technology to transfer the force between the pipe and the sliding body, the deformation and failure characteristics of the pipe are analyzed. Zhou et al. [26] created a vertical pipe-landslide model for pipelines and landslides, adopted Moore–Coulomb and Ramberg–Osgood constitutive models for landslides and pipelines, respectively, and analyzed the deformation and failure characteristics of pipelines under different factors.

From the above studies, it can be seen that many previous studies have been done on the response state of pipelines under landslides and the pipe–soil interaction. However, when laying pipelines in landslide-prone areas, the influence of different buried depths of pipelines, different sliding displacements, and the cross-cutting relationship between pipelines and landslide sliding directions on the mechanical response of pipelines has still not been studied in depth. Therefore, in this paper, FLAC3D numerical simulation software is used to establish an ideal model of completely coupled tube–soil interaction of pipeline oblique crossing landslide. It has important theoretical and practical significance to provide the basis for the risk assessment and prevention of pipeline landslides. The novelty of this work is as follows:

(1) The stress mechanism of pipelines at different intersection angles of landslides and pipelines.
(2) The influence of buried depths and sliding displacements on the stress of gas pipeline oblique crossing landslide.

2. Influence of the Buried Depth of Pipeline and Displacement of Sliding Body on Pipeline Oblique Crossing Landslide

FLAC3D can be used to simulate the mechanical properties and analyze the plastic flow of soil, rock, and other materials, which is widely used in the field of geotechnical engineering [27–29]. Therefore, in this study, FLAC3D6.0 was used to establish an ideal model of pipeline oblique crossing landslide [30,31], and further analyze the influence of buried depth, the displacement of the sliding body, and buried position on stress and deformation under the effect of a landslide.

2.1. Establishment of Model

This paper mainly studies the mechanical response of pipeline oblique crossing landslide. According to the relationship between landslides and pipelines, and considering the boundary effect of the model, a pipeline-landslide ideal model with a landslide width of 60 m, a landslide height of 45 m, a sliding body thickness of 2–10 m, and a slope of 30–45° was established (Figure 2). The intersection angle between the pipeline and landslide is 45°, the buried depth of the pipeline is taken as seven cases: 1 m, 1.5 m, 2 m, 2.5 m, 3 m, 3.5 m, and 4 m, respectively, and the maximum displacement of the sliding body is taken as four cases: 1 m, 2 m, 3 m, and 4 m, respectively, for analysis. In the process of modeling and simulation, the parameters related to the pipeline remain unchanged.

The landslide is divided into sliding body and sliding bed and divided by a triangular mesh element. In the modeling process, the mesh of the pipeline and the sliding body around the pipe are encrypted. In the calculation process, the Moore–Coulomb model is adopted for the landslide, and the velocity boundary conditions with normal constraints are used around and at the bottom of the model, and the slope body adopts the free boundary. The pipeline model is built based on the landslide model using the SHELL element in FLAC3D6.0. Due to the large stiffness difference between the pipeline and the sliding body, a pipe–soil interface is established between the pipeline and the sliding body to reduce the large error caused by the stiffness difference (Figure 3). In the calculation process, the elastic-plastic constitutive model is adopted and applies the gravity load.

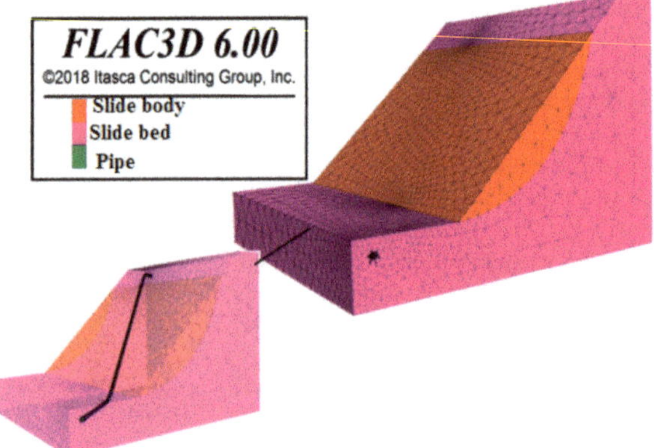

Figure 2. Ideal piping-landslide model with complete coupling.

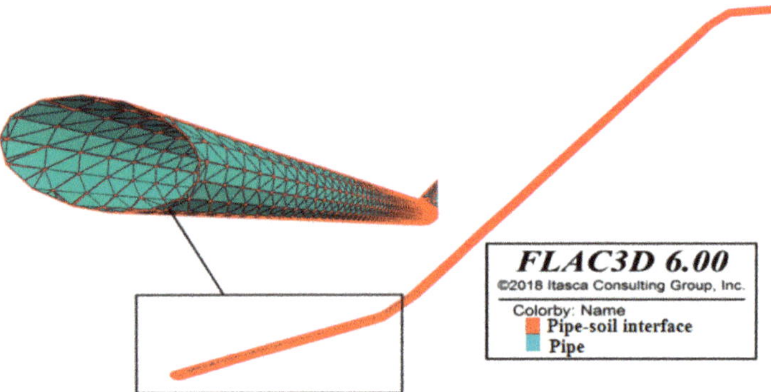

Figure 3. Pipe–soil interface models.

2.2. Geotechnical Parameters

The physical and mechanical parameters of the landslide are shown in Table 2, while the parameters of the pipeline and pipe–soil interface are shown in Tables 3 and 4.

Table 2. Physical and mechanical parameters of the landslide (from experimental data).

Material	Natural Gravity (kN/m^3)	Cohesion (kPa)	Internal Friction Angle (°)	Elastic Modulus (MPa)	Poisson's Ratio
Sliding body	16.75	7.5	10.1	20	0.30
Sliding bed	25.00	11.6	18.0	35	0.28

Table 3. Parameters of the pipeline [11].

Material	Density (kg/m^3)	Elastic Modulus (MPa)	Poisson's Ratio	Steel Grade	Yield Strength (MPa)
pipeline	7850	210,000	0.3	X70	483

Table 4. Parameters of the pipe–soil interface [11].

Material	Normal Stiffness (MPa/m)	Shear Stiffness (MPa/m)	Internal Friction Angle (°)	Cohesion (kPa)	Tensile Strength (MPa)
pipe-soil interface	4038	4038	15	0	0

2.3. Influence of Buried Depth and Displacement of Sliding Body on Force of Pipeline

In order to study the influence of buried depth and sliding displacement on the mechanical response of the pipeline, the corresponding pipe-soil coupling models with seven different buried depths were established respectively when the maximum sliding body was 1 m, 2 m, 3 m, and 4 m. During the analysis, the intersection angle between the pipeline and the landslide was 45°.

(1) Mechanical response of pipeline with the displacement of sliding body

In order to analyze the variation of the mechanical response of the pipeline with the displacement of the sliding body, a pipe-soil coupling model with a constant buried depth of 2 m was randomly selected to analyze the simulation results.

(2) Mechanical response of pipeline with buried depth

Figure 4 shows the combined displacement cloud diagram of the sliding body under different maximum displacement conditions when the buried depth of the pipeline is 2 m. Figure 5 shows the combined displacement cloud diagram of the sliding body under different buried depths when the displacement of the sliding body is 2 m. It shows that the overall distribution characteristics of combined displacement are consistent under the conditions of different maximum sliding displacements and different buried depths of the pipeline because the buried depth of the pipeline does not change much, and the relative plane position relationship between the pipeline and landslide is unchanged. However, due to the existence of the pipeline, the continuity distribution characteristics of landslide displacement were broken, and the sliding displacement in the buried position of the pipeline was the smallest, which increased from the center of the pipeline to both sides.

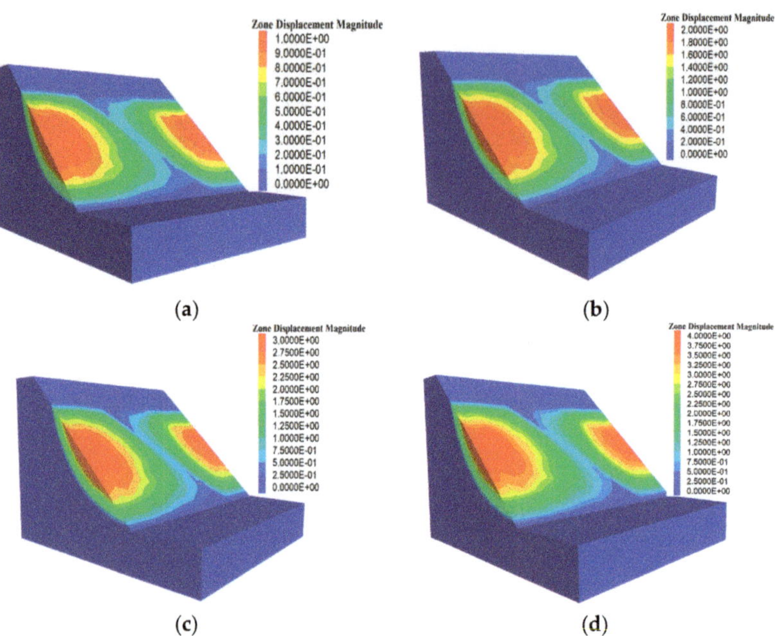

Figure 4. Combined displacement cloud diagram of the sliding body under different displacement conditions when the buried depth of the pipeline is 2 m: (**a**) the maximum displacement of the sliding body is 1 m; (**b**) the maximum displacement of the sliding body is 2 m; (**c**) the maximum displacement of the sliding body is 3 m; and (**d**) the maximum displacement of the sliding body is 4 m.

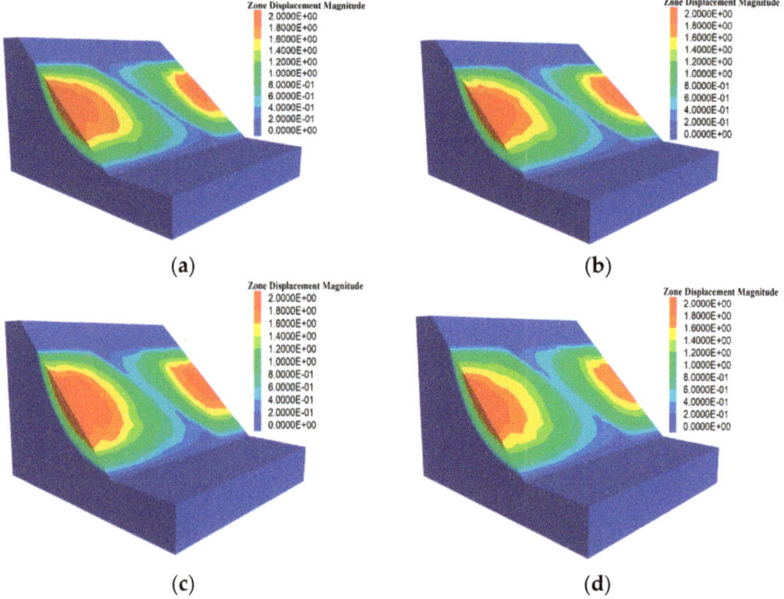

Figure 5. *Cont.*

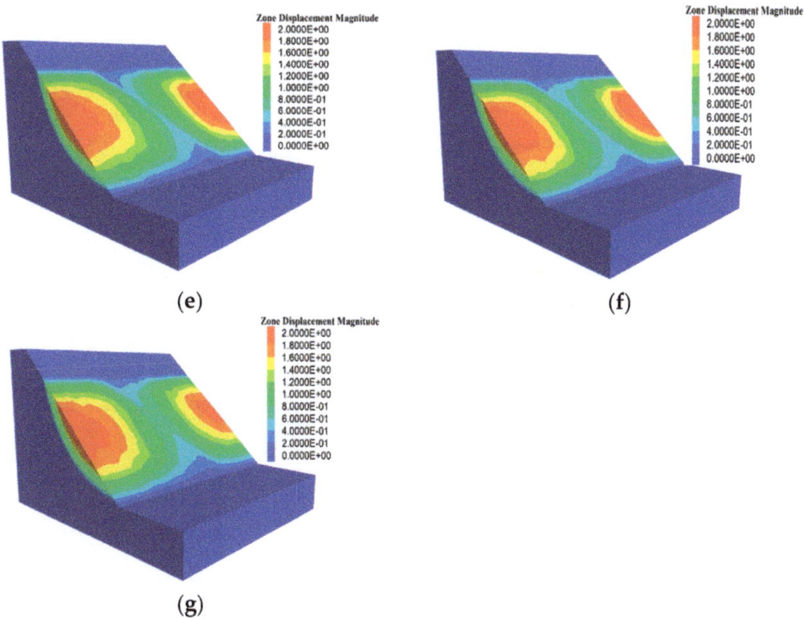

Figure 5. Combined displacement cloud diagram of the sliding body under different buried depth conditions when the displacement of the sliding body is 2 m: (**a**) the buried depth is 1 m; (**b**) the buried depth is 1.5 m; (**c**) the buried depth is 2m; (**d**) the buried depth is 2.5 m; (**e**) the buried depth is 3 m; (**f**) the buried depth is 3.5 m; (**g**) the buried depth is 4 m.

Figure 6 shows the combined displacement cloud diagram of the pipeline under different maximum displacement conditions of the sliding body when the buried depth of the pipeline is 2 m, and the shear stress distribution diagram is in the XZ direction. Figure 7 shows the combined displacement cloud diagram of the pipeline under different buried depth conditions when the sliding displacement is 2 m, and the shear stress distribution diagram is in the XZ direction. It can be seen from the cloud diagram that the overall distribution characteristics of the pipeline combined displacement and shear stress cloud diagram do not change under the conditions of different maximum displacement of the sliding body and different buried depths of the pipeline. The size distribution of pipeline displacement is consistent with that of the displacement of the sliding body, that is, the position of sliding displacement is large, the pipeline displacement is relatively large, and the position of sliding displacement is small, the pipeline displacement is relatively small. It can be seen from the shear stress distribution in the XZ direction that stress concentration is obvious at the shear outlet and the trailing edge of the landslide, and the two stress directions are opposite. The shear stress at the trailing edge of the landslide is opposite to the direction of the free face, while the shear outlet of the landslide shows the stress shear in the same direction as that of the free face under the effect of landslide pushing. The absolute value of stress at the shear outlet is the largest, which should be regarded as the key location.

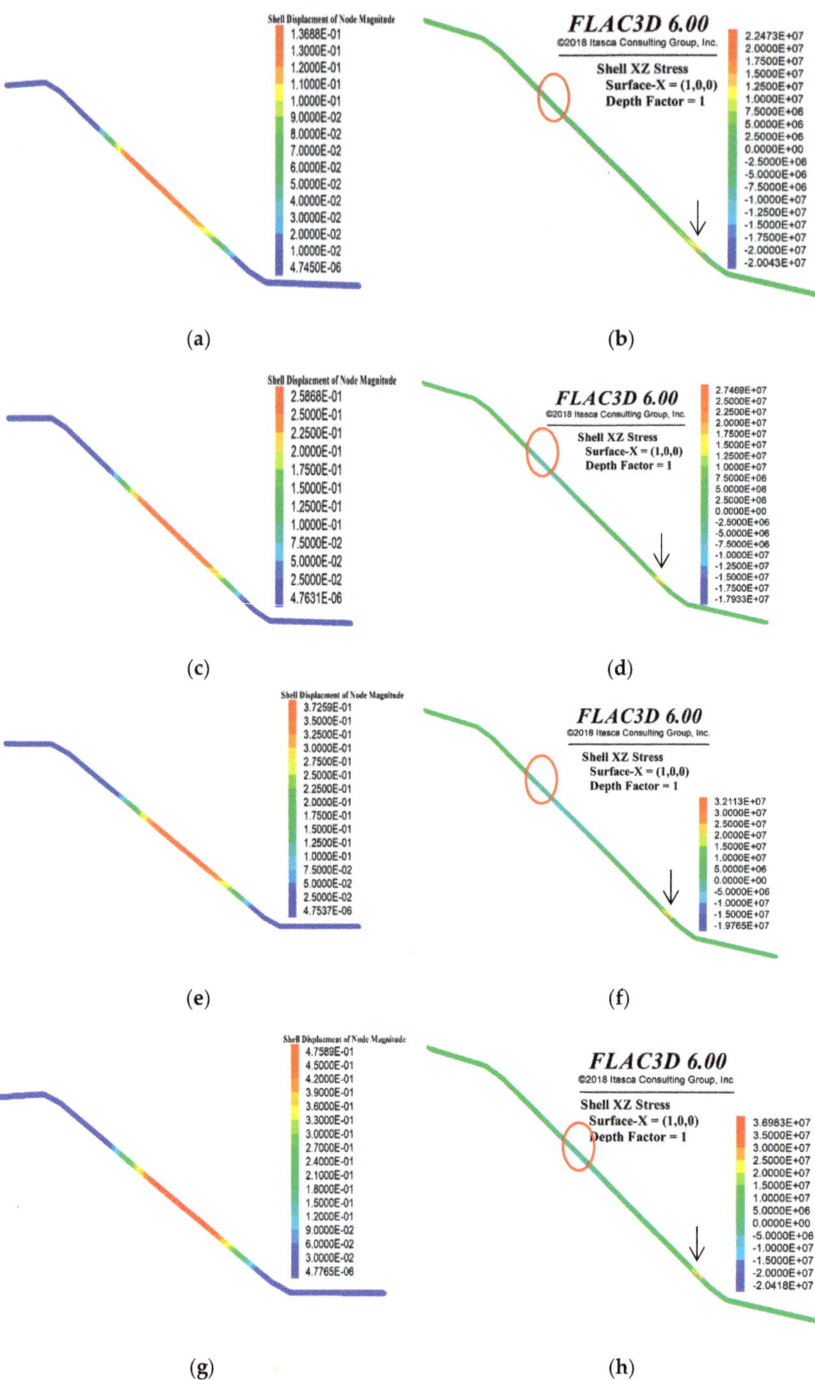

Figure 6. Combined displacement and stress cloud diagram of the pipeline under different displacement conditions when the buried depth of the pipeline is 2 m: (**a**) combined displacement of the pipeline when

the displacement of the sliding body is 1 m; (**b**) stress of the pipeline when the displacement of the sliding body is 1 m; (**c**) combined displacement of the pipeline when the displacement of the sliding body is 2 m; (**d**) stress of the pipeline when the displacement of the sliding body is 1 m; (**e**) combined displacement of the pipeline when the displacement of the sliding body is 3 m; (**f**) stress of the pipeline when the displacement of the sliding body is 3 m; (**g**) combined displacement of the pipeline when the displacement of the sliding body is 4 m; (**h**) stress of the pipeline when the displacement of the sliding body is 4 m.

Tables 5 and 6 show the maximum displacement and stress of the pipeline under different displacements of the sliding body and different buried depths of the pipeline, respectively, and Figure 8 shows the corresponding curves. The displacement and stress of the pipeline are significantly affected by the displacement of the sliding body, and the displacement and stress of the pipeline increase significantly with the increase of the displacement of the sliding body. The influence of pipeline-buried depth on displacement and stress of the pipeline is obvious under the effect of the landslide. With the increase of pipeline-buried depth, the displacement of pipeline shows an overall decrease, while stress shows an overall increase. However, with the increase of the buried depth of the pipeline, due to the difference of stress and strain in different parts of the sliding body, the pushing effect on the pipeline is different. Furthermore, the displacement and stress of the pipeline fluctuate with the increase of the buried depth of the pipeline. When the buried depth of the pipeline is 3–3.5 m, the displacement and stress are close to the peak, indicating that the buried depth is of great risk. Furthermore, the pipeline is in a failure state when the displacement of the sliding body is greater than 1 m and the buried depth of the pipeline is greater than 1.5 m, so the displacement of the sliding body should be strengthened to give an early warning in time.

In order to analyze the variation of the mechanical response of the pipeline with the buried depth, a pipe-soil coupling model with a constant sliding displacement of 2 m was randomly selected to analyze the simulation results.

Table 5. Maximum displacement of the pipeline under different sliding displacements and different buried depths (mm).

Sliding Displacement (m) Buried Depths (m)	1	1.5	2	2.5	3	3.5	4
1 m	146.12	139.94	136.88	122.92	132.70	121.11	127.27
2 m	281.74	272.37	258.68	236.25	263.42	230.18	240.19
3 m	409.13	395.06	372.59	346.85	382.22	338.37	350.82
4 m	540.89	515.52	475.89	443.67	490.89	439.10	459.60

Table 6. Maximum stress of the pipeline under different sliding displacements and different buried depths (Pa).

Sliding Displacement (m) Buried Depths (m)	1	1.5	2	2.5	3	3.5	4
1 m	6.51×10^7	5.15×10^8	7.81×10^8	7.35×10^8	8.94×10^8	9.96×10^9	8.6×10^8
2 m	3.02×10^8	8.87×10^8	1.29×10^9	1.36×10^9	1.25×10^9	1.53×10^9	1.18×10^9
3 m	7.84×10^8	1.33×10^9	1.58×10^9	1.55×10^9	1.72×10^9	1.68×10^9	1.59×10^9
4 m	9.61×10^8	1.53×10^9	1.87×10^9	1.95×10^9	2.03×10^9	2.26×10^9	2.09×10^9

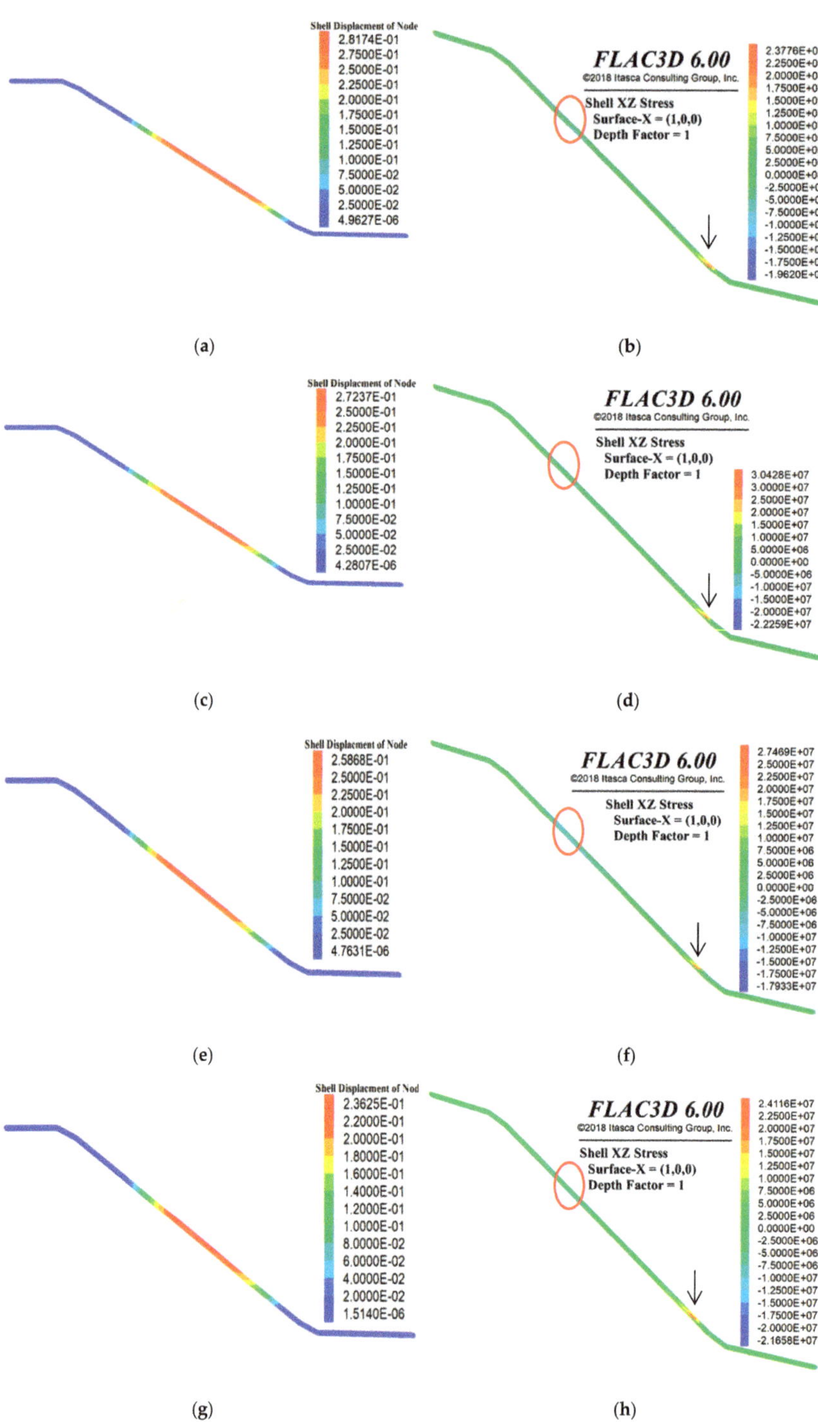

(**a**)

(**b**)

(**c**)

(**d**)

(**e**)

(**f**)

(**g**)

(**h**)

Figure 7. *Cont.*

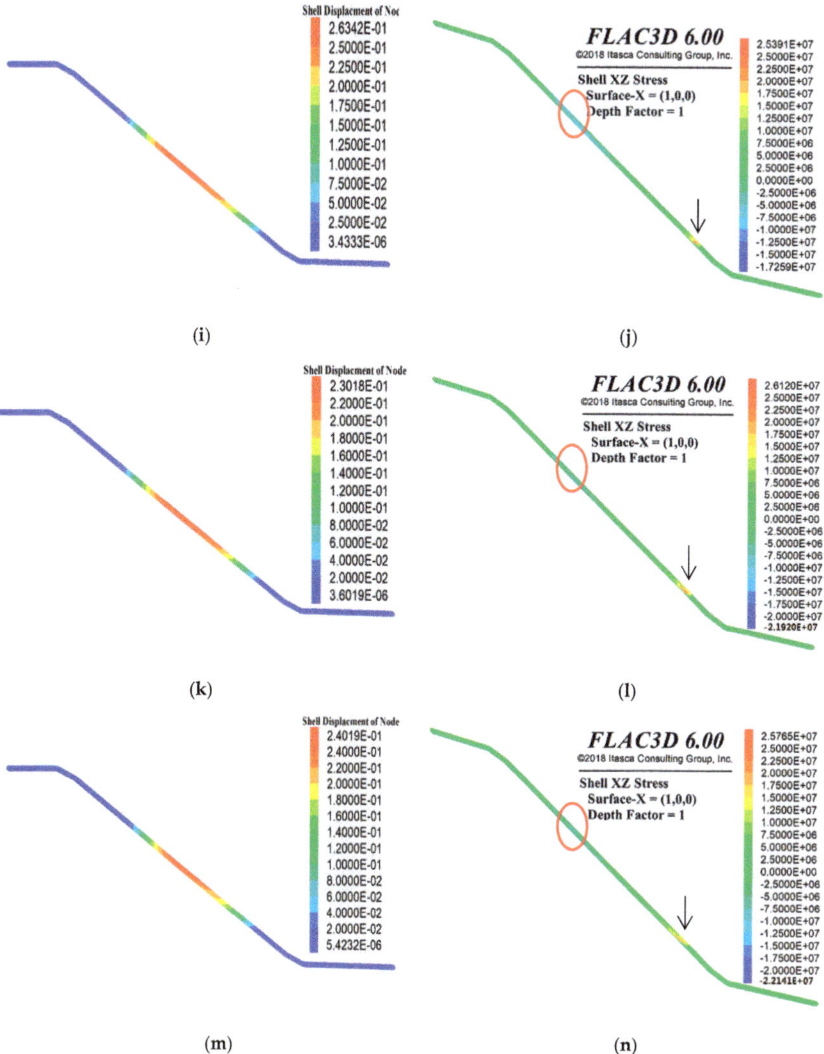

Figure 7. Combined displacement and stress cloud diagram of the pipeline under different buried depths when the displacement of the sliding body is 2 m: (**a**) combined displacement of the pipeline when the buried depth is 1 m; (**b**) stress of the pipeline when the buried depth is 1 m; (**c**) combined displacement of the pipeline when the buried depth is 1.5 m; (**d**) stress of the pipeline when the buried depth is 1.5 m; (**e**) combined displacement of the pipeline when the buried depth is 2 m; (**f**) stress of the pipeline when the buried depth is 2 m; (**g**) combined displacement of the pipeline when the buried depth is 2.5 m; (**h**) stress of the pipeline when the buried depth is 2.5 m; (**i**) combined displacement of the pipeline when the buried depth is 3 m; (**j**) stress of the pipeline when the buried depth is 3 m; (**k**) combined displacement of the pipeline when the buried depth is 3.5 m; (**l**) stress of the pipeline when the buried depth is 3.5 m; (**m**) combined displacement of the pipeline when the buried depth is 4 m; (**n**) stress of the pipeline when the buried depth is 4 m.

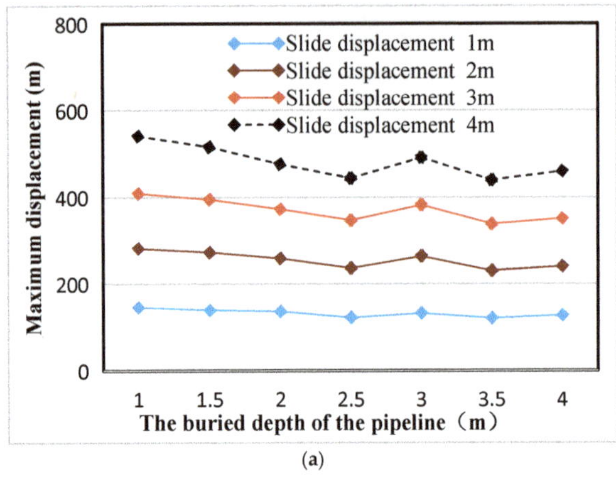

(a)

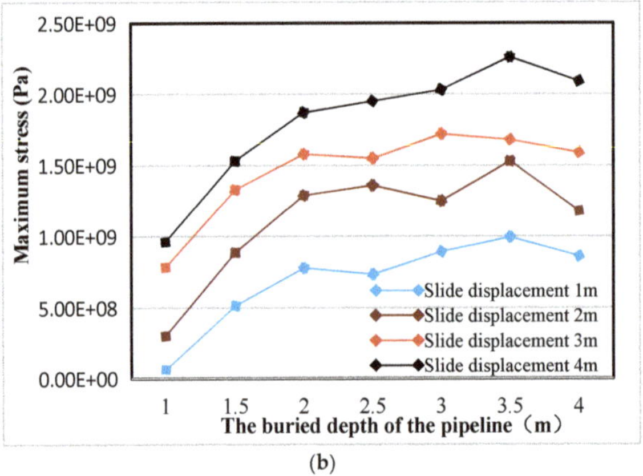

(b)

Figure 8. Maximum displacement and stress of the pipeline under different buried depths and different sliding displacements: (**a**) maximum displacement and (**b**) maximum stress.

3. Effect of Intersection Angle between Pipeline and Landslide on Mechanical Response of Pipeline

The stress characteristics are not consistent for different intersection angles of the pipeline and landslide. Therefore, different intersection angles of the pipeline and landslide will affect the deformation and failure of the pipeline. This paper studies the variation law of pipeline displacement under different intersection angles. Seven cases of the intersection angle between the pipeline and the landslide are 0°, 15°, 30°, 45°, 60°, 75° and 90°, the buried depth of the pipeline is 2 m, and the maximum displacement of the sliding body is 4 m for analysis.

Figure 9a–n show the combined displacement cloud diagram of the sliding body and pipeline under different intersection angles of the pipeline and landslide (0°, 15°, 30°, 45°, 60°, 75° and 90°). Due to the different intersection angles between the pipeline and the landslide, the displacement field and the size of the sliding body vary. When the intersection angle is 0°, as shown in Figure 9a, the displacement field of the pipeline centered around on both sides of the symmetrical distribution; with the increase of intersection angle, the displacement becomes a more uneven distribution. The symmetry of displacement distribution on both sides of the pipeline becomes worse and worse, and the symmetric distribution of displacement on both sides of the sliding body centered on the pipeline appears until the intersection angle increases to 90°, as shown in Figure 9m. The distribution of pipeline displacement varies with the intersection angle between the pipeline and landslide. When the intersection angle is 0–45°, the pipeline displacement distribution is similar, whereas when the intersection angle is 45–90°, the pipeline displacement distribution is not similar. It can be seen that a 45° intersection angle between the pipeline and landslide is the cut-off point for force change of the pipeline. The main reason is that when the intersection angle is greater than 45°, the pipeline no longer intersects the shear outlet of the landslide, but intersects the boundary on both sides of the landslide, and the force characteristics change significantly.

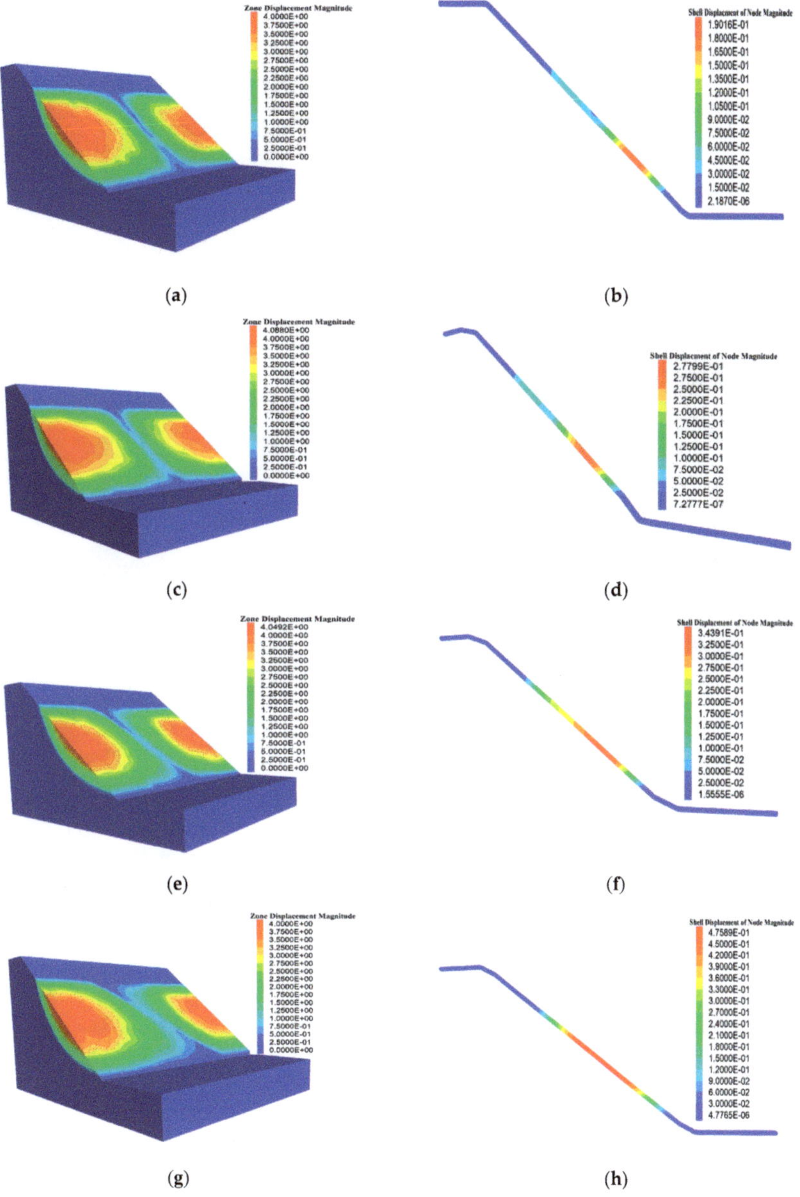

Figure 9. *Cont.*

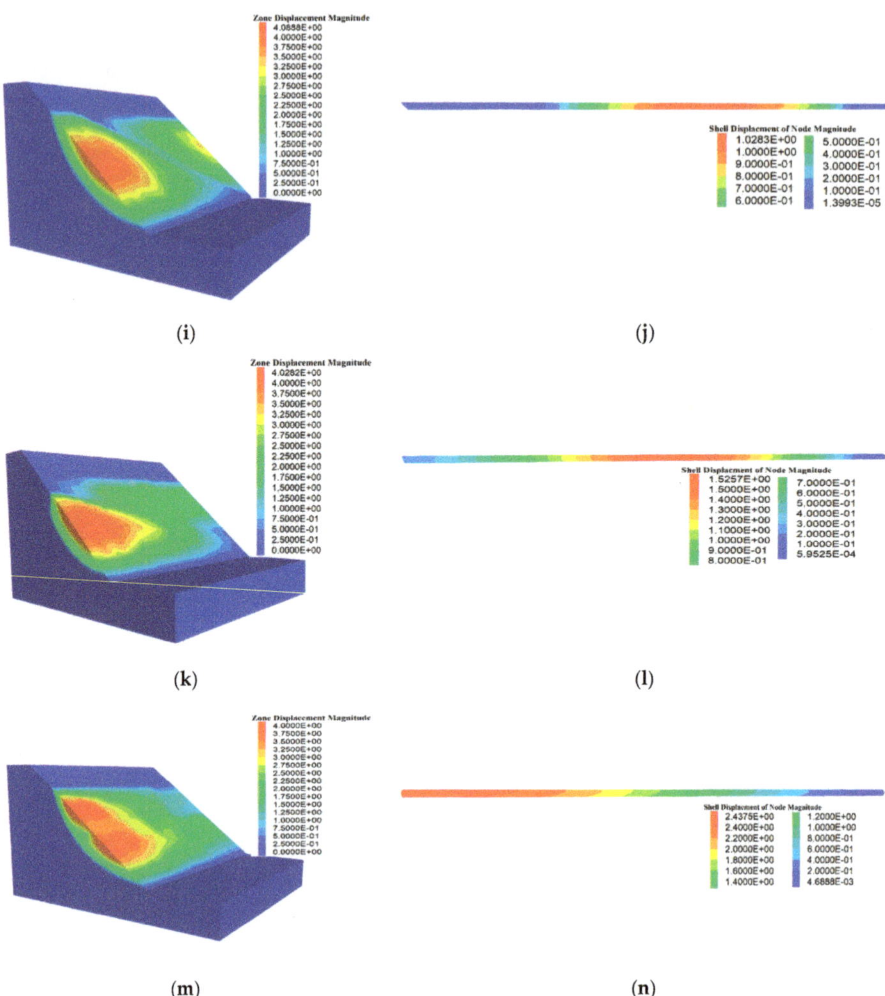

(**i**) (**j**)

(**k**) (**l**)

(**m**) (**n**)

Figure 9. Combined displacement cloud diagram of the sliding body and pipeline under different intersection angles of the pipeline and landslide: (**a**) combined displacement of the sliding body when the intersection angle is 0°; (**b**) combined displacement of the pipeline when the intersection angle is 0°; (**c**) combined displacement of the sliding body when the intersection angle is 15°; (**d**) combined displacement of the pipeline when the intersection angle is 15°; (**e**) combined displacement of the sliding body when the intersection angle is 30°; (**f**) combined displacement of the pipeline when the intersection angle is 30°; (**g**) combined displacement of the sliding body when the intersection angle is 45°; (**h**) combined displacement of the pipeline when the intersection angle is 45°; (**i**) combined displacement of the sliding body when the intersection angle is 60°; (**j**) combined displacement of the pipeline when the intersection angle is 60°; (**k**) combined displacement of the sliding body when the intersection angle is 75°; (**l**) combined displacement of the pipeline when the intersection angle is 75°; (**m**) combined displacement of the sliding body when the intersection angle is 90°; (**n**) combined displacement of the pipeline when the intersection angle is 90°.

As shown in Table 7 and Figure 10, the maximum displacement of the pipeline increased with the increase of the intersection angle between the pipeline and the landslide. When the intersection angle between the pipeline and landslide direction is 0°, the pipeline displacement is the minimum, which is 190.16 mm. When the intersection angle between the pipeline and landslide direction is 90°, the pipeline displacement reaches the maximum, which is 2437.5 mm. When the intersection angle is 0°–45°, the increase rate of pipeline displacement is small, and when the intersection angle is 45°–90°, the increase rate of pipeline displacement is large.

Table 7. Maximum displacement of the pipeline under different intersection angles between the pipeline and the landslide.

Intersection Angle (°)	0	15	30	45	60	75	90
Maximum displacement (mm)	190.16	277.99	343.91	475.89	1028.3	1525.7	2437.5

In order to analyze the deformation and stress characteristics of the pipeline in detail, the stress value at the maximum displacement of the pipeline under different intersection angles is extracted. As shown in Table 8 and Figure 11, the maximum stress of the pipeline increases with the increase of the intersection angle between the pipeline and the landslide. When the intersection angle between the pipeline and landslide is 90°, that is, the pipeline and landslide slide vertically, the maximum stress value of the pipeline reaches 2.16×10^9 Pa. Similar to the variation characteristics of displacement, when the intersection angle is 0°–45°, the increase rate of maximum stress is small, and when the intersection angle is 45°–90°, the increase rate of stress is large. Therefore, by comprehensive analysis of the characteristics of displacement and stress of the pipeline, it can be seen that the smaller the intersection angle between the pipeline and landslide, the safer the pipeline is. When the intersection angle between the pipeline and landslide direction is greater than 45°, the threat to the pipeline increases.

Table 8. Maximum stress of the pipeline under different intersection angles between the pipeline and the landslide.

Intersection Angle (°)	0	15	30	45	60	75	90
Maximum stress (Pa)	4.97×10^7	9.72×10^7	1.2×10^8	2.08×10^8	6.08×10^8	9.49×10^8	2.16×10^9

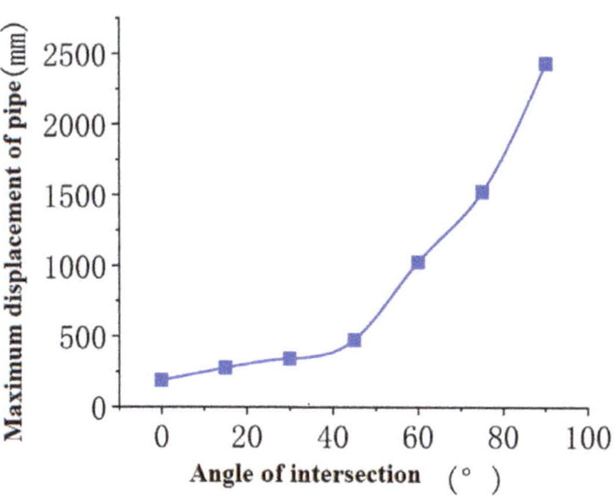

Figure 10. Maximum displacement of the pipeline at different intersection angles.

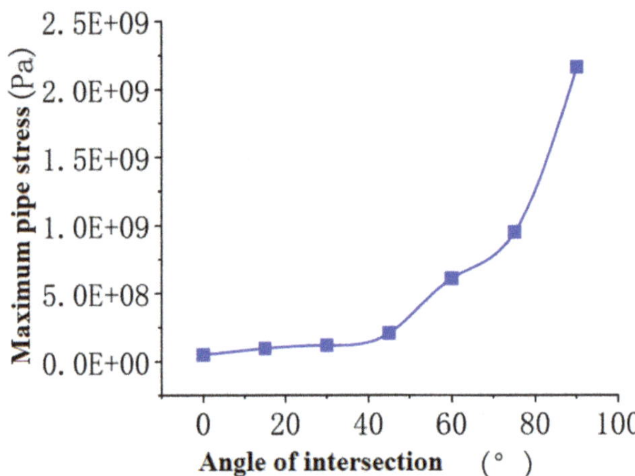

Figure 11. Maximum stress of the pipeline at different intersection angles.

4. Typical Case of Pipeline Oblique Crossing Landslide

Taking the landslide of the Second Branch of Yunnan Oil Transport (MJ053) pipeline of China National Pipeline Network as a typical case, the complete coupling model of pipeline-landslide is established using FLAC3D. The stability of the landslide and the displacement and stress of the pipeline oblique crossing landslide are analyzed.

4.1. Landslide Characteristics

The landslide was located at Jianshui section (MJ053) of Baikun Pipeline in Jianshui County, Yunnan Province, and the pipeline crossed the landslide area diagonally (Figure 12). The geomorphic unit of the site is hilly, and the micro-geomorphic features are high in the southeast and low in the northwest. The elevation of the landslide is 1347 m–1363 m, and it presents a "slow-steep" step-like landform longitudinally. The middle and lower part of the front edge is relatively slow, ranging from 2° to 8°. The upper slope of the landslide is relatively large, ranging from 15° to 45°. The main sliding direction of the landslide is 312°, the landslide is curved in plane form, the average length of the landslide is about 90 m, the average width is about 80 m, the total area is about 7200 m^2, the average thickness of the landslide is about 3 m, and the total volume is about 2.16×10^4 m^3, which is a shallow soil landslide. The surface deformation in the landslide area is mainly characterized by fracture development (Figure 13). A total of 8 fractures (L1 to L8) were developed in the landslide area, with a maximum width of 0.1 cm–10 cm, and local dislocation occurred to form the landslide back wall of 0.1 m to 1.5 m (Figure 14).

According to the field drilling data, the stratigraphic lithology exposed in the landslide area mainly consists of plain fill (Q_4^{ml}), clay (Q_4^{al+pl}), and argillaceous siltstone (N) (Figure 15). The hydrogeological conditions are simple. According to the occurrence conditions and migration forms of groundwater, it can be divided into two categories: loose rock pore water and bedrock fissure water. There is no spring in the landslide area. The stable groundwater depth measured in the borehole in the rainy season is 0.50~0.80 m, and the stable groundwater depth measured in the dry season is 1.50~3.00 m.

Figure 12. Photo of landslide.

Figure 13. Serious deformation of the original drainage ditch.

Figure 14. Cracks on trailing edge of landslide.

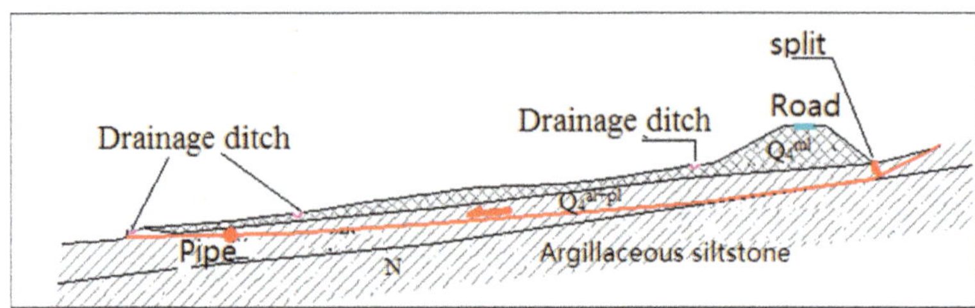

Figure 15. Typical engineering geological section of landslide.

4.2. Numerical Simulation Analysis of Pipeline-Landslide

(1) Modeling

According to the landslide survey report, the model size is 149 m long, 113 m wide, and 44 m high after considering the boundary conditions. The sliding body, sliding zone, and bedrock are divided by triangular grid elements. In order to improve the calculation accuracy and the accuracy of numerical simulation, the pipeline grid, the sliding body around the pipeline, and the sliding zone grid are locally encrypted. The model consists of 38,902 triangular nodes and 217,257 triangular elements, as shown in Figure 16.

The pipeline is located in the middle part of the landslide, and the buried depth of the pipeline is about 2.0 m on average. In the process of modeling and calculation, the influence of pipe diameter and other parameters on the mechanical response state of pipe is ignored. The pipeline type is X70 tubular steel, with a wall thickness of 14.6 mm. The SHELL element in FLAC3D was used to establish the fully coupled pipe-soil model. In order to improve the accuracy of the numerical calculation, a pipe-soil contact surface is established on the surface of the pipeline to provide the sliding between the pipeline and the sliding body.

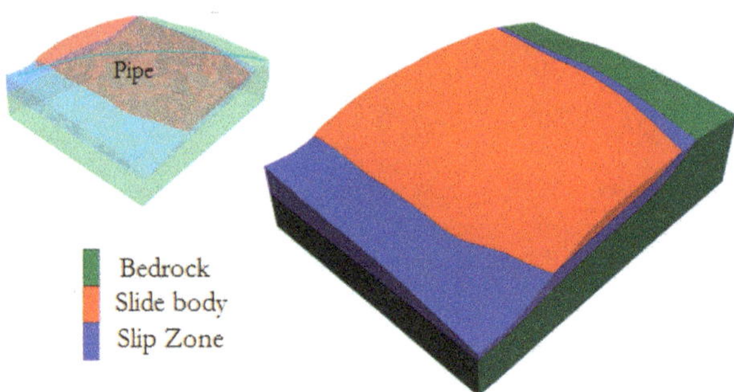

Figure 16. Landslide–pipeline interaction model.

(2) Geotechnical parameters

The calculated parameters are obtained according to the site survey report and after several inversion trials. The calculated parameters are shown in Tables 9–11.

Table 9. Geotechnical parameters of the landslide under natural conditions.

Material	Natural Gravity (kN/m^3)	Cohesion (kPa)	Internal Friction Angle (°)	Elastic Modulus (MPa)	Poisson's Ratio
Sliding body	16.75	7.5	10.1	15.0	0.30
Sliding zone	25.00	9.2	12.4	23.0	0.28
Bedrock	28.00	11.6	22.0	35.0	0.25

Table 10. Mechanical parameters of the pipeline.

Material	Density (kg/m³)	Elastic Modulus (MPa)	Poisson's Ratio	Steel Grade	Yield Strength (MPa)
pipeline	7850	210,000	0.3	X70	483

Table 11. Parameters of pipe–soil interface.

Pipe-Soil Interface	Cohesion (kPa)	Internal Friction Angle (°)	Dilatancy Angle (°)	Normal Stiffness, Kn (MPa/m)	Shear Stiffness, Ks (MPa/m)	Tensile Strength kPa
value	0	0	0	4038.5	4038.5	0

(3) Pipeline-landslide stability analysis

The landslide stability was calculated by the strength reduction method using FLAC3D6.0. As shown in Figure 17, the stability coefficient of the landslide is 0.971, which is in an unstable state under natural working conditions. At this time, the surface sliding body has been transformed from the creeping stage to the sliding, and the deep sliding body is in the critical stage from creeping to sliding. According to the survey report, the maximum displacement of the landslide is 500 mm, which is in good agreement with the numerical analysis result of 453.74 mm. Based on the in situ deformation and numerical simulation results, it can be seen that the landslide is in an unstable state under natural conditions, and it is necessary to take comprehensive engineering measures to control the landslide to reduce its threat to the oil pipeline.

Figure 17. Landslide stability.

(4) Stress analysis of pipeline

According to the calculation results of landslide stability, the stability coefficient in the natural condition is 0.971, which is in an unstable state, and the slope deformation is obvious. At this time, the pipeline is pushed by the soil around the pipeline to produce a large displacement. As shown in Figure 18, the maximum displacement of the pipeline under the landslide is 156.87 mm, distributed at the side edge of the landslide, which is basically consistent with the surface displacement of the sliding body. It can be seen that the pipeline oblique crossing landslide is mainly affected by the landslide pushing, resulting in deformation and displacement. The site survey report shows that the actual displacement of the pipeline is about 150 mm, which is close to the numerical simulation results.

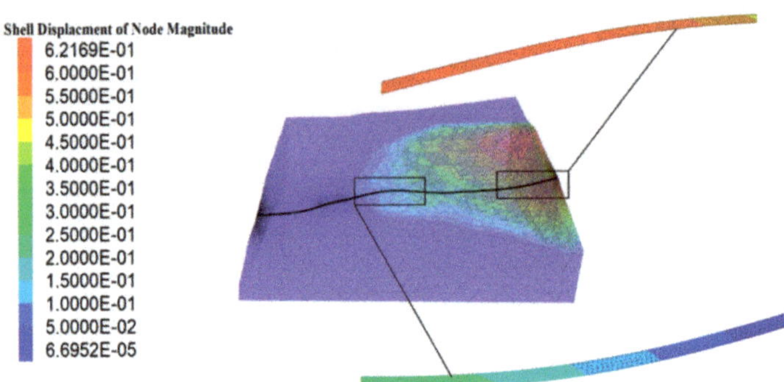

Figure 18. The displacement of the pipeline.

Figure 19 shows the vertical stress variation cloud diagram of the pipeline under the action of the landslide. The results show that the pipeline is subjected to the pushing or tension and compression of the sliding body, resulting in the tensile stress and compressive stress concentration. The maximum tensile stress of the pipeline is 2.3454×10^7 Pa at the maximum displacement of the sliding body, and the maximum compressive stress of the pipeline is 1.2751×10^7 Pa at the small displacement of the pipeline. The tensile strength of the oil pipeline is 9.86×10^6 Pa. At this time, the maximum tensile stress of the pipeline is greater than the ultimate tensile strength, and the local damage of the pipeline occurs, which is consistent with the on-site pipeline state.

In order to further analyze the variation of pipeline displacement in the process of the landslide, seven monitoring points are set on the pipeline model to monitor the displacement variation of the pipeline. The arrangement of monitoring points is shown in Figure 20. According to the numerical simulation results, the pipeline displacement monitoring data is extracted, and the pipeline displacement change curve is obtained after processing (Figure 21). With the development of the landslide, the pipeline displacement shows an overall increasing trend. The pipeline displacement increases significantly where the displacement of the sliding body is large, and the pipeline displacement changes little where the displacement of the sliding body is small.

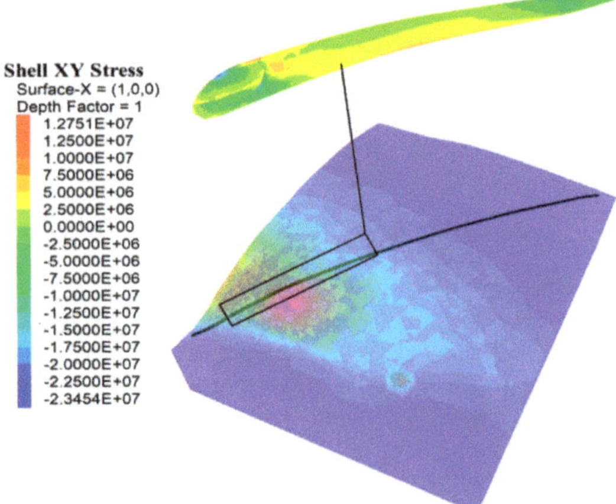

Figure 19. Vertical pipeline stress.

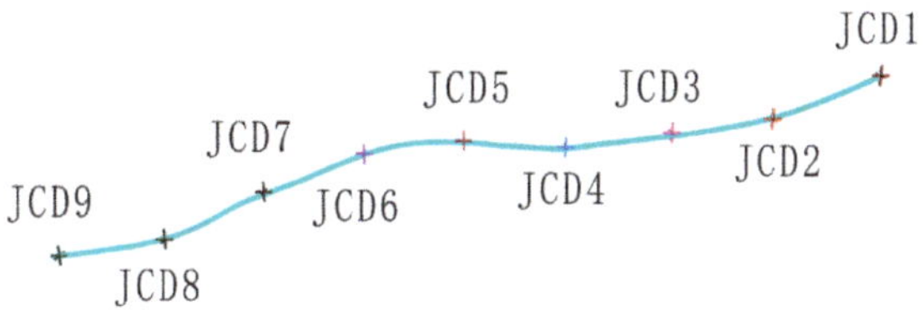

Figure 20. Monitoring point arrangement.

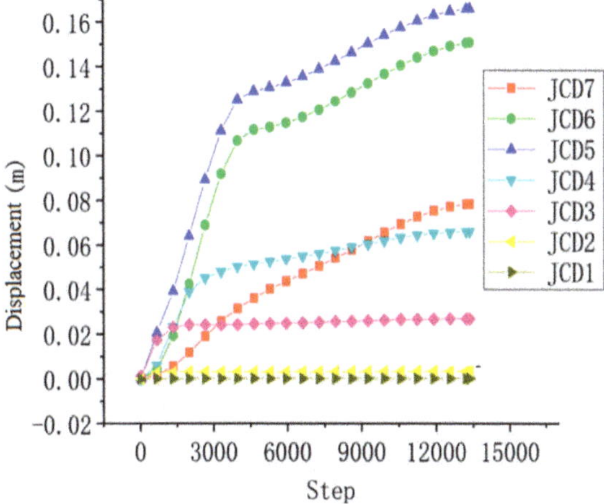

Figure 21. Displacement monitoring curve.

According to the above numerical simulation, the stability coefficient of the landslide is less than 0.971, and it is in an unstable state. Part of the soil in the landslide area has slipped obviously, leading to obvious deformation and displacement of the pipeline. Referring to the specification, it can be seen that the pipeline displacement is greater than the maximum allowable displacement value of the specification, and the stress is in the limit equilibrium state. Therefore, the landslide directly threatens the pipeline safety, and it is necessary to treat the pipeline landslide.

5. Conclusions

The ideal model of pipe-soil coupling interaction of oil and gas pipeline oblique crossing landslide is established using FLAC3D. The stress and deformation characteristics of pipelines under different sliding displacements, different buried depths, and different intersection angles between landslides and pipelines were analyzed, and a typical case of pipeline oblique crossing landslide was used for analysis. The main conclusions are as follows:

(1) The displacement field distribution of the sliding body is influenced by the existence of the pipeline. Due to the existence of the pipeline, the continuous distribution characteristics of landslide displacement are broken, and the displacement of the sliding body in the buried position of the pipeline is the smallest, increasing from the center of the pipeline to both sides. The overall distribution characteristics of displacement field of the sliding body are consistent under different maximum displacements of the sliding body and different buried depths of the pipeline. However, with the increase of the intersection angle between the pipeline and the landslide, the displacement distribution of the sliding body becomes more and more uneven, and the symmetry of the displacement distribution on both sides of the pipeline becomes worse and worse.

(2) The stress distribution characteristics of pipeline oblique crossing landslides have its particularity. The stress concentration is obvious at the shear outlet and the trailing edge of the landslide, and the two stress directions are opposite. The shear stress at the trailing edge of the landslide is opposite to the direction of the free face, while the shear outlet of the landslide shows the stress shear in the same direction as that of the free face under the effect of landslide pushing. The absolute value of stress at the shear outlet is the largest, which should be regarded as the key location.

(3) The displacement and stress of pipeline oblique crossing landslides are significantly affected by the displacement of the sliding body and the buried depth of the pipeline. The displacement and stress of the pipeline always increase with the increase of the maximum displacement of the sliding body and decrease with the increase of the buried depth of the pipeline.

(4) The intersection angle of the pipeline and landslide has a significant influence on the stress of the pipeline. With the increase of the intersection angle between the pipeline and landslide, the maximum displacement and stress of the pipeline increase. The smaller the intersection angle between the pipeline and landslide, the safer the pipeline is. Moreover, there is a significant difference between the pipeline displacement field when the intersection angle is $0°$–$45°$ and when the intersection angle is $45°$–$90°$. The main reason is that when the intersection angle is greater than $45°$, the pipe no longer intersects the landslide shear outlet, but intersects the boundary on both sides of the landslide, and the stress characteristics change significantly.

In short, there are still many problems in pipeline-landslide disasters that need in-depth study. On the basis of the above research, it is the main direction of future research to put forward some targeted prevention measures and incorporate the research results into pipeline safety evaluation.

Author Contributions: F.-Y.A., T.-H.C., C.Y., Y.-F.W. and S.-Q.Y. contributed equally to this work. All authors have read and agreed to the published version of the manuscript.

Funding: This research was funded by the National Natural Science Foundation of China, grant number 42267020, the Key Research and Development Program of Yunnan Province in 2022, grant number 202203AC100003.

Institutional Review Board Statement: Not applicable.

Informed Consent Statement: Not applicable.

Data Availability Statement: The data used to support the findings of this study are included within the article.

Acknowledgments: The authors acknowledge the PipeChina's help and support.

Conflicts of Interest: The authors declare no conflict of interest, financial, or otherwise.

References

1. Qiuyang, L.I.; Minghua, Z.H.; Zhang, B.; Wen, W.E.; Lele, W.A.; Xueqin, Z.H.; Lin, C.H. Current construction status and development trend of global oil and gas pipelines in 2020. *Oil Gas Storage Transp.* **2021**, *40*, 1330–1337.
2. Interstate Natural Gas Association of America. Interstate Natural Gas Pipeline Efficiency [EB/OL]. Available online: http://www.ingaa.org/11885/Reports/10927.aspx (accessed on 1 January 2020).
3. EGIG. *The 10th Report of the Gas Pipeline Incidents of European Gas Pipeline Incident Data Group*; R.0395[R]; EGIG: Groningen, The Netherlands, 2018; Volume 17, pp. 10–30.
4. Gao, P. New progress in China's oil and gas pipeline construction in 2021. *Int. Pet. Econ.* **2022**, *30*, 12–19.
5. Qingwen, M.; Chenghua, W.; Jiming, K. Dynamical mechanisms of effects of landslides on long distance oil and gas pipelines. *Wuhan Univ. J. Nat. Sci.* **2006**, *11*, 820–824. [CrossRef]
6. Novotný, J. Stability problems in road and pipeline constructions and their mitigation—Examples from Sakhalin and Azerbaijan. *J. Mt. Sci.* **2011**, *8*, 307–313. [CrossRef]
7. Borfecchia, F.; De Canio, G.; De Cecco, L.; Giocoli, A.; Grauso, S.; La Porta, L.; Martini, S.; Pollino, M.; Roselli, I.; Zini, A. Mapping the earthquake-induced landslide hazard around the main oil pipeline network of the Agri Valley (Basilicata, southern Italy) by means of two GIS-based modelling approaches. *Nat. Hazards* **2015**, *81*, 759–777. [CrossRef]

8.   Zhang, S.-Z.; Li, S.-Y.; Chen, S.-N.; Wu, Z.-Z.; Wang, R.-J.; Duo, Y.-Q. Stress analysis on large-diameter buried gas pipelines under catastrophic landslides. *Pet. Sci.* **2017**, *14*, 579–585. [CrossRef]

9.   Zhu, C.-J. Treatment of landslide in Baiyangping section of Sichuan gas pipeline project. *Oil-Gasfield Surf. Eng.* **2010**, *29*, 4–5.

10.   Zhang, J. Study on Landslide Stability and Treatment Monitoring in the Jiange Section of the Lanzhou-Chengdu-Chongqing Oil Pipeline. Master's Thesis, Southwest Petroleum University, Chengdu, China, 2017.

11.   Xi, S. Study on Critical Criterion and Sensitive Section of Deformation and Failure of Buried Pipeline in Landslide Area. Ph.D. Thesis, China University of Geosciences, Beijing, China, 2018.

12.   Borodavkin, P.P. *Buried Pipeline*; Feng Liang Version; Petroleum Industry Press: Beijing, China, 1980.

13.   Lin, D.; Xu, K.F.; Huang, R.Q.; Zhu, Y.; Luo, M. Landslides Classification of Pipeline for Transporting Oil and Gas. *Welded Pipe Tube* **2009**, *21*, 66–68.

14.   Rajani, B.B.; Robertson, P.K.; Morgenstern, N.R. Simplified design methods for pipelines subject to transverse and longitudinal soil movements. *Can. Geotech. J.* **1995**, *32*, 309–323. [CrossRef]

15.   O'Rourke, M.J.; Liu, X.; Flores-Berrones, R. Steel Pipe Wrinkling due to Longitudinal Permanent Ground Deformation. *J. Transp. Eng.* **1995**, *121*, 443–451. [CrossRef]

16.   Cocchetti, G.; di Prisco, C.; Galli, A.; Nova, R. Soil-pipeline interaction along unstable slopes:a coupled three-dimensional approach. Part 1:Theoretical fonnulation. *Can. Geotech. J.* **2007**, *46*, 1289–1304. [CrossRef]

17.   Cocchetti, G.; di Prisco, C.; Galli, A. Soil-pipeline interaction along unstable slopes:a coupled three-dimensional approach. Part 2: Numerical analyses. *Can. Geotech. J.* **2009**, *46*, 1305–1321. [CrossRef]

18.   Yun, L.; Kang, L. Reliability analysis of high pressure buried pipeline underlandslide. *Appl. Mech. Mater.* **2014**, *501–504*, 1081–1086. [CrossRef]

19.   Audibert, J.M.; ENyman, K.J. Soil restraint against hori zontal mo-tion of pipes. *J. Geotech. Eng. Divi-Sion* **1977**, *103*, 1119–1142. [CrossRef]

20.   Calvetti, F.; Di Prisco, C.; Nova, R. Experimental and numeri-cal analysis of soil-pipe interactior. *J. Geotech. Geoenviron. Eng.* **2004**, *130*, 1292–1299. [CrossRef]

21.   Georgiadis, M. Landslide drag forces on pipelines. *Soils Found.* **1991**, *31*, 156–161. [CrossRef]

22.   Yoshizaki, K.; Sakanoue, T. Experimental study on Soil-pipeline interaction using eps back fill. In *New Pipeline Technologies Security and Safety Baltimore*; ASCE: Reston, VA, USA, 2003; pp. 1126–1134.

23.   Karimain, S.A. Response of Buried Steel Pipelines Subjected to Longitudinal and Transverse Ground Movement. Ph.D. Thesis, University of Tehran, Teheran, Iran, 2003.

24.   Yatabe, H.; Fukuda, N.; Masuda, T.; Toyoda, M. Analytical Study of Appropriate Designfr High-Grade Induction Bend Pipes Subjected to Large Ground Deformation. *J. Offshore Mech. Arct. Eng.* **2004**, *126*, 376–383. [CrossRef]

25.   Vans, A.J. Three-Dimensional Finite Element Analysis of Longitudinal Support in Buried Pipes. Master's Thesis, Utah State University, Logan, UT, USA, 2004.

26.   Zhou, X. Study on Effects of Different Factors on Pipelines Risk Under Landslide. *Ind. Saf. Environ. Prot.* **2012**, *38*, 42–44.

27.   Fa-You, A.; Hei, M.-C.; Yan, S.-Q.; Zhang, P. Investigating the Effect of the Rock-Socketed Depth of the Hinged Cable-Anchored Pile on the Earthquake Response Characteristics of Supporting Structures. *Adv. Civ. Eng.* **2022**, *2022*, 2249654. [CrossRef]

28.   Wu, J.-J.; Li, Y.; Cheng, Q.-G.; Wen, H.; Liang, X. A simplified method for the determination of vertically loaded pile-soil interface parameters in layered soil based on FLAC3D. *Front. Struct. Civ. Eng.* **2015**, *10*, 103–111. [CrossRef]

29.   Peng, S.; Liao, W.; Liu, E. Pipe–soil interaction under the rainfall-induced instability of slope based on soil strength reduction method. *Energy Rep.* **2020**, *6*, 1865–1875. [CrossRef]

30.   Liu, E.; Li, D.; Li, W.; Liao, Y.; Qiao, W.; Liu, W.; Azimi, M. Erosion simulation and improvement scheme of separator blowdown system—A case study of Changning national shale gas demonstration area. *J. Nat. Gas Sci. Eng.* **2021**, *88*, 103856. [CrossRef]

31.   Liu, E.; Wang, X.; Zhao, W.; Su, Z.; Chen, Q. Analysis and Research on Pipeline Vibration of a Natural Gas Compressor Station and Vibration Reduction Measures. *Energy Fuels* **2020**, *35*, 479–492. [CrossRef]

 *sustainability*

Article

Seismotectonics of Shallow-Focus Earthquakes in Venezuela with Links to Gravity Anomalies and Geologic Heterogeneity Mapped by a GMT Scripting Language

Polina Lemenkova * and **Olivier Debeir**

Laboratory of Image Synthesis and Analysis (LISA), École Polytechnique de Bruxelles (Brussels Faculty of Engineering), Université Libre de Bruxelles (ULB), Building L, Campus de Solbosch, ULB—LISA CP165/57, Avenue Franklin D. Roosevelt 50, 1000 Brussels, Belgium
* Correspondence: polina.lemenkova@ulb.be; Tel.: +32-471860459

Abstract: This paper presents a cartographic framework based on algorithms of GMT codes for mapping seismically active areas in Venezuela. The data included raster grids from GEBCO, EGM-2008, and vector geological layers from the USGS. The data were iteratively processed in the console of GMT, converted by GDAL, formatted, and mapped for geophysical data visualisation; the QGIS was applied for geological mapping. We analyzed 2000 samples of the earthquake events obtained from the IRIS seismic database with a 25-year time span (1997–2021) in order to map the seismicity. The approach to linking geological, topographic, and geophysical data using GMT scripts aimed to map correlations among the geophysical phenomena, tectonic processes, geological setting, seismicity, and earthquakes. The practical application of the GMT scripts consists in automated mapping for the visualization of geological risks and hazards in the mountainous region of the Venezuelan Andes. The proposed method integrates the approach of GMT scripts with state-of-the-art GIS techniques, which demonstrated its effectiveness as a tool for mapping spatial datasets and rapid data processing in an iterative regime. In this context, using GMT and GIS to find similarities between the regional earthquake distribution and the geological and topographic setting is essential for hazard risk assessment. This study can serve as a basis for predictive seismic analysis in geologically vulnerable regions of Venezuela. In addition to a technical demonstration of GMT algorithms, this study also contributes to geological and geophysical mapping and seismic hazard assessments in South America. We present the full scripts used for mapping in a GitHub repository.

Keywords: risk; hazard; sustainability; earthquake; seismicity; cartography; GMT; geophysics

PACS: 91.10.Da; 91.10.-v; 91.10.Jf; 91.10.Op; 91.30.Dk; 93.85.Pq; 91.70.-c

MSC: 86Axx; 86A04; 86A15; 86A30; 86A60; 86-XX; 86-04; 86-08

JEL Classification: Y91; Q00; Q01; Q2; Q20; Q24; Q3; Q35; Q5; Q50; Q51; Q54; Q55; Q56; C6; C61; C63

Citation: Lemenkova, P.; Debeir, O. Seismotectonics of Shallow-Focus Earthquakes in Venezuela with Links to Gravity Anomalies and Geologic Heterogeneity Mapped by a GMT Scripting Language. *Sustainability* **2022**, *14*, 15966. https://doi.org/10.3390/su142315966

Academic Editors: Chong Xu and Jian Chen

Received: 4 November 2022
Accepted: 26 November 2022
Published: 30 November 2022

Publisher's Note: MDPI stays neutral with regard to jurisdictional claims in published maps and institutional affiliations.

1. Introduction

1.1. Background and Motivation

Nowadays, geological hazards are acknowledged as some of the most disastrous and high-risk events in mountainous regions. These involve a series of negative consequences for nature (landslides, tsunami, rock falls, land mass movements) and complex social problems (destruction of houses and infrastructure, injuries and losses of human lives) caused by seismic risks [1–3]. Early prediction of seismic risks in geologically vulnerable regions has become more valuable in a large variety of time-critical geospatial applications focused on evaluating geological hazards for the mitigation of earthquake hazards at the regional scale. The integration of geological and geophysical data is an effective approach for risk assessment in mountainous regions, for the prevention of damages and danger, and for serving for early warning, long-term monitoring, management, and mitigation of seismic hazards.

Recently, advances in cartographic development and mapping have resulted in the realization of an important application of spatial data analysis: the prediction and prognosis of seismic activities or imminent earthquake events from events observed in the past [4–6]. With enough data on earthquake events, these methods enable an accurate prognosis of seismicity with the aim of mitigating hazard risks [7,8]. Early detection of seismic risks can be solved by evaluating and visualizing different geophysical determinants that affect disaster fatalities at the country level, namely, seismic hazard strength, population density, distribution of locations, demographic and socioeconomic data, ands geological background layers [9–12]. For instance, a systematic examination of the correlations and links among various geophysical and geological determinants underlying disasters is useful for the prevention of socioeconomic risks in geologically unstable zones [13,14]. In earthquake analysis, the capability of predicting earthquake events, their magnitude, and their focal depth is highly desirable for real-time risk assessment [15,16].

To build effective geological modeling approaches for seismic hazard prediction, real-time mapping applications are beneficial for earthquake prognoses. However, obtaining an automated approach to cartographic dataset processing for seismic prediction is usually difficult, resulting in biases toward a particular GIS software. The approach presented in this study aims to solve this problem by introducing an integrative study oof scripts developed in the Generic Mapping Tools (GMT) scripting toolset [17], in addition to the traditional mapping. This framework, which is different from the conventional state-of-the-art GIS methods, presents a rapid workflow of mapping based on the data processing from the console using command line and codes written in the GMT syntax. Although advanced methods of geophysical mapping of such seismically active regions as South America present a very important task, the scripting of cartography remains a new topic. To the best of our knowledge, the majority of similar studies used the conventional GIS rather than scripts for the visualization of geophysical data [18].

While conventional methods of GIS enable the plotting the maps in a traditional regime, this often requires manual operations and is not suitable for automated mapping. In such a way, it limits the types of cartographic activities that can be used for operative geological analyses and seismic prediction. In addition, data handling is limited by the restricted compatibility of formats accepted in GIS. In contrast, scripts present a rapid and repeatable procedure of automated mapping that can be integrated into real-time risk assessment and hazard analysis. As a response to these needs, GMT presents a more flexible cartographic approach and versatile strategy of spatial data processing, and it is aimed at seismic risk assessment, geological exploration, and management by using multi-source datasets.

1.2. Contemporaneity and Objectives

Integrated mapping using various multi-source data has the objective of visualizing a complex interplay among the geological, topographic, tectonic, and geophysical parameters. Regional effects of these settings cause the Caribbean region and Venezuela to be exposed to a high risk of earthquakes, volcanos, and associated tsunami hazards. For instance, the tectonic structures of the deep basins and erosional troughs affect the submarine geomorphology, which can be interpreted using geophysical and gravimetric measurements. Earthquake-related aftershocks, hazards, and landslides result in damage to the geotechnical infrastructure and bridges [19], demolished buildings, loss of human lives [20], and aseismic slips and creeps [21]. Such consequences affect both nature and the social sectors of Venezuela and require regular monitoring and mapping for risk and vulnerability assessment.

The visualization of the information on these processes is important for seismic prognosis and earthquake modeling. Recent research demonstrated that geological and geophysical mapping creates a basis for natural hazard risk assessment [22–26]. Therefore, the data derived from hazard risk mapping can be used as background information and for the recognition of complex geological actions and processes. Previous cartographic investigations were mostly carried out by using traditional GIS in geological and environmental studies [27–30]. However, none of them focused on an integrated approach that would combine various cartographic techniques for the geophysical mapping of Venezuela.

Scripting techniques applied to geological mapping have not been adequately presented in the existing literature in comparison with the traditional GIS. At the same time, scripts employed for visualization of complex geophysical and geological phenomena enable automation and increase the precision of the processing of cartographic data. The use of machine-based methods enables one to accurately map and highlight correlations between geological and geophysical phenomena. With this aim, GMT scripts in combination with QGIS and its plugins were applied for the geophysical mapping of Venezuela in order to visualize the regional seismicity in connection with the geological structure and geophysical setting.

The cartographic objective of this study is to demonstrate the functionality of GMT scripts in combination with GIS techniques. The geological aim is to highlight the correlations between the geological setting, topography, geophysical anomalies, and regional seismicity through the analysis of the distribution, magnitude, and depth of earthquakes. To this end, this study applied GMT cartographic scripting techniques and QGIS in order to process high-resolution open-source raster grids and vector data. We used a variety of datasets to build the project with observed geological and geophysical processes and structures. The primary context of this work is the analysis of seismic events by using the visualization of earthquake locations in the region.

2. Study Area

This study focuses on Venezuela, one the of the most seismically active regions of South America. The spatial extent of the study area was 59.5°–74° W, 0°–12.5° N (Figure 1).

Figure 1. Topographic map of the region of Venezuela. Mapping: GMT. Source: authors.

Its location in the Andes exposes Venezuela to a high risk of seismic hazards. Geologically, the most important factors that have a significant influence on the seismicity in Venezuela and its surroundings include lithospheric plate subduction and the associated complex geological processes of the oblique collision of the Caribbean arc and a zone of active deformation in the eastern offshore Trinidad area [31]. This is reflected in the formation of the Venezuelan Andes as a N50°E-oriented mountain belt extending from the Colombian border in the SW to the Caribbean Sea in the NE with the associated Boconó and Valera thrust faults [32]. Seismotectonic activities in Venezuela can be characterized by a spatio-temporal composition of the constituent geophysical and geological forces, their evolutional processes, and their complex interactions. Such linkages are related to the actual location and distribution of seismic sources of the earthquakes in the Andean region with regard to the tectonics and overall dynamics of the Earth's crust in South America.

The high frequency and magnitude of earthquakes in Venezuela are associated with complex geological and geophysical settings; Figure 2. In addition, the tectonic interactions of the lithospheric plates trigger seismic activity and tectonic-related hazards that affect the lives of people and infrastructures [33–35]. The Caribbean–South American plate boundary is a very wide tectonically active zone [36,37] with recorded seismicity at various depths. Thus, anomalous series of earthquakes took place on the east coast of Trinidad with a reported depth of the main shock of 53 km [38]. A seismicity deeper than 40 km was reported on the southern end of the Lesser Antilles subduction zone and the active island arc [39].

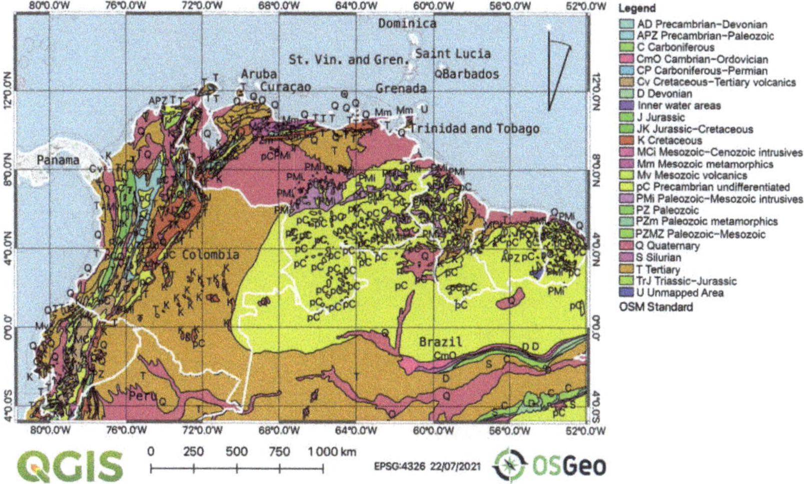

Figure 2. Geological map of the region of Venezuela. Mapping: QGIS. Source: authors.

Venezuela is one of the most countries that is most vulnerable to earthquakes in the Amazon Basin. The geological hazards in this region are exacerbated by social vulnerability due to the anthropogenic exposure, with a frequently injured and affected population living in the high mountainous regions of Andes, which increases the population who is at risk during earthquakes. Furthermore, catastrophic earthquakes produce a chain effect that triggers tsunamis, which amplifies the damage to the environment and human society. For example, it is reported that since the 19th century, tsunamis in the Caribbean Sea have resulted in the loss of several thousands of lives and the exposure of many people to risk and damage from landslides [40].

Venezuela is located in the north of the South American continent, where a collision of the South American and Caribbean tectonic plates causes complex regional geological patterns (Figure 2). The geodynamics of the interacting plates result in the formation of thrust belts and foreland basins in northern Venezuela that are shaped by the interplay of the geological layers. Outcrops of Precambrian (pC), Quaternary (Q), Triassic (T), and Paleozoic–Mesozoic intrusives (PMi) dominate in the central regions of the country, while regional outcrops of Cretaceous (K) and Paleozoic metamorphics (PZm) and Mesozoic–Cenozoic intrusives (MCi) are found in the Falcón Basin (Figure 2).

The northwestern and northeastern regions of Venezuela are prone to earthquakes [41,42] due to the tectonic activity of the lithospheric plates [43,44]. According to the Incorporated Research Institutions for Seismology (IRIS) database, the repetitive occurrence of seismic events, such as volcanism and earthquakes, in northwestern Venezuela and Trinidad and Tobago was recorded as a series of events over the last 25 years. The ongoing effects of earthquakes in Venezuela [45–50] and coastal regions of Trinidad and Tobago [51,52] have made geophysical mapping an important task for seismic hazard mapping, earthquake prognosis, and risk assessment [53].

The region of the Caribbean Sea is notable for its complex geologic setting (Figure 3), where several tectonic plates experience collision, interaction, and subduction—these are the Caribbean, Cocos, Nazca, and South American plates. This results in a high seismic activity in the region. The influence of tectonic processes is reflected in a variety of topographic forms, geological structures, and geomorphic settings of Venezuela, as discussed earlier [54–57]. The geodynamic heterogeneity of the upper mantle has resulted in a specific mountain geomorphology with notable features, such as the extent of the Tobago Trough, the Lara–Falcón Basin, and the crustal structure of the Cordillera de Mérida in southwestern Venezuela, which are important segments of the Andean orogeny [58]. The distribution of earthquakes in this region is strongly controlled by the tectonic and geophysical setting, i.e., closeness to the margins of the colliding South American and Caribbean tectonic plates. Other important factors include the mountain geomorphology and topographic variability, which are shaped by the geophysical–geological forces that affect the variations in depth and magnitude of seismic events.

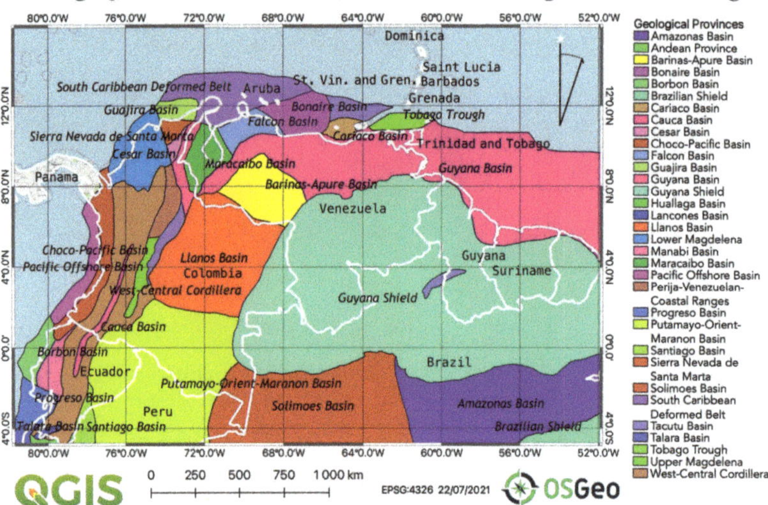

Figure 3. Geological provinces in the region of Venezuela. Mapping: QGIS. Source: authors.

Active tectonic movements cause the formation of orogenic mountain belts and topographies. In addition, regional bathymetry is complicated by the deep-sea trenches and troughs in the margins of the Pacific Ocean and the Caribbean Sea along the subduction path [59,60]. The complexity of the geological history is especially notable in NW

Venezuela—in the Falcón Basin—where interactions between the lithospheric plates and subducting Caribbean slab cause associated seismic activity and geological processes, such as an outcrop of igneous intrusive bodies, crustal thinning, decreased Moho depth, gravity anomalies, and rifting [61]. Previous studies [62] identified crustal thinning beneath the Falcón Basin along the western extension of the Oca–Ancón fault system, which was interpreted as a back-arc basin, as well as suture zones between the Proterozoic and Paleozoic provinces (Ouachita–Marathon-related suture) and Paleozoic and Meso-Cenozoic terranes (peri-Caribbean suture), where the seismic velocity changes.

3. Methodology

3.1. Data

Topographic, geophysical, seismic, and geological data were taken from open sources in order to present a multi-disciplinary cartographic analysis of Venezuela. The topographic data were collected from the open-access repository of the General Bathymetric Chart of the Oceans (GEBCO) in NetCDF, and they were collected in raster format with grid cells [63]. The GEBCO presents the most comprehensive data on the topography and bathymetry of the Earth with a 15 arc second resolution [64]. GEBCO data are widely used in the geosciences due to their robustness, i.e., their high resolution, precision, and reliability as a cartographic data source [65,66].

Geological data (Figures 2 and 3) were collected from the United States Geological Survey (USGS). The geological maps were processed as .shp vector layers by using the open-source QGIS software due to its compatibility with the ArcGIS data format. The visualization of these maps followed the traditional mapping process of QGIS [67]. In the first stage, vector layers of the geological provinces and outcrops were visualized within a graphical user interface (GUI). In the second stage, the study area was designated by using an overlay of the Digital Chart of the World (DCW), which is a comprehensive 1:1,000,000 scale vector base map uploaded into the QGIS project.

The geoid map was visualized based on data collected from EGM2008 [68]. The free-air gravity maps and vertical gradient were based on open-source datasets available from the public repositories of the Scripps Institution of Oceanography and described in [69]. The seismic dataset used to plot maps of earthquakes was obtained from the Incorporated Research Institutions for Seismology (IRIS) database by using the spatial extent of Venezuela and the neighboring areas of Trinidad, Tobago, and the surroundings. This dataset covered the earthquake events in the region of Venezuela for the period from 1997 to 2021 and included earthquakes with magnitudes from 1.9 to 7.3 M_L on the Richter scale.

3.2. Methods

Maps showing the topographic, seismic, and geophysical setting of Venezuela (Figures 1 and 4–7) were made by using the scripting techniques of GMT [17] by following the existing methodology described by [70,71]. The methodological workflow included five algorithms in GMT codes, which are presented in Appendix A of this article. Prior to data visualization and mapping, the scripts were written using GMT syntax and organized in the Xcode environment. As mentioned in the previous subsection, we processed the high-resolution data by using GMT. Conventionally, the language of GMT was used in several modules for computing and processing raster data in order to plot cartographic elements. The area of interest was first selected and cut from the GEBCO grid by using the following snippet of code: *gmtgrdcutGEBCO_2019.nc − R286/300.5/0/12.5 − Gve_relief.nc.*

After the first step, all other substantial cartographic elements were added into a script by using the following GMT modules: psbasemap, grdimage, psscale, pstext, psclip, and grdcontour. Specifically, these included images that were visualized in selected color palettes, grids, clipped insertions of regions into a global map, coastlines, borders, rivers, bar legends on the maps, scale bars, text annotations, time stamps, etc. The technique of scripting was applied by using existing detailed descriptions [72]. For example, the 'grdcontour' module was used to compute, model, and visualize topographic isolines by using quantitative DEM data as follows: '$gmtgrdcontourve\_relief1.nc - R - J - C1000 - Wthinner, darkbrown - O - K >> \$ps$'. The purpose of the geoid visualization (Figure 4) in relation to the geophysical differentiation was to investigate how the gravity values responded to the effects of topographic elevations and bathymetric depressions.

While GEBCO data were gathered in NetCDF format ($GEBCO\_2019.nc$), the geoid data first needed to be converted from the .adf format. This was done with the following code of GMT: '$gmtgrdconvertn00w90/w001001.adf geoid\_02.grd$'. The extent and range of the data were examined through the Geospatial Data Abstraction Library (GDAL) and interpreted for the color palette as follows: '$gmtmakecpt - C33_blue\_red.cpt - T - 55/25 > colors.cpt$'. The geoid was modeled by using the following code: '$gmtgrdimagegeoid\_02.grd - Ccolors.cpt - R286/300.5/0/12.5 - JM6.5i - P - Xc - I + a15 + ne0.75 - K > \$ps$'. The visualized data are shown in Figure 4.

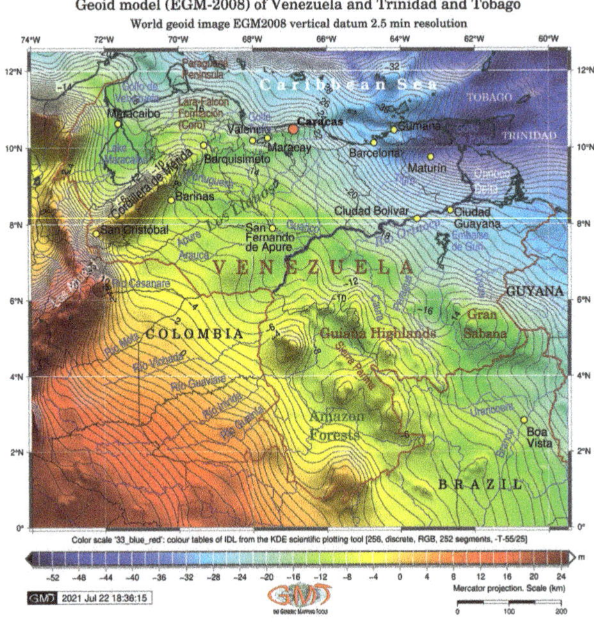

Figure 4. Geoid model of the region of Venezuela. Mapping: GMT. Source: authors.

The gravity grids and the geophysical data (Figures 5 and 6) were also converted from the initial raw IMG format into the GRD format, which is compatible with GMT. The code was executed as follows: '$gmtimg2grdgrav\_27.1.img - R286/300.5/0/12.5 - Ggrav\_VE.grd - T1 - I1 - E - S0.1 - V$' (for Figure 5). The same procedure was repeated for Figure 6 (vertical gradient of gravity) by following the examples described in existing work [73]. Annotations were added to all of the maps by using the 'pstext' module of GMT, as in the following example: '$gmtpstext - R - J - N - O - K - F + f10p, 0, black + jLB + a - 0 >> \$ps << EOF294.208.24CiudadBolvarEOF$'. The data were converted into the 'ngdc' format, visualized with the 'psxy' module, and mapped, as shown in Figure 5.

The earthquakes in Figure 7 (map of seismicity) were plotted by using the 'psxy' module of GMT with the following code: '$gmtpsxy - R - Jquakes\_VE.ngdc - Wfaint - i4, 3, 6, 6s0.05 - h3 - Scc - Csteps.cpt - O - K >> \$ps$'. Here, 'i4,3,6,6s0.05' indicates the numbers of the columns in the table ($quakes\_VE.ngdc$) that were converted from the original CSV from IRIS. The volcanoes were visualized with the following code: '$gmtpsxy - R - Jvolcanoes.gmt - St0.4c - Gred - Wthinnest - O - K >> \$ps$'. The complex legend showing the magnitude of events was plotted by using the following code: '$gmtpslegend -$

R − J − Dx1.5/ − 3.0 + w17.8c + o − 2.0/0.1c − F+pthin+ithinner+gwhite -O -K « FIN » $ps H 10 Helvetica Seismicity: earthquakes magnitude (M) from 1.9 to 7.3 N 9 S 0.3c c 0.3c red 0.01c 0.5c M (7.1-7.3) <...> FIN'.

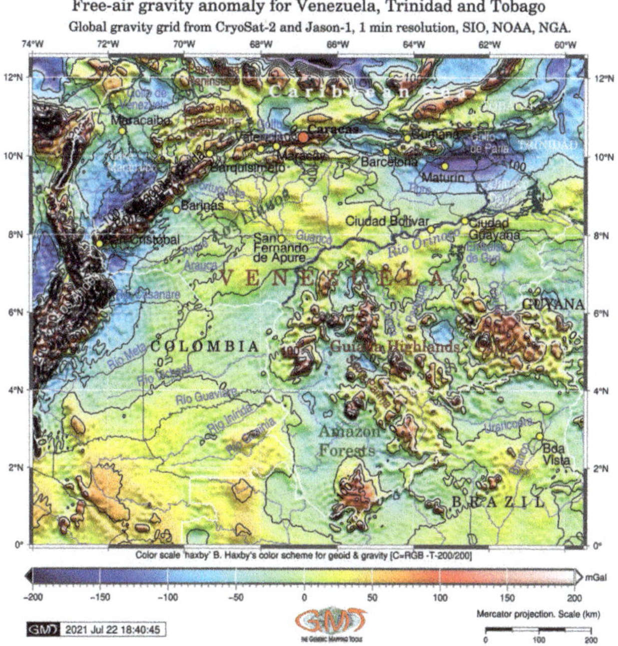

Figure 5. Free-air gravity model of the region of Venezuela. Mapping: GMT. Source: authors.

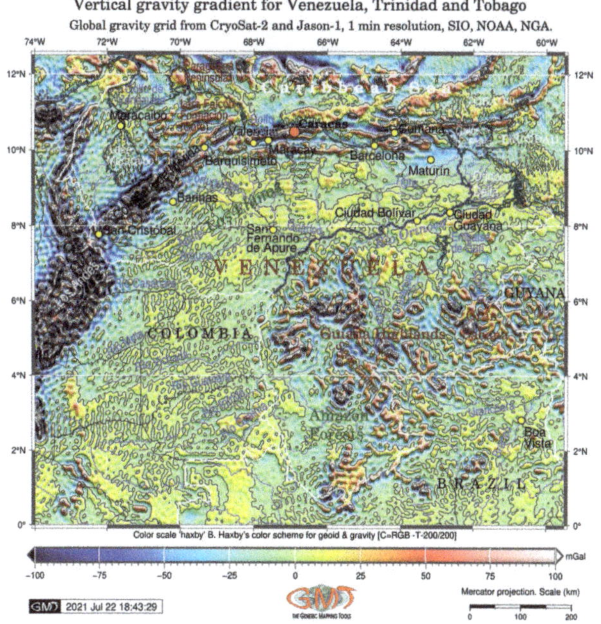

Figure 6. Vertical gradient of gravity in the region of Venezuela. Mapping: GMT. Source: authors.

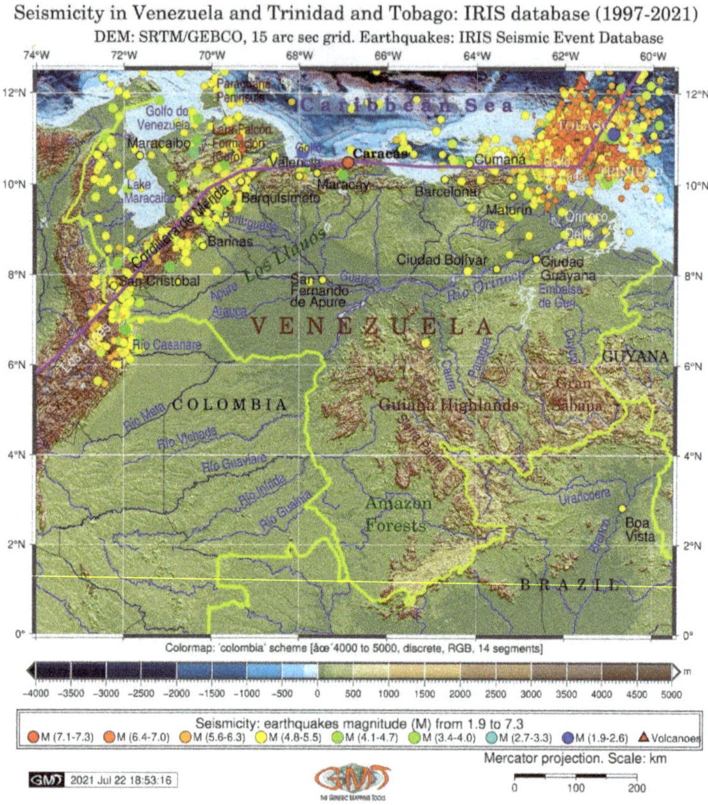

Figure 7. Seismic map of earthquakes in the region of Venezuela. Mapping: GMT. Source: authors.

In practice, we fused the data from seismic events to the topographic grid to show the locations of the events on the map with their attributes, which showed magnitudes ranging from 1.9 to 7.3. In our maps, we demonstrated that the density of the earthquake events visibly increased along the lines of the boundaries of lithospheric plates (thick purple line on the map in Figure 7), with a maximal concentration of events in the Andes (Cordillera de Merida) and the region of Trinidad and Tobago.

The full algorithms of the GMT codes are available in the authors' GitHub repository: https://github.com/paulinelemenkova/Mapping_Venezuela_GMT_Scripts, accessed on 1 November 2022.

4. Results

Figure 1 was plotted by using the GEBCO raster grid, which was selected due to its quality and reliability. A precise topographic map is essential to a fundamental understanding of the geophysics and tectonics of a region. Topographic maps enable the analysis of earthquake locations and the evaluation of the associated effects on ocean and terrestrial geomorphology. Therefore, the analysis of topography is necessary for the mapping of seismicity, risk assessment, and tsunami wave propagation. Here, the topographic and bathymetric elevations of the study area encompassing Venezuela and the islands of Trinidad and Tobago ranged from −4948 to 5536 m according to the GEBCO grid.

Figures 2 and 3 indicate the spatial extent of the geological provinces and outcrops in Venezuela. The map in Figure 2 shows that most of the territory of Venezuela is covered by the Guyana Shield (aquamarine color) in the southeastern parts of the country. The Barinas–Apure sedimentary basin is located in Llanos and the region of the Andean foothills in western Venezuela [74]. The Barinas–Apure basin (yellow color on the map) includes oil fields, which comprise the third most important area of production in Venezuela after the Maracaibo and Eastern Venezuela basins according to the petroleum accumulation. The Falcón sedimentary basin, which produces indigenous Miocene oil from structural and stratigraphic accumulations [75], is colored in slate blue on the map.

The South Caribbean deformed belt (purple color in Figure 2) represents a submarine prism formed between the subducting tectonic plates in the Colombian and Venezuelan basins and the arc terranes along the northern rim of South America [76]. Its southwestern part, the Sinu fold belt, forms an accretionary wedge in the area of subduction of the Caribbean Plate under the South American Plate [77]. Other important geological provinces include the Guyana Basin, which stretches along the passive margin of northeastern South America, and the Maracaibo Basin (green, Figure 2). The Cariaco Basin (beige color in Figure 2) is an East–West-stretching basin [78] situated in the Gulf of Cariaco, which is located on the continental shelf of the Caribbean Sea, off the eastern coast of the surroundings of Barcelona, Figure 1).

The Tobago Trough (chartreuse green color in Figure 2) is a modern marine forearc basin of the Lesser Antilles arc and is located above the sedimentary succession. It is at least 10 km thick and is a probable oceanic basement that was formed by sediments from various stratigraphic sequences—Quaternary, early Pleistocene–late Miocene, Miocene, and late Eocene. In addition, it may also include mid-Cretaceous or older intrusive suites [79]. The distribution of the Quaternary sediments covering most of the northwest of the country (dark pink color in Figure 3) corresponds with previous studies on Plio-Quaternary extension in the Venezuelan Andes. Thus, this shows the extensional structures corresponding to elongated tilted blocks, with the geometry and kinematics of the structures corresponding to the earlier syn-orogenic extension [80].

Figure 4 shows the variations in the geoid heights over the study area. The highest values for the geoid (up to 25 m, bright red colors) are visible in the southwestern region of the map at the border with Colombia. However, the majority of the territory of Venezuela covers the extent from −50 to 8 m, with the highest values being in the mountains, showing a clearly visible correlation with topographic elevations. The undulations of the geoid in the Guyana Highlands correspond to the higher topographic elevations on the shield, with table-like 'tepuis' mountains. In general, the marine areas of the Caribbean Sea have lower geoid heights (blue-colored areas in Figure 4), which correspond well with the bathymetric depressions reflected in the geophysical setting of the Earth.

The comparison of the geoid mode in Figure 4 with the topographic and geological maps in Figures 1 and 2 shows the elevated values of the geoid that are notable in the region of the Guyana Shield. The Guyana Shield is a Precambrian geological formation on the NE coast of South America and is an area of special importance in the historical geology of Venezuela due to the high potential for mineral exploration [81]. Aside from the geological features, the environmental importance of this unique region consists in the highest biodiversity in the world, as it includes the numerous endemic species [82] of the tropical forests of Amazonia [83].

Figure 5 shows the variations in the gravity over the study area. The notable minimum of the free-air gravity coincides with distribution of the Guyana Basin, the southeastern part of Venezuela, and Trinidad and Tobago. In particular, this points to the tectonic basement depression formed during the Jurassic rifting, as well as Lake Maracaibo, an important geological region of Venezuela. The northeastern coasts of Lake Maracaibo are occupied by the Bolivar Coastal Field, the largest oil field in South America. With a basin area of 50,000 km^2 and oil production of 30×109 bbl , it is the second most prolific hydrocarbon basin in the world after the Middle Eastern hydrocarbon basins [84].

Maracaibo Lake has a wide variety of oil reservoir rocks with structural traps in the tectonic units, e.g., normal or inverted faults on the South American plate [85]. The oil formed in the Maracaibo Basin is mostly heavy black oil with asphaltene deposition [86,87]. The gravity in Lake Maracaibo reaches below −200 mGal in correlation with the bathymetric depression (Figure 1) and the geological extent of the Maracaibo Basin (Figure 2). The highest values for gravity reach +200 mGal and correspond well with the topographic elevation of the Cordillera de Mérida, Columbian Andes, Guiana Highlands, and Sierra Parima. The Amazon and Orinoco Flow Basins mostly have gentle fluctuations in gravity, with values in the range between −25 and 30 mGal (Figure 5).

Figure 6 refers to a vertical gradient of normal gravity as an approximation in which a gravity measurement point, which is almost never placed on the reference ellipsoid's surface, is adjusted according to geophysical corrections [88,89]. The details regarding the distribution of vertical gravity values over Venezuela, Trinidad, and Tobago showed that the lowest values (below −100 mGal) mostly correspond to the extreme amplitude of topographic changes, e.g., the abrupt borders between the mountains/hills and the margins of the bathymetric depression, while the highest values (over 100 mGal) were detected on the tops of the mountains, which corresponded well with the local geomorphology of the region.

The distribution of the earthquakes and the seismicity of Venezuela, Trinidad, and Tobago (Figure 7) depend on the proximity of the area to a border of lithospheric plates with active margins, such as the South American and Caribbean plates, which are depicted in the map with purple thick lines. The second factor includes active geological processes (faults) and the geomorphology. Hence, the Cordillera de Mérida had the majority of the detected earthquakes after the active zone of Trinidad and Tobago. Most of the earthquakes mapped in Venezuela belonged to the 'shallow' category, that is, they had a depth between 0 and 70 km. Earthquakes at depths from 74 to 155 km were located offshore of Sucre, Venezuela, and those that were detected in Trinidad and Tobago were predominantly shallow, not exceeding 70 km. The deepest earthquakes in the inspected database were located in northern Columbia (182–200 km).

5. Conclusions

In this paper, we proposed two algorithms for the geophysical mapping of Venezuela, which included GMT-based scripts and traditional plotting by using QGIS, with the aim of analyzing the geohazard risk of Venezuela. Investigations and risk assessments of geohazards require advanced methods of mapping that are based on the programming of algorithms, as demonstrated in this paper. The operative use of correctly and effectively plotted maps, along with tabular data, supports reasonable and substantiated recommendations aimed at the prevention and mitigation of earthquake hazard risks in seismically active regions, such as the South American Andes. The algorithms based on GMT enabled improved performance of the cartographic workflow in comparison with that of the conventional approaches.

We demonstrated the technical advantages of the approaches to programming in cartography, which included scripting, as this is a rapidly developing branch of geoinformation. Due to the high degree of automation, scripting is the best technical cartographic solution for seismic mapping, as it enables rapid machine-based processing of data, which is beneficial for prognosis, predictive mapping, risk assessment, and hazard visualization in a real-time regime. Automation enables the rapid processing of large datasets and the effective use of multi-source heterogeneous data. Aside from that, the known advantages of scripting in cartography include machine-based graphics, which result in an aesthetically refined plotting by GMT.

Scripting cartographic methods for GMT were addressed to evaluate the geophysical and geological correspondences in a seismically active region of Venezuela and the Caribbean Sea basin. We used high-resolution geophysical, topographic, and geological datasets, such as GEBCO, EGM-2008 for geoid, gravity grids, and seismic data obtained from IRIS. The results corresponded with and supported those of previous studies on the geological and geophysical methods of gravity measurements, seismic observations, and earthquake monitoring in Venezuela and the surroundings [90,91].

This study presented a machine-based mapping workflow that contributed to the regional studies of the eastern Caribbean and Venezuela. The GMT code snippets have been presented in a GitHub repository as a demonstration of the technical possibilities of this toolset. The results showed optimized and improved cartographic performance and demonstrated the advantages of the automating processing of geophysical datasets. Multi-source geophysical, geologic, seismic, and topographic data were used for a comparative analysis of the geophysical processes. The tectonic evolution of the Caribbean Sea basin largely affected its geological setting, which could be tracked in the relevant geophysical data, indicating a high seismicity and high earthquake risk in the Andes.

Since the eastern Caribbean is a tectonically complex area that includes a subduction zone, island arc, and active volcanoes [92], a regional analysis and visualization of the varied settings were performed. Specifically, high seismicity was detected in the northwestern region of Venezuela and the northeastern region of Trinidad and Tobago, which is consistent with the geological structure associated with the tectonic setting and the related movements of the subduction of the Caribbean Plate. Although continued research is needed in order to improve earthquake monitoring in the region of Venezuela, this study presents a contribution to the seismic study of the Caribbean and South America, as it shows the earthquakes that were detected according to the IRIS database based on a 25-year time span (1997–2021).

A better understanding of the geological and tectonic factors affecting seismicity and regulating the appearance, magnitude, and focal depth of earthquakes is critical for seismically active regions, such as the Caribbean and Venezuela. Practical applications of geological risk assessment in South America include the use of information by public administrations in city management and urban planning for environmental management and the mitigation and management of hazard risk in regions prone to geological hazards, such as Venezuela.

The main contributions that we have made and the contemporaneity of the presented study are highlighted as follows:

1. Region: The region of Venezuela is at high risk of seismicity and earthquakes because it is located in the zone of the collision of the South American and Caribbean tectonic plates and the Andean orogeny. This requires spatial analysis of the regions that are at risk.
2. Geohazard: The risk assessment of geophysical hazards was based on the effective integration and visualization of multi-source data, which provided detailed insights into the environmental and geological settings of Venezuela and supported complex investigations in seismically active regions of South America.
3. Data: Geological disasters are related to high seismicity, exposure to hazards, and vulnerability of people at risk. This requires complex and detailed studies that summarize these factors and visualize regional geophysical settings, as shown in this paper.
4. Methods: The combination of scripting for GMT and GIS is a solid foundation for cartographic data analysis. An efficient method for processing multi-format data ensures hazard mapping and geophysical and geological visualization with the aim of preventing and evaluating seismic hazards.

In this paper, we approached the problem of automated mapping in the geophysical sciences and the estimation of the level of seismicity in the region of Venezuela for risk assessment studies as a contribution from the new perspective of data analysis. The scripting approach was mainly designed for the integration of data and the presented maps for studies of risk assessment, though it is also valuable as a general-purpose mapping technique. The limitations of GMT may be constrained in specific tasks, such as the analysis of remote sensing data or image processing, where general-purpose languages, such as Python and R, perform better.

The methodology that was presented and explained here for plotting maps with codes from a console can be potentially extended to other regions of the world, since the approach to data analysis remains identical. It is possible to reuse the presented codes for other seismically active regions of the world. The only modifications of the method would include the spatial extent of the study area, that is, the cartographic coordinates for plotting the maps and for downloading the data from the IRIS catalogue.

The implementation of the GMT scripts for other regions implies that one only needs to access datasets for a given region and to adjust the codes to this regional extent in order to map the data fairly well. We consider this cartographic scripting approach as a promising extension of our method for similar future work on seismicity and risk assessment. We demonstrated that the performance and visualization that were obtained are effective and that the level of workflow automation is high. The experimental results verify the usefulness of our approach for both of the following research objectives: the evaluation of the cartographic performance of GMT and the mapping of seismicity in the region of Venezuela and the northern Andes.

Author Contributions: Supervision, conceptualization, methodology, software, resources, funding acquisition, and project administration, O.D.; writing—original draft preparation, methodology, software, data curation, visualization, formal analysis, validation, writing—review and editing, and investigation, P.L. All authors have read and agreed to the published version of the manuscript.

Funding: This project was supported by the Federal Public Planning Service Science Policy or Belgian Science Policy Office, Federal Science Policy—BELSPO (B2/202/P2/SEISMOSTORM).

Institutional Review Board Statement: Not applicable.

Informed Consent Statement: Not applicable.

Data Availability Statement: The GitHub repository contains the GMT scripts used for the mapping in this study: https://github.com/paulinelemenkova/Mapping_Venezuela_GMT_Scripts, accessed on 1 November 2022.

Acknowledgments: The authors thank the three anonymous reviewers for their careful reading, critical suggestions, and useful comments that helped improve an earlier version of this manuscript.

Conflicts of Interest: The authors declare no conflict of interest.

Abbreviations

The following abbreviations are used in this manuscript:

CSV	Comma-separated values
DCW	Digital Chart of the World
EGM	Earth Gravitational Models
GEBCO	General Bathymetric Chart of the Oceans
GIS	Geographic Information System
GMT	Generic Mapping Tools
GUI	Graphical User Interface
QGIS	Quantum GIS
IRIS	Incorporated Research Institutions for Seismology
NetCDF	Network Common Data Form
NGDC	National Geophysical Data Center
USGS	United States Geological Survey

Appendix A. GMT Scripts

Appendix A.1. GMT Script for Topographic Mapping

Listing A1: GMT code used to plot Figure 1 (topographic map).

```sh
#!/bin/sh
# Purpose: shaded relief grid raster map from the GEBCO dataset (here: Venezuela)
# GMT modules: gmtset, gmtdefaults, grdcut, makecpt, grdimage, psscale, grdcontour,
    psbasemap, gmtlogo, psconvert
# http://soliton.vm.bytemark.co.uk/pub/cpt-city/arendal/tn/arctic.png.index.html
# GMT set up
gmt set FORMAT_GEO_MAP=dddF \
# Overwrite defaults of GMT
gmtdefaults -D > .gmtdefaults
#chsh -s /bin/bash
```

```
10 chsh -s /bin/zsh
11 gmt grdcut GEBCO_2019.nc -R286/300.5/0/12.5 -Gve_relief.nc
12 gmt grdcut ETOPO1_Ice_g_gmt4.grd -R286/300.5/0/12.5 -Gve_relief1.nc
13 gdalinfo ve_relief.nc -stats
14 # Minimum=-4947.963, Maximum=5535.625, Mean=162.220, StdDev=788.097
15 # create mask of vector layer from the DCW of country's polygon
16 gmt pscoast -R286/300.5/0/12.5 -Dh -M -EVE > ve.txt
17 # Make color palette
18 gmt makecpt -Carctic.cpt > pauline.cpt
19 # Generate a file
20 ps=Topography_VE.ps
21 # Make background transparent image
22 gmt grdimage ve_relief.nc -Cpauline.cpt -R286/300.5/0/12.5 -JM6i -P -I+a15+ne0.75 -t20 -
      Xc -K > $ps
23 # Add isolines
24 gmt grdcontour ve_relief1.nc -R -J -C500 -W0.1p -O -K >> $ps
25 # Add coastlines, borders, rivers
26 gmt pscoast -R -J -P \
27     -Ia/thinner,blue -Na -N1/thickest,darkred -W0.1p -Df -O -K >> $ps
28 # CLIPPING
29 # 1. Start: clip the map by mask to only include country
30 gmt psclip -R286/300.5/0/12.5 -JM6.0i ve.txt -O -K >> $ps
31 # 2. create map within mask
32 # Add raster image
33 gmt grdimage ve_relief.nc -Cpauline.cpt -R286/300.5/0/12.5 -JM6.0i -I+a15+ne0.75 -Xc -P
      -O -K >> $ps
34 # Add isolines
35 gmt grdcontour ve_relief1.nc -R -J -C1000 -Wthinner,darkbrown -O -K >> $ps
36 # Add coastlines, borders, rivers
37 gmt pscoast -R -J \
38     -Ia/thinner,blue -Na -N1/thicker,tomato -W0.1p -Df -O -K >> $ps
39 # 3: Undo the clipping
40 gmt psclip -C -O -K >> $ps
41 # Add color barlegend
42 gmt psscale -Dg286/-1.0+w15.2c/0.4c+h+o0.0/0i+ml -R -J -Cpauline.cpt \
43     -Bg500a1000f100+l"Color scale 'arctic': global bathymetry/topography relief [-5000
        to 4000, mixed, RGB, 110 segments]" \
44     -I0.2 -By+lm -O -K >> $ps
45 # Add grid
46 gmt psbasemap -R -J \
47     --MAP_FRAME_AXES=WEsN \
48     --FORMAT_GEO_MAP=ddd:mm:ssF \
49     -Bpx4f2a2 -Bpyg4f2a2 -Bsxg4 -Bsyg2 \
50     -B+t"Topographic map of Venezuela, Trinidad and Tobago" -O -K >> $ps
51 # Add scalebar, directional rose
52 gmt psbasemap -R -J \
53     -Lx12.7c/-2.2c+c50+w200k+l"Mercator projection. Scale: km"+f \
54     -UBL/-5p/-65p -O -K >> $ps
55 # Texts
56 gmt pstext -R -J -N -O -K \
57 -F+jTL+f10p,21,darkred+jLB+a-0 -Gwhite@60 >> $ps << EOF
58 290.1 11.8 Peninsula
59 EOF
60 # insert map
61 # Countries codes: ISO 3166-1 alpha-2. Continent codes AF (Africa), AN (Antarctica), AS
      (Asia), EU (Europe), OC (Oceania), NA (North America), or~SA (South America). -EEU+
      ggrey
62 gmt psbasemap -R -J -O -K -DjBL+w3.3c+stmp >> $ps
63 read x0 y0 w h < tmp
64 gmt pscoast --MAP_GRID_PEN_PRIMARY=thinner,white -Rg -JG293/6N/$w -Da -Gbrown -A5000 -Bg
      -Wfaint -ESA+gpeachpuff -EVE+gyellow -Slightsteelblue3 -O -K -X$x0 -Y$y0 >> $ps
65 #gmt pscoast -Rg -JG12/5N/$w -Da -Gbrown -A5000 -Bg -Wfaint -ECM+gbisque -O -K -X$x0 -
      Y$y0 >> $ps
66 gmt psxy -R -J -O -K -T   -X-${x0} -Y-${y0} >> $ps
67 # Add GMT logo
68 gmt logo -Dx6.2/-2.9+o0.1i/0.1i+w2c -O -K >> $ps
69 # Add subtitle
70 gmt pstext -R0/10/0/15 -JX10/10 -X0.5c -Y5.0c -N -O \
71     -F+f10p,Helvetica,black+jLB >> $ps << EOF
72 2.3 13.3 SRTM/GEBCO 15 arc sec resolution global terrain model grid
73 EOF
74 # Convert to image file using GhostScript
75 gmt psconvert Topography_VE.ps -A0.5c -E720 -Tj -Z
```

Appendix A.2. GMT Script for Geoid Mapping

Listing A2: GMT code used to plot Figure 4 (geoid modeling).

```
1 #!/bin/sh
2 # Purpose: geoid of Venezuela
```

```
 3  # GMT modules: gmtset, gmtdefaults, grdcut, makecpt, grdimage, psscale, grdcontour,
        psbasemap, gmtlogo, psconvert
 4  # http://soliton.vm.bytemark.co.uk/pub/cpt-city/kst/tn/33_blue_red.png.index.html
 5  # GMT set up
 6  gmt set FORMAT_GEO_MAP=dddF \
 7  # Overwrite defaults of GMT
 8  gmtdefaults -D > .gmtdefaults
 9  gmt grdconvert n00w90/w001001.adf geoid_02.grd
10  gdalinfo geoid_02.grd -stats
11  # Minimum=-70.856, Maximum=32.958, Mean=-24.768, StdDev=19.081
12  # Generate a color palette table from grid
13  gmt makecpt -C33_blue_red.cpt -T-55/25 > colors.cpt
14  # Generate a file
15  ps=Geoid_VE.ps
16  gmt grdimage geoid_02.grd -Ccolors.cpt -R286/300.5/0/12.5 -JM6.5i -P -Xc -I+a15+ne0.75 -
        K > $ps
17  # Add shorelines
18  gmt grdcontour geoid_02.grd -R -J -C1.0 -A2.0+f8p,0,black -Wthinner,dimgray -O -K >> $ps
19  # Add grid
20  gmt psbasemap -R -J \
21      --MAP_FRAME_AXES=WEsN \
22      --FORMAT_GEO_MAP=ddd:mm:ssF \
23      -Bpx4f2a2 -Bpyg4f2a2 -Bsxg4 -Bsyg2 \
24      --MAP_TITLE_OFFSET=0.8c \
25      --FONT_ANNOT_PRIMARY=7p,0,black \
26      --FONT_LABEL=7p,25,black \
27      --FONT_TITLE=13p,25,black \
28      -B+t"Geoid model (EGM-2008) of Venezuela and Trinidad and Tobago" -O -K >> $ps
29  # Add legend
30  gmt psscale -Dg286/-1.0+w16.5c/0.4c+h+o0.0/0i+ml+e -R -J -Ccolors.cpt \
31      -Bg2f0.2a4+l"Color scale '33_blue_red': colour tables of IDL from the KDE scientific
        plotting tool [256, discrete, RGB, 252 segments, -T-55/25]" \
32      -I0.2 -By+lm -O -K >> $ps
33  # Add scale, directional rose
34  gmt psbasemap -R -J \
35      --FONT=7p,0,black \
36      --FONT_ANNOT_PRIMARY=6p,0,black \
37      --MAP_TITLE_OFFSET=0.1c \
38      --MAP_ANNOT_OFFSET=0.1c \
39      -Lx14.7c/-2.3c+c50+w200k+l"Mercator projection. Scale (km)"+f \
40      -UBL/-5p/-65p -O -K >> $ps
41  # Add coastlines, borders, rivers
42  gmt pscoast -R -J -P -Ia/thinnest,blue -Na -N1/thick,brown -Wthick,darkslategray -Df -O
        -K >> $ps
43  # Texts
44  gmt pstext -R -J -N -O -K \
45      -F+jTL+f10p,21,darkred+jLB+a-0 >> $ps << EOF
46  290.0 10.5 (Coro)
47  EOF
48  # Add GMT logo
49  gmt logo -Dx7.0/-2.9+o0.1i/0.1i+w2c -O -K >> $ps
50  # Add subtitle
51  gmt pstext -R0/10/0/15 -JX10/10 -X0.1c -Y5.9c -N -O \
52      -F+f10p,25,black+jLB >> $ps << EOF
53  3.0 13.6 World geoid image EGM2008 vertical datum 2.5 min resolution
54  EOF
55  # Convert to image file using GhostScript
56  gmt psconvert Geoid_VE.ps -A1.0c -E720 -Tj -Z
```

Appendix A.3. GMT Script for Free-Air Gravity Mapping

Listing A3: GMT code used to plot Figure 5 (free-air gravity modeling).

```
 1  #!/bin/sh
 2  # Purpose: mapping free-air gravity anomaly of Venezuela
 3  # GMT modules: gmtset, gmtdefaults, img2grd, makecpt, grdimage, psscale, grdcontour,
        psbasemap, gmtlogo, psconvert, pscoast
 4  # http://soliton.vm.bytemark.co.uk/pub/cpt-city/pj/4/index.html
 5  # GMT set up
 6  gmt set FORMAT_GEO_MAP=dddF \
 7      MAP_FRAME_PEN=dimgray \
 8  # Overwrite defaults of GMT
 9  gmtdefaults -D > .gmtdefaults
10  gmt img2grd grav_27.1.img -R286/300.5/0/12.5 -Ggrav_VE.grd -T1 -I1 -E -S0.1 -V
11  gdalinfo grav_VE.grd -stats
12  # Minimum=-218.668, Maximum=942.457, Mean=5.694, StdDev=64.118
13  # Generate a color palette table from grid
14  gmt makecpt -Chaxby -T-200/200 > colors.cpt
15  # Generate a file
16  ps=Grav_VE.ps
```

```
17 gmt grdimage grav_VE.grd -Ccolors.cpt -R286/300.5/0/12.5 -JM6.0i -P -I+a15+ne0.75 -Xc -K
       > $ps
18 # Add isolines
19 gmt grdcontour grav_VE.grd -R -J -C50 -A100 -Wthinner -O -K >> $ps
20 # Add grid
21 gmt psbasemap -R -J \
22     --MAP_FRAME_AXES=WEsN \
23     --FORMAT_GEO_MAP=ddd:mm:ssF \
24     -Bpx4f2a2 -Bpyg4f2a2 -Bsxg4 -Bsyg2 \
25     -B+t"Free-air gravity anomaly for Venezuela, Trinidad and Tobago" -O -K >> $ps
26 # Add legend
27 gmt psscale -Dg286/-1.0+w15.0c/0.4c+h+o0.0/0i+ml+e -R -J -Ccolors.cpt \
28     --FONT_LABEL=7p,Helvetica,black \
29     --FONT_ANNOT_PRIMARY=7p,0,black \
30     --FONT_TITLE=8p,25,black \
31     -Bg50f5a50+l"Color scale 'haxby' B. Haxby's color scheme for geoid & gravity [C=RGB
       -T-200/200]" \
32     -I0.2 -By+l"mGal" -O -K >> $ps
33 # Add scale, directional rose
34 gmt psbasemap -R -J \
35     -Lx14.0c/-2.3c+c50+w200k+l"Mercator projection. Scale (km)"+f \
36     -UBL/-5p/-70p -O -K >> $ps
37 # Add coastlines, borders, rivers
38 gmt pscoast -R -J -P -Ia/thinnest,blue -Na -N1/thick,white -Wthin,darkslategray -Df -O -
       K >> $ps
39 # Texts
40 gmt pstext -R -J -N -O -K \
41 -F+jTL+f14p,32,darkgreen+jLB+a-330 >> $ps << EOF
42 290.8 7.8 Los Llanos
43 EOF
44 # Add GMT logo
45 gmt logo -Dx7.0/-2.9+o0.1i/0.1i+w2c -O -K >> $ps
46 # Add subtitle
47 gmt pstext -R0/10/0/15 -JX10/10 -X0.0c -Y4.8c -N -O \
48     -F+f10p,25,black+jLB >> $ps << EOF
49 1.0 13.6 Global gravity grid from CryoSat-2 and Jason-1, 1 min resolution, SIO, NOAA,
       NGA.
50 EOF
51 # Convert to image file using GhostScript
52 gmt psconvert Grav_VE.ps -A0.5c -E720 -Tj -Z
```

Appendix A.4. GMT Script for Vertical Gradient of Gravity

Listing A4: GMT code used to plot Figure 5 (free-air gravity modeling).

```
 1 #!/bin/sh
 2 # Purpose: geophysical mapping of Venezuela (vertical free-air gravity gradient)
 3 # GMT modules: gmtset, gmtdefaults, img2grd, makecpt, grdimage, psscale, grdcontour,
       psbasemap, gmtlogo, psconvert, pscoast
 4 # http://soliton.vm.bytemark.co.uk/pub/cpt-city/pj/4/index.html
 5 # GMT set up
 6 gmt set FORMAT_GEO_MAP=dddF \
 7 # Overwrite defaults of GMT
 8 gmtdefaults -D > .gmtdefaults
 9 gmt img2grd curv_27.1.img -R286/300.5/0/12.5 -Ggrav_v_VE.grd -T1 -I1 -E -S0.1 -V
10 gdalinfo grav_v_VE.grd -stats
11 # Minimum=-488.036, Maximum=870.350, Mean=0.358, StdDev=43.643
12 # Generate a color palette table from grid
13 gmt makecpt -Chaxby -T-100/100 > colors.cpt
14 # Generate a file
15 ps=Grav_VE_v.ps
16 gmt grdimage grav_v_VE.grd -Ccolors.cpt -R286/300.5/0/12.5 -JM6.0i -P -I+a15+ne0.75 -Xc
       -K > $ps
17 # Add isolines
18 gmt grdcontour grav_v_VE.grd -R -J -C100 -Wthinnest -O -K >> $ps
19 # Add grid
20 gmt psbasemap -R -J \
21     --MAP_FRAME_AXES=WEsN \
22     --FORMAT_GEO_MAP=ddd:mm:ssF \
23     -Bpx4f2a2 -Bpyg4f2a2 -Bsxg4 -Bsyg2 \
24     -B+t"Vertical gravity gradient for Venezuela, Trinidad and Tobago" -O -K >> $ps
25 # Add legend
26 gmt psscale -Dg286/-1.0+w15.0c/0.4c+h+o0.0/0i+ml+e -R -J -Ccolors.cpt \
27     -Bg25f2.5a25+l"Color scale 'haxby' B. Haxby's color scheme for geoid & gravity [C=
       RGB -T-200/200]" \
28     -I0.2 -By+l"mGal" -O -K >> $ps
29 # Add scale, directional rose
30 gmt psbasemap -R -J \
31     -Lx14.0c/-2.3c+c50+w200k+l"Mercator projection. Scale (km)"+f \
32     -UBL/-5p/-70p -O -K >> $ps
```

```
33  # Add coastlines, borders, rivers
34  gmt pscoast -R -J -P -Ia/thinnest,blue -Na -N1/thick,white -Wthin,darkslategray -Df -O -
        K >> $ps
35  # Texts
36  gmt pstext -R -J -N -O -K \
37  -F+jTL+f12p,21,white+jLB+a-315 >> $ps << EOF
38  286.7 5.8 Los Andes
39  EOF
40  # Add GMT logo
41  gmt logo -Dx7.0/-2.9+o0.1i/0.1i+w2c -O -K >> $ps
42  # Add subtitle
43  gmt pstext -R0/10/0/15 -JX10/10 -X0.0c -Y4.8c -N -O \
44      -F+f10p,25,black+jLB >> $ps << EOF
45  1.0 13.6 Global gravity grid from CryoSat-2 and Jason-1, 1 min resolution, SIO, NOAA,
        NGA.
46  EOF
47  # Convert to image file using GhostScript
48  gmt psconvert Grav_VE_v.ps -A0.5c -E720 -Tj -Z
```

Appendix A.5. GMT Script for Seismic Mapping

Listing A5: GMT codes used to plot Figure 7 (seismic mapping).

```
1   #!/bin/sh
2   # Purpose: seismicity map of Venezuela (earthquake distribution according to the IRIS
        database (1997-2021) of seismic events)
3   # GMT modules: gmtset, gmtdefaults, grdcut, makecpt, grdimage, psscale, grdcontour,
        psbasemap, gmtlogo, psconvert
4   # http://soliton.vm.bytemark.co.uk/pub/cpt-city/wkp/shadowxfox/tn/colombia.png.index.
        html
5   # GMT set up
6   gmt set FORMAT_GEO_MAP=dddF
7   gmtdefaults -D > .gmtdefaults
8   # Extract a topographic subset of Venezuela and surroundings:
9   gmt grdcut ETOPO1_Ice_g_gmt4.grd -R286/300.5/0/12.5 -Gve_relief1.nc
10  gmt grdcut GEBCO_2019.nc -R286/300.5/0/12.5 -Gve_relief.nc
11  gdalinfo -stats ve_relief.nc
12  # Min=-4947.963 Max=5535.625
13  # Make color palette
14  gmt makecpt -Ccolombia.cpt > pauline.cpt
15  gmt makecpt -Cseis -T1.9/7.3/0.5 -Z > steps.cpt
16  ps=Seis_VE.ps
17  # Make raster image
18  gmt grdimage ve_relief.nc -Cpauline.cpt -R286/300.5/0/12.5 -JM6.5i -I+a15+ne0.75 -Xc -P
        -K > $ps
19  # Add legend
20  gmt psscale -Dg286/-1.0+w16.5c/0.4c+h+o0.0/0i+ml+e -R -J -Cpauline.cpt \
21      -Bg500f50a500+l"Colormap..." \
22      -I0.2 -By+lm -O -K >> $ps
23  # Add isolines
24  gmt grdcontour ve_relief1.nc -R -J -C500 -Wthinnest,brown -O -K >> $ps
25  # Add coastlines, borders, rivers
26  gmt pscoast -R -J -P \
27      -Ia/thinner,blue -Na -N1/thickest,olivedrab1 -Wthin,lightcyan2 -Df -O -K >> $ps
28  # Add grid
29  gmt psbasemap -R -J \
30      --MAP_FRAME_AXES=WEsN \
31      --FORMAT_GEO_MAP=ddd:mm:ssF \
32      -Bpx4f2a2 -Bpyg4f2a2 -Bsxg4 -Bsyg2 \
33      -B+t"Seismicity in Venezuela and Trinidad and Tobago: IRIS database (1997-2021)" -O
        -K >> $ps
34  # Add scale, directional rose
35  gmt psbasemap -R -J \
36      -Lx14.0c/-3.6c+c50+w200k+l"Mercator projection. Scale: km"+f \
37      -UBL/-5p/-110p -O -K >> $ps
38  # Add earthquake points
39  # separator in numbers of table: dot (.), not comma ! (British style)
40  gmt psxy -R -J quakes_VE.ngdc -Wfaint -i4,3,6,6s0.05 -h3 -Scc -Csteps.cpt -O -K >> $ps
41  # Add geological lines and points
42  gmt psxy -R -J volcanoes.gmt -St0.4c -Gred -Wthinnest -O -K >> $ps
43  # fabric and magnetic lineation picks fracture zones
44  gmt psxy -R -J GSFML_SF_FZ_KM.gmt -Wthicker,goldenrod1 -O -K >> $ps
45  gmt psxy -R -J GSFML_SF_FZ_RM.gmt -Wthicker,pink -O -K >> $ps
46  gmt psxy -R -J ridge.gmt -Sf0.5c/0.15c+l+t -Wthick,red -Gyellow -O -K >> $ps
47  gmt psxy -R -J ridge.gmt -Sc0.05c -Gred -Wthickest,red -O -K >> $ps
48  # tectonic plates
49  gmt psxy -R -J TP_Caribbean.txt -L -Wthickest,purple -O -K >> $ps
50  gmt psxy -R -J TP_Cocos.txt -L -Wthickest,purple -O -K >> $ps
51  gmt psxy -R -J TP_Nazca.txt -L -Wthickest,purple -O -K >> $ps
52  gmt psxy -R -J TP_South_Am.txt -L -Wthickest,purple -O -K >> $ps
```

```
53  # Texts
54  gmt pstext -R -J -N -O -K \
55     -F+f10p,17,black+jLB+a-0 >> $ps << EOF
56  293.20 10.58 Caracas
57  EOF
58  # Processed likewise for similar texts
59  gmt pslegend -R -J -Dx1.5/-3.0+w17.8c+o-2.0/0.1c \
60       -F+pthin+ithinner+gwhite \
61       --FONT=8p,black -O -K << FIN >> $ps
62  H 10 Helvetica Seismicity: earthquakes magnitude (M) from 1.9 to 7.3
63  N 9
64  S 0.3c c 0.3c red 0.01c 0.5c M (7.1-7.3)
65  S 0.3c c 0.3c tomato 0.01c 0.5c M (6.4-7.0)
66  S 0.3c c 0.3c orange 0.01c 0.5c M (5.6-6.3)
67  S 0.3c c 0.3c yellow 0.01c 0.5c M (4.8-5.5)
68  S 0.3c c 0.3c chartreuse1 0.01c 0.5c M (4.1-4.7)
69  S 0.3c c 0.3c chartreuse1 0.01c 0.5c M (3.4-4.0)
70  S 0.3c c 0.3c cyan3 0.01c 0.5c M (2.7-3.3)
71  S 0.3c c 0.3c blue 0.01c 0.5c M (1.9-2.6)
72  S 0.3c t 0.3c red 0.03c 0.5c Volcanoes
73  FIN
74  # Add GMT logo
75  gmt logo -Dx7.0/-4.5+o0.1i/0.1i+w2c -O -K >> $ps
76  # Add subtitle
77  gmt pstext -R0/10/0/15 -JX10/10 -X0.5c -Y5.9c -N -O \
78       -F+f11p,25,black+jLB >> $ps << EOF
79  1.0 13.6 DEM: SRTM/GEBCO, 15 arc sec grid. Earthquakes: IRIS Seismic Event Database
80  EOF
81  # Convert to image file using GhostScript
82  gmt psconvert Seis_VE.ps -A1.7c -E720 -Tj -Z
```

References

1. Laffaille, J.; Ferrer, C.; Laffaille, K. Venezuela: The Construction of Vulnerability and Its Relation to the High Seismic Risk. *Dev. Earth Surf. Process.* **2009**, *13*, 99–114. [CrossRef]

2. Schmitz, M.; Hernández, J.J.; Rocabado, V.; Domínguez, J.; Morales, C.; Valleé, M.; García, K.; Sánchez-Rojas, J.; Singer, A.; Oropeza, J.; et al. The Caracas, Venezuela, Seismic Microzoning Project: Methodology, results, and implementation for seismic risk reduction. *Prog. Disaster Sci.* **2020**, *5*, 100060. [CrossRef]

3. Weber, J.C.; Saleh, J.; Balkaransingh, S.; Dixon, T.; Ambeh, W.; Leong, T.; Rodriguez, A.; Miller, K. Triangulation-to-GPS and GPS-to-GPS geodesy in Trinidad, West Indies: Neotectonics, seismic risk, and geologic implications. *Mar. Pet. Geol.* **2011**, *28*, 200–211. [CrossRef]

4. Khan, A.; Gupta, S.; Gupta, S.K. Multi-hazard disaster studies: Monitoring, detection, recovery, and management, based on emerging technologies and optimal techniques. *Int. J. Disaster Risk Reduct.* **2020**, *47*, 101642. [CrossRef]

5. Mesgar, M.A.A.; Jalilvand, P. Vulnerability Analysis of the Urban Environments to Different Seismic Scenarios: Residential Buildings and Associated Population Distribution Modelling through Integrating Dasymetric Mapping Method and GIS. *Procedia Eng.* **2017**, *198*, 454–466. . [CrossRef]

6. Nath, S.K. An initial model of seismic microzonation of Sikkim Himalaya through thematic mapping and GIS integration of geological and strong motion features. *J. Asian Earth Sci.* **2005**, *25*, 329–343. [CrossRef]

7. Kwag, S.; Choi, E.; Hahm, D.; Eem, S.; Ju, B.S. On improving seismic risk and cost for nuclear energy facility based on multi-objective optimization considering seismic correlation. *Energy Rep.* **2022**, *8*, 7230–7241. [CrossRef]

8. Kwag, S.; Choi, E.; Eem, S.; Ha, J.G.; Hahm, D. Toward improvement of sampling-based seismic probabilistic safety assessment method for nuclear facilities using composite distribution and adaptive discretization. *Reliab. Eng. Syst. Saf.* **2021**, *215*, 107809. [CrossRef]

9. Liu, H.; Cui, X.; Yuan, D.; Wang, Z.; Jin, J.; Wang, M. Study of Earthquake Disaster Population Risk Based on GIS A Case Study of Wenchuan Earthquake Region. *Procedia Environ. Sci.* **2011**, *11*, 1084–1091. 2011 2nd International Conference on Challenges in Environmental Science and Computer Engineering (CESCE 2011). [CrossRef]

10. Zheng, X.; Feng, C.; Ishiwatari, M. Examining the Indirect Death Surveillance System of The Great East Japan Earthquake and Tsunami. *Int. J. Environ. Res. Public Health* **2022**, *19*, 12351. [CrossRef]

11. Peng, C.; Jiang, P.; Ma, Q.; Su, J.; Cai, Y.; Zheng, Y. Chinese Nationwide Earthquake Early Warning System and Its Performance in the 2022 Lushan M6.1 Earthquake. *Remote Sens.* **2022**, *14*, 4269. [CrossRef]

12. Adu-Gyamfi, B.; Shaw, R. Risk Awareness and Impediments to Disaster Preparedness of Foreign Residents in the Tokyo Metropolitan Area, Japan. *Int. J. Environ. Res. Public Health* **2022**, *19*, 11469. [CrossRef] [PubMed]

13. Yousuf, M.; Bukhari, S.K.; Bhat, G.R.; Ali, A. Understanding and managing earthquake hazard visa viz disaster mitigation strategies in Kashmir valley, NW Himalaya. *Prog. Disaster Sci.* **2020**, *5*, 100064. [CrossRef]

14. Kamata, H.; Seto, S.; Suppasri, A.; Sasaki, H.; Egawa, S.; Imamura, F. A study on hypothermia and associated countermeasures in tsunami disasters: A case study of Miyagi Prefecture during the 2011 great East Japan earthquake. *Int. J. Disaster Risk Reduct.* **2022**, *81*, 103253. [CrossRef]

15. Shao, Z.; Wu, Y.; Ji, L.; Diao, F.; Shi, F.; Li, Y.; Long, F.; Zhang, H.; Wang, W.; Wei, W.; et al. Assessment of strong earthquake risk in the Chinese mainland from 2021 to 2030. *Earthq. Res. Adv.* **2022**, *2*, 100177. [CrossRef]

16. Yaghmaei-Sabegh, S.; Pavel, F.; Shahvar, M.; Qadri, S.T. Empirical frequency content models based on intermediate-depth earthquake ground-motions. *Soil Dyn. Earthq. Eng.* **2022**, *155*, 107173. [CrossRef]

17. Wessel, P.; Luis, J.F.; Uieda, L.; Scharroo, R.; Wobbe, F.; Smith, W.H.F.; Tian, D. The Generic Mapping Tools version 6. *Geochem. Geophys. Geosyst.* **2019**, *20*, 5556–5564. [CrossRef]

18. Castillo, A.; López-Almansa, F.; Pujades, L.G. Seismic risk analysis of urban non-engineered buildings: application to an informal settlement in Mérida, Venezuela. *Nat. Hazards* **2011**, *59*, 891–916. [CrossRef]

19. Salcedo, D.A. Behavior of a landslide prior to inducing a viaduct failure, Caracas–La Guaira highway, Venezuela. *Eng. Geol.* **2009**, *109*, 16–30. [CrossRef]

20. Rodríguez, L.; Diederix, H.; Torres, E.; Audemard, F.; Hernández, C.; Singer, A.; Bohórquez, O.; Yepez, S. Identification of the seismogenic source of the 1875 Cucuta earthquake on the basis of a combination of neotectonic, paleoseismologic and historic seismicity studies. *J. S. Am. Earth Sci.* **2018**, *82*, 274–291. [CrossRef]

21. Pousse Beltran, L.; Pathier, E.; Jouanne, F.; Vassallo, R.; Reinoza, C.; Audemard, F.; Doin, M.P.; Volat, M. Spatial and temporal variations in creep rate along the El Pilar fault at the Caribbean-South American plate boundary (Venezuela), from InSAR. *J. Geophys. Res. Solid Earth* **2016**, *121*, 8276–8296. [CrossRef]

22. Lyu, H.M.; Shen, J.S.; Arulrajah, A. Assessment of Geohazards and Preventative Countermeasures Using AHP Incorporated with GIS in Lanzhou, China. *Sustainability* **2018**, *10*, 304. [CrossRef]

23. Karymbalis, E.; Andreou, M.; Batzakis, D.V.; Tsanakas, K.; Karalis, S. Integration of GIS-Based Multicriteria Decision Analysis and Analytic Hierarchy Process for Flood-Hazard Assessment in the Megalo Rema River Catchment (East Attica, Greece). *Sustainability* **2021**, *13*, 10232. [CrossRef]

24. Lemenkova, P. Handling Dataset with Geophysical and Geological Variables on the Bolivian Andes by the GMT Scripts. *Data* **2022**, *7*, 74. [CrossRef]

25. He, B.; Bai, M.; Shi, H.; Li, X.; Qi, Y.; Li, Y. Risk Assessment of Pipeline Engineering Geological Disaster Based on GIS and WOE-GA-BP Models. *Appl. Sci.* **2021**, *11*, 9919. [CrossRef]

26. Valkanou, K.; Karymbalis, E.; Papanastassiou, D.; Soldati, M.; Chalkias, C.; Gaki-Papanastassiou, K. Assessment of Neotectonic Landscape Deformation in Evia Island, Greece, Using GIS-Based Multi-Criteria Analysis. *ISPRS Int. J. Geo-Inf.* **2021**, *10*, 118. [CrossRef]

27. González-Longatt, F.; Medina, H.; González, J.S. Spatial interpolation and orographic correction to estimate wind energy resource in Venezuela. *Renew. Sustain. Energy Rev.* **2015**, *48*, 1–16. [CrossRef]

28. Hackley, P.; Urbani, F.; Karlsen, A.W.; Garrity, C.P. Geologic Shaded Relief Map of Venezuela. U.S. Geological Survey, Open File Report 2005–1038. 2006. Available online: https://pubs.usgs.gov/of/2005/1038/index.htm (accessed on 1 November 2022).

29. Schmitz, M.; Ramírez, K.; Mazuera, F.; Ávila, J.; Yegres, L.; Bezada, M.; Levander, A. Moho depth map of northern Venezuela based on wide-angle seismic studies. *J. S. Am. Earth Sci.* **2021**, *107*, 103088. [CrossRef]

30. Sun, C.G.; Chun, S.H.; Ha, T.G.; Chung, C.K.; Kim, D.S. Development and application of a GIS-based tool for earthquake-induced hazard prediction. *Comput. Geotech.* **2008**, *35*, 436–449. [CrossRef]

31. Escalona, A.; Mann, P. Tectonics, basin subsidence mechanisms, and paleogeography of the Caribbean-South American plate boundary zone. Thematic Set on: Tectonics, basinal framework, and petroleum systems of eastern Venezuela, the Leeward Antilles, Trinidad and Tobago, and offshore areas. *Mar. Pet. Geol.* **2011**, *28*, 8–39. [CrossRef]

32. Dhont, D.; Monod, B.; Hervouët, Y.; Backé, G.; Klarica, S.; Choy, J.E. 3D geological modeling of the Trujillo block: Insights for crustal escape models of the Venezuelan Andes. *J. Am. Earth Sci.* **2012**, *39*, 245–251. [CrossRef]

33. Masy, J.; Niu, F.; Levander, A.; Schmitz, M. Mantle flow beneath northwestern Venezuela: Seismic evidence for a deep origin of the Mérida Andes. *Earth Planet. Sci. Lett.* **2011**, *305*, 396–404. [CrossRef]

34. Pérez, O.J.; Wesnousky, S.G.; De La Rosa, R.; Márquez, J.; Uzcátegui, R.; Quintero, C.; Liberal, L.; Mora-Páez, H.; Szeliga, W. On the interaction of the North Andes plate with the Caribbean and South American plates in northwestern South America from GPS geodesy and seismic data. *Geophys. J. Int.* **2018**, *214*, 1986–2001. [CrossRef]

35. Suárez, G.; Nábělek, J. The 1967 Caracas Earthquake: Fault geometry, direction of rupture propagation and seismotectonic implications. *J. Geophys. Res.* **1990**, *95*, 17459–17474. [CrossRef]

36. Audemard, F.A. Néotectonique, Sismotectonique et aléa Sismique du nord-ouest du Vénézuela (Système de Failles d'Oca-Ancón). Ph.D. Thesis, University of Montpellier I, Montpellier, France, 1993; 369p.

37. Backé, G.; Dhont, D.; Hervouët, Y. Spatial and temporal relationships between compression, strike-slip and extension in the Central Venezuelan Andes: Clues for Plio-Quaternary tectonic escape. *Tectonophysics* **2006**, *425*, 25–53. [CrossRef]

38. Marshall, J.L.; Russo, R.M. Relocated aftershocks of the March 10, 1988 Trinidad earthquake: Normal faulting, slab detachment and extension at upper mantle depths. *Tectonophysics* **2005**, *398*, 101–114. [CrossRef]

39. Reinoza, C.; Jouanne, F.; Audemard, F.A.; Schmitz, M.; Beck, C. Geodetic exploration of strain along the El Pilar Fault in northeastern Venezuela. *J. Geophys. Res. Solid Earth* **2015**, *120*, 1993–2013. [CrossRef]

40. Harbitz, C.B.; Glimsdal, S.; Bazin, S.; Zamora, N.; Løvholt, F.; Bungum, H.; Smebye, H.; Gauer, P.; Kjekstad, O. Tsunami hazard in the Caribbean: Regional exposure derived from credible worst case scenarios. *Cont. Shelf Res.* **2012**, *38*, 1–23. [CrossRef]

41. Audemard, F.A. Revised seismic history of the El Pilar fault, Northeastern Venezuela, from the Cariaco 1997 earthquake and recent preliminary paleoseismic results. *J. Seismol.* **2007**, *11*, 311–326. [CrossRef]

42. Pérez, O.J.; Sanz, C.; Lagos, G. Microseismicity, tectonics and seismic potential in southern Caribbean and northern Venezuela. *J. Seismol.* **1997**, *1*, 15–28. [CrossRef]

43. Barr, K.G.; Robson, G.R. Seismic Delays in the Eastern Caribbean. *Geophys. J. Int.* **1963**, *7*, 342–349. [CrossRef]

44. Hayes, G.P.; Mcnamara, D.E.; Seidman, L.; Roger, J. Quantifying potential earthquake and tsunami hazard in the Lesser Antilles subduction zone of the Caribbean region. *Geophys. J. Int.* **2014**, *196*, 510–521. [CrossRef]

45. Audemard, F.A. Surface rupture of the Cariaco July 09, 1997 earthquake on the El Pilar fault, northeastern Venezuela. *Tectonophysics* **2006**, *424*, 19–39. [CrossRef]

46. Baumbach, M.; Grosser, H.; Romero Torres, G.; Rojas Gonzales, J.L.; Sobiesiak, M.; Welle, W. Aftershock pattern of the July 9, 1997 Mw=6.9 Cariaco earthquake in Northeastern Venezuela. *Tectonophysics* **2004**, *379*, 1–23. [CrossRef]

47. Colón, S.; Audemard, F.A.; Beck, C.; Avila, J.; Padrón, C.; De Batist, M.; Paolini, M.; Leal, A.F.; Van Welden, A. The 1900 Mw 7.6 earthquake offshore north–central Venezuela: Is La Tortuga or San Sebastián the source fault? *Mar. Pet. Geol.* **2015**, *67*, 498–511. [CrossRef]

48. Pousse-Beltran, L.; Vassallo, R.; Audemard, F.; Jouanne, F.; Oropeza, J.; Garambois, S.; Aray, J. Earthquake geology of the last millennium along the Boconó Fault, Venezuela. *Tectonophysics* **2018**, *747–748*, 40–53. [CrossRef]

49. Rial, J.A. The Caracas, Venezuela, earthquake of July 1967: A multiple-source event. *J. Geophys. Res. Solid Earth* **1978**, *83*, 5405–5414. [CrossRef]

50. Yamazaki, Y.; Audemard, F.; Altez, R.; Hernández, J.; Orihuela, N.; Safina, S.; Schmitz, M.; Tanaka, I.; Kagawa, H. Estimation of the seismic intensity in Caracas during the 1812 earthquake using seismic microzonation methodology. *Revista Geográfica Venezolana* **2005**, *46*, 199–216.

51. Deville, E.; Guerlais, S.H. Cyclic activity of mud volcanoes: Evidences from Trinidad (SE Caribbean). *Mar. Pet. Geol.* **2009**, *26*, 1681–1691. [CrossRef]

52. Latchman, J.L.; Ambeh, W.B.; Lynch, L.L. Attenuation of seismic waves in the Trinidad and Tobago area. *Tectonophysics* **1996**, *253*, 111–127. [CrossRef]

53. Bucci, M.G.; Tuttle, M.P. Liquefaction Susceptibility and Hazard Mapping: Insights From Case Histories of Earthquake-Induced Liquefaction. *Ref. Modul. Earth Syst. Environ. Sci.* **2022**, *2*, 563–582. [CrossRef]

54. Diebold, J.B.; Stoffa, P.L.; Buhl, P.; Truchan, M. Venezuela Basin crustal structure. *J. Geophys. Res. Solid Earth* **1981**, *86*, 7901–7923. [CrossRef]

55. Fajardo, A.; Aubourg, C.; Nivière, B.; Regard, V.; Uzcátegui, R. Neotectonic evolution of Northeastern Venezuela. Geomorphological and seismic analysis in the Monagas Fold and Thrust Belt. *J. S. Am. Earth Sci.* **2020**, *104*, 102867. [CrossRef]

56. Lemenkova, P. GRASS GIS for topographic and geophysical mapping of the Peru-Chile Trench. *Forum Geogr.* **2020**, *19*, 143–157. [CrossRef]

57. Pérez, O.J.; Aggarwal, Y.P. Present-day tectonics of the southeastern Caribbean and northeastern Venezuela. *J. Geophys. Res. Solid Earth* **1981**, *86*, 10791–10804. [CrossRef]

58. Avila-García, J.; Schmitz, M.; Mortera-Gutierrez, C.; Bandy, W.; Yegres, L.; Zelt, C.; Aray-Castellano, J. Crustal structure and tectonic implications of the southernmost Merida Andes, Venezuela, from wide-angle seismic data analysis. *J. S. Am. Earth Sci.* **2022**, *116*, 103853. [CrossRef]

59. Gough, D.I.; Heirtzler, J.R. Magnetic Anomalies and Tectonics of the Cayman Trough. *Geophys. J. Int.* **1969**, *18*, 33–49. [CrossRef]

60. Lemenkova, P. Geomorphology of the Puerto Rico Trench and Cayman Trough in the Context of the Geological Evolution of the Caribbean Sea. *Ann. Univ. Mariae-Curie-Sklodowska, Sect. Geogr. Geol. Mineral. Petrogr.* **2020**, *75*, 115–141. [CrossRef]

61. Bezada, M.J.; Schmitz, M.; Jácome, M.I.; Rodríguez, J.; Audemard, F.; Izarra, C. Crustal structure in the Falcón Basin area, northwestern Venezuela, from seismic and gravimetric evidence. *J. Geodyn.* **2008**, *45*, 191–200. [CrossRef]

62. Mazuera, F.; Schmitz, M.; Escalona, A.; Zelt, C.; Levander, A. Lithospheric structure of northwestern Venezuela from wide-angle seismic data: Implications for the understanding of continental margin evolution. *J. Geophys. Res. Solid Earth* **2019**, *124*, 13124–13149. [CrossRef]

63. Group, G.C. GEBCO 2020 Grid, 2020. Available online: https://www.bodc.ac.uk/data/published_data_library/catalogue/10.5285/a29c5465-b138-234d-e053-6c86abc040b9/ (accessed on 1 November 2022).

64. Schenke, H., General Bathymetric Chart of the Oceans (GEBCO). In *Encyclopedia of Earth Sciences Series*; Chapter Encyclopedia of Marine Geosciences; Springer: Dordrecht, The Netherlands, 2016; pp. 268–269. [CrossRef]

65. Mcmichael-Phillips, J. The Nippon Foundation-GEBCO Seabed 2030 Project: The Most Ambitious Seafloor Mapping Initiative in History. *Mar. Technol. Soc. J.* **2021**, *55*, 25–28. [CrossRef]

66. Cortina, G.C.Z.; Wigley, R.; Pe'eri, S. The GEBCO and NOAA Chart Adequacy Workshop. In *Abstracts of the International Cartographic Conference (ICA)*; ICA: Bern, Switzerland 2019; Chapter 1, p. 51. [CrossRef]

67. Lemenkova, P. Tanzania Craton, Serengeti Plain and Eastern Rift Valley: mapping of geospatial data by scripting techniques. *Est. J. Earth Sci.* **2022**, *71*, 61–79. [CrossRef]

68. Pavlis, N.K.; Holmes, S.; Kenyon, S.C.; Factor, J.K. The development and evaluation of the Earth Gravitational Model 2008 (EGM2008). *J. Geophys. Res.* **2012**, *117*, B04406. [CrossRef]

69. Sandwell, D.T.; Müller, R.D.; Smith, W.H.F.; Garcia, E.; Francis, R. New global marine gravity model from CryoSat-2 and Jason-1 reveals buried tectonic structure. *Science* **2014**, *7346*, 65–67. [CrossRef]

70. Lemenkova, P. Console-Based Mapping of Mongolia Using GMT Cartographic Scripting Toolset for Processing TerraClimate Data. *Geosciences* **2022**, *12*, 140. [CrossRef]

71. Lemenkova, P. Mapping Climate Parameters over the Territory of Botswana Using GMT and Gridded Surface Data from TerraClimate. *ISPRS Int. J. Geo-Inf.* **2022**, *11*, 473. [CrossRef]

72. Lemenkova, P. Topography of the Aleutian Trench south-east off Bowers Ridge, Bering Sea, in the context of the geological development of North Pacific Ocean. *Baltica* **2021**, *34*, 27–46. [CrossRef]

73. Lemenkova, P. The visualization of geophysical and geomorphologic data from the area of Weddell Sea by the Generic Mapping Tools. *Stud. Quat.* **2021**, *38*, 19–32. [CrossRef]

74. Callejón, A.F.; von der Dick, H.H. Numerical identification of microseeps insurface soil gases of western Venezuela, and its significance for hydrocarbon exploration. In *Surface Exploration Case Histories: Applications of Geochemistry, Magnetics, and Remote Sensing*; AAPG Studies in Geology 48 and SEG Geophysical References Series; AAPG: Tulsa, OK, USA, 2002; Volume 11, pp. 381–392.

75. De Rodriguez, G. Fundamental Geological Characteristics of the Venezuelan Oil Basins. In Proceedings of the 3rd World Petroleum Congress, The Hague, The Netherlands, 28 May–6 June 1951 .

76. Mann, P.; Kroehler, M.; Escalona, A.; Magnani, B.; Christeson, G. Subduction Along the South Caribbean Deformed Belt: Age of Initiation and Backthrust Origin. In *American Geophysical Union*; Fall Meeting 2007; AGU: Washington, DC, USA, 2007; Chapter Abstract ID. T11D-02, p. 2.

77. Rodríguez, I.; Bulnes, M.; Poblet, J.; Masini, M.; Flinch, J. Structural style and evolution of the offshore portion of the Sinu Fold Belt (South Caribbean Deformed Belt) and adjacent part of the Colombian Basin. *Mar. Pet. Geol.* **2021**, *125*, 104862. [CrossRef]

78. Schubert, C. Origin of the Cariaco Basin, Southern Caribbean Sea. *Mar. Geol.* **1982**, *47*, 345–360. [CrossRef]

79. Speed, R.; Torrini, R.; Smith, P.L. Tectonic evolution of the Tobago Trough forearc basin. *J. Geophys. Res.* **1989**, *94*, 2913–2936. [CrossRef]

80. Dhont, D.; Backé, G.; Hervouët, Y. Plio-Quaternary extension in the Venezuelan Andes: Mapping from SAR JERS imagery. *Tectonophysics* **2005**, *399*, 293–312. [CrossRef]

81. Gibbs, A.K.; Barron, C.N. *Geology of the Guiana Shield*; Clarendon Press: Oxford, UK, 1993; Volume 22, 246p.

82. Hollowell, T.; Reynolds, R.P. Checklist of the Terrestrial Vertebrates of the Guiana Shield. In *Bulletin of the Biological Society of Washington*; Biological Society of Washington: Washington, DC, USA, 2005; Chapter 13, pp. 1–106.

83. Hammond, D.S. *Tropical Forests of the Guiana Shield*; CABI Publishing: Wallingford, UK, 2005.

84. Escalona, A.; Mann, P. An overview of the petroleum system of Maracaibo Basin. *AAPG Bull.* **2006**, *90*, 657–678. [CrossRef]

85. Bockmeulen, H.; Barker, C.; Dickey, P.A. Geology and geochemistry of crude oils, Bolivar coastal fields, Venezuela. *AAPG Bull.* **1983**, *67*, 242–270. [CrossRef]

86. Escobar, M.; Márquez, G.; Inciarte, S.; Rojas, J.; Esteves, I.; Malandrino, G. The organic geochemistry of oil seeps from the Sierra de Perijá eastern foothills, Lake Maracaibo Basin, Venezuela. *Org. Geochem.* **2011**, *42*, 727–738. [CrossRef]

87. Memon, A.; Borman, C.; Mohammadzadeh, O.; Garcia, M.; Tristancho, D.J.R.; Ratulowski, J. Systematic evaluation of asphaltene formation damage of black oil reservoir fluid from Lake Maracaibo, Venezuela. *Fuel* **2017**, *206*, 258–275. [CrossRef]

88. Li, X.; Götze, H.J. Ellipsoid, geoid, gravity, geodesy, and geophysics. *Geophysics* **2001**, *66*, 1660–1668. . [CrossRef]

89. Yilmaz, M.; Kozlu, B. The Comparison of Gravity Anomalies based on Recent High-Degree Global Models. *Afyon Kocatepe Univ. J. Sci. Eng.* **2018**, *18*, 981–990. [CrossRef]

90. Guenther, F.B. Relations between Pasadena magnitude and vertical ground acceleration from long distance earthquakes, derived by gravity earthtide measurements at Caracas, Venezuela. *Phys. Earth Planet. Inter.* **1980**, *21*, 261–266. [CrossRef]

91. Schmitz, M.; Avila, J.; Bezada, M.; Vieira, E.; Yáñez, M.; Levander, A.; Zelt, C.A.; Jácome, M.I.; Magnani, M.B. Crustal thickness variations in Venezuela from deep seismic observations. *Tectonophysics* **2008**, *459*, 14–26. [CrossRef]

92. Ambeh, W.B.; Lynch, L.L. Coda Q in the Eastern Caribbean, West Indies. *Geophys. J. Int.* **1993**, *112*, 507–516. [CrossRef]

Article

Effect of Land Use and Drainage System Changes on Urban Flood Spatial Distribution in Handan City: A Case Study

Beibei Liu [1], Chaowei Xu [2], Jiashuai Yang [2], Sen Lin [1] and Xi Wang [2,*]

[1] National Disaster Reduction Center of China, Ministry of Emergency Management of China, Beijing 100124, China

[2] College of Urban and Environmental Sciences, Peking University, Beijing 100871, China

* Correspondence: pkuwangx@pku.edu.cn

Abstract: This study simulated urban flooding under various land use and drainage system conditions and described the process of historical ground–underground construction and its influence on spatial variations in waterlogging, taking Handan City as an example. The obtained results can provide support for urban water security and sustainable urban water resource management. The land use change, represented by the expansion of sealed surfaces, has a positive impact on the distribution and the volume of flood in Handan City, while the drainage system has the opposite effect. The flooding distribution changes over decades reveal that flooding risk is reduced in most areas by improved drainage conditions but exacerbated in impervious areas and riversides due to increasing impermeable areas, the rapid draining of pipes, and poor outlet conditions. This study demonstrates how the dual changes in land use and drainage pipeline networks affect urban flooding distribution; we suggest considering land use and the extension of drainage pipelines in future construction.

Keywords: land use change; drainage system; urban water security; SWMM; Handan

Citation: Liu, B.; Xu, C.; Yang, J.; Lin, S.; Wang, X. Effect of Land Use and Drainage System Changes on Urban Flood Spatial Distribution in Handan City: A Case Study. *Sustainability* **2022**, *14*, 14610. https://doi.org/10.3390/su142114610

Academic Editors: Chong Xu and Jian Chen

Received: 22 September 2022
Accepted: 4 November 2022
Published: 7 November 2022

Publisher's Note: MDPI stays neutral with regard to jurisdictional claims in published maps and institutional affiliations.

1. Introduction

Against the background of rapid economic growth in China, the rate of urbanization in China increased dramatically from 26% in 1990 to 64.72% in 2021 [1]. Rapid urbanization causes many environmental problems, among which urban water security is becoming more and more notable [2]. Taking the 2021 Henan floods as an example, China's Henan province suffered from a catastrophic flooding disaster from 17 to 23 July 2021 as a result of heavy rainfall. Especially on 20 July, Zhengzhou, the provincial capital of Henan, experienced a severe river flood and concurrent urban waterlogging. A record amount of rain fell within one hour, reaching 201.9 mm, breaking the previous record in China, which had stood since 1951. The flood disaster caused 380 deaths and led to a direct economic loss of CNY 41 billion (USD 5.67 billion) after investigation and evaluation in Zhengzhou [3]. Urbanization has transformed the processes of the urban hydrological cycle, which, in turn, has intensified the frequency and intensity of extreme hydrological events and exerted a far-reaching impact on urban water security [4,5]. Therefore, it is essential to evaluate the influence of urbanization on urban floods and identify the impact factors.

Urbanization has affected processes of the urban hydrological cycle in the following ways: (a) expanding impermeable surfaces, such as sidewalks and buildings, may increase surface runoff and reduce groundwater recharge [6–9]; (b) lowering surface roughness by reducing the vegetation distribution and flattening the surface may accelerate the surface runoff process [10,11]; (c) hardening river bottoms may accelerate the river confluence process [12,13]; (d) building drainage pipelines may accelerate pipeline confluence [14,15]. Hence, the dual changes in ground (such as expanding impermeable surfaces and lowering surface roughness) and underground construction (such as hardening river bottoms and building drainage pipeline networks) must be considered in urban stormwater management.

Some researchers have attempted to assess urban waterlogging risk under different land use scenarios, which would assist in land use planning and policymaking [16]. Nonetheless, it is still challenging to build high-resolution models to predict urban development patterns [17], especially in rapidly developing areas. Additionally, historical records regarding the urbanization process and urban flood conditions are often limited [14], which restricts the research to exploring the impacts of the urbanization process on floods in rapidly developing areas [18,19]. Some researchers have pointed out that drainage conditions can strongly influence rainfall–runoff processes in urban areas [20–22]. In addition to land use, the urban drainage network system is regarded as one of the main factors in urban flooding [23,24]. According to existing research, an undeveloped drainage system is more likely to result in flooding in cities [25–28]. Due to poor drainage conditions, torrential rain may result in severe temporal and regional waterlogging [29–31]. Previous studies have shown that urban construction and expansion exert pressure on the current drainage system, particularly underground networks, thereby increasing the risk of frequent flooding [32–34]. However, more research is needed to investigate the relationships between changes in ground–underground conditions (i.e., the detailed spatial distribution of land use classifications, drainage systems, and related indicators) and urban floods in historical urbanization. Consequently, it is crucial to determine the effect of surface and underground construction on flood risk, which requires records of land use and drainage system construction in the long term.

This study addresses the combined effect of land use change and drainage system change on urban spatial variations in waterlogging in Handan City, a medium-sized city in China. With five rivers flowing through, Handan provides simple conditions for urban flooding simulations. The SWMM was used to simulate floods under different land use and drainage network conditions because this model can mechanistically quantify the impact of surface land use and underground pipe networks on urban waterlogging; moreover, it has the advantages of being open source and performing with high calculation speeds. The control variables method was used in this study. Firstly, based on previous research [35,36], this study simulated urban floods by considering the sole impact of land use change on spatial variations in waterlogging and flooding volume. Land use in 1987, 1997, 2007, and 2017 was interpreted according to satellite remote sensing data from Google Maps. Secondly, the sole impact of drainage system change on waterlogging was assessed, and drainage system data from different years were collected from the local drainage network database. Finally, this study probed the comprehensive influence of dual changes in land use and drainage pipeline network on the spatial distribution of urban waterlogging.

2. Materials and Methods

2.1. Study Area

Handan City is located in southern Hebei Province, China (Figure 1a). At the end of 2021, the city had a total population of 9.36 million, ranking second in Hebei; GDP totaled CNY 411.4 billion (USD 56.94 billion), ranking fourth in Hebei [37]. Handan lies in the North China Plain, in a continental semi-arid climate zone, with strong monsoonal influence. The mean daily temperatures range from $-0.9\ ^\circ C$ in January to $27.3\ ^\circ C$ in July, while the annual mean temperature is $14.3\ ^\circ C$. The average annual precipitation is 502 mm, and the rainy season generally occurs in July and August.

Surrounded by the ring road, the study area was the central area of Handan City (Figure 1b). The area of this region is 91.7 km^2, and the geographical coordinates are 36°20′–36°44′ N and 114°03′–114°40′ E. The topography of Handan City is generally high in the south and west and low in the north and east (Figure 1c). With its typical northern climate characteristics and medium-sized area, the downtown Handan area represents a typical medium-sized, rapidly developing city in northern China.

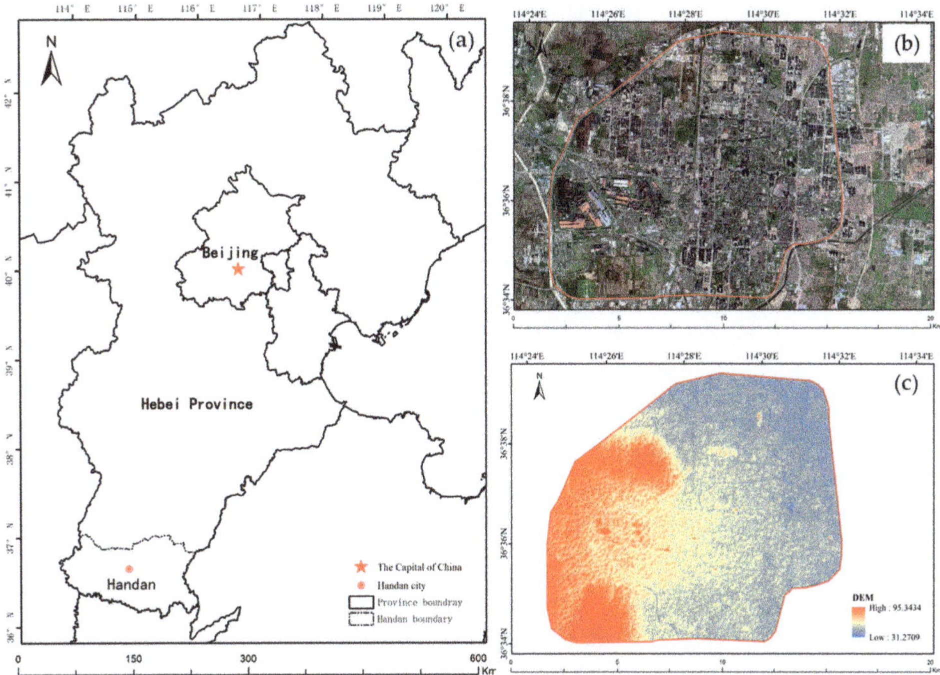

Figure 1. Study area. (**a**) Location of the study area; (**b**) remote sensing view of the study area; and (**c**) digital elevation model (DEM) of the study area.

The study area is a high-risk area for urban waterlogging problems. A typical example is the "7·18 Handan Heavy Rainstorm event" in 2016, which caused the collapse of 36,000 houses, affecting 2.04 million people, with 108 reported dead and missing, and a direct economic loss of CNY 18.9 billion (USD 2.62 billion). Urban flooding management is of great significance for Handan City in the prevention and mitigation of natural disasters. However, only a few researchers have performed studies related to urban stormwater management in Handan [38–43].

2.2. Data

In order to simulate the rainfall runoff of the study area, the input data included the following: (1) weather data providing the input rainfall data; (2) land use data providing the infiltration parameters for the runoff generation process and the roughness parameters of different land covers for the overland flow concentration process; and (3) drainage network data providing the concentration paths and their physical parameters for the concentration process. Additionally, to validate the simulation, social media data were used to represent the observed data.

2.2.1. Weather Data

The design storm curve in this study was provided by the water conservancy bureau of Handan, ascertained from the Chicago hyetograph method developed by Tongji University [44].

Equation (1) is a common form of the design rainstorm intensity formula:

$$i = \frac{A_1(1 + BlgT)}{(t + b)^c} \tag{1}$$

where i is the mean rainstorm intensity, mm/min; T is the design rainfall return period, a; t is the rainfall duration, min; and A_1, b, B, and c are parameters related to local rainstorm characteristics and need to be solved. Among them, A_1 is a parameter representing the rainfall intensity; B is a dimensionless parameter representing the variation in the rainfall intensity; b and c are constants independent of the return period.

The design drainage frequency is 50%, which can provide reasonable safety for traffic and pedestrians at a reasonable cost. The formula only applies to rainfall continuing for less than 3 h, and the recurrence interval for storm sewer design is 2 years, according to the local design standard for wastewater engineering; therefore, a typical design rainfall process lasting for 1 h with a return period of 2 years was selected. Equation (2) is the design rainstorm intensity formula from the Handan Hydrology Bureau with experimental parameter values. The experienced value of A_1 is 6.025 mm/min, B is 0.86, b is 7.22, and c is 0.528. Thus, Equation (2) is obtained.

$$i = \frac{6.025(1 + 0.86 lgT)}{(t + 7.22)^{0.528}} \tag{2}$$

2.2.2. Land Use Data

The land use data of Handan in 1987, 1997, 2007, and 2017 were interpreted using remote sensing data from Google Maps. The spatial resolution of the downloaded remote sensing images of the study area was up to 1 m. For these remote sensing data, eCognition software was used to perform supervised classification, and the corresponding classification results and remote sensing images were corrected and coordinate-transformed. Based on the land surface properties of Handan City, the land use types were classified into 6 classes, i.e., farmland, grassland, water, woodland, bare land, and impervious surface (Figure 2).

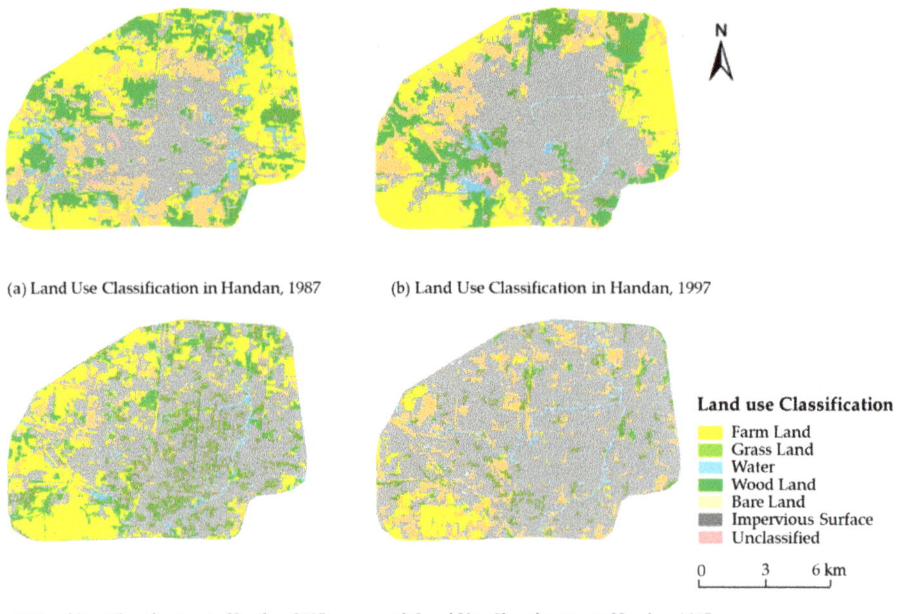

(a) Land Use Classification in Handan, 1987   (b) Land Use Classification in Handan, 1997

(c) Land Use Classification in Handan, 2007   (d) Land Use Classification in Handan, 2017

Figure 2. The land use classification distribution of Handan in (**a**) 1987, (**b**) 1997, (**c**) 2007, and (**d**) 2017.

Land use change is a major issue in rapid globalization and urbanization, particularly in rapidly developing countries and regions throughout Asia. After experiencing developments in the economy, population growth, and the development of land policy, the extent of land use change in Handan from 1987 to 2017 is shown in Table 1, with the matrix depicted in Table 2.

Table 1. Percentage of classified land use types in Handan from 1987 to 2017 (unit: %).

Land Use Classification	Year			
	1987	**1997**	**2007**	**2017**
Farm land	29.91	24.02	17.49	5.45
Grass land	14.56	9.63	3.12	0.71
Water	4.64	1.78	1.66	1.19
Wood land	2.26	5.02	4.66	2.79
Bare land	16.00	14.09	14.65	13.29
Impervious surface	32.44	44.73	58.41	76.57
Unclassified	0.21	0.74	0.02	0.00

Table 2. Land use conversion matrix from 1987 to 2017 (km^2).

Land Use		2017						
		Farm Land	**Grass Land**	**Water**	**Wood Land**	**Bare Land**	**Impervious Surface**	**Total**
1987	Farm land	2.40	0.28	0.24	0.76	3.73	20.02	27.42
	Grass land	1.06	0.09	0.15	0.31	2.51	9.24	13.35
	Water	0.18	0.05	0.23	0.15	0.66	2.98	4.25
	Wood land	0.27	0.02	0.01	0.05	0.27	1.45	2.07
	Bare land	0.57	0.09	0.08	0.32	2.09	11.52	14.67
	Impervious surface	0.52	0.12	0.37	0.95	2.93	24.84	29.74
	unclassified	0	0	0.01	0.02	0	0.17	0.19
	Total	5	0.65	1.09	2.56	12.19	70.22	91.70

Table 1 summarizes the changes in different types of land use over the past 30 years. The overall trend shows that the decreases in farmland and grassland are 24.46% and 13.85%, respectively, and the increase in impermeable area is 44.13%. Over the past three decades, the impermeable surface has increased significantly, from 32.44% to 76.57%.

From the land use transfer matrix (Table 2), the types of land use with the greatest changes are farmland, grassland, and impermeable surface. The increasing impermeable areas are mainly farmland (20.02 km^2), bare land (11.52 km^2), and grassland (9.24 km^2). Moreover, of all the converted land use types, the conversion rates to impermeable surface are as follows: 72.99% of farmland, 69.19% of grassland, 70.06% of water, 70.16% of woodland, and 78.53% of bare land in 1987 converted to imperious surface by 2017. It can be inferred that buildings and pavements with rapid urbanization are occupying all other types of land use.

2.2.3. Drainage Network Data

The drainage pipe network data provide the pipeline concentration parameters (length, shape, depth, and Manning's roughness coefficient of the conduit). When considering the ground and underground concentration progress, rivers are taken as open channels incorporated into the drainage network for calculation because solidification of the river bottom was completed. The pipeline concentration parameters of the river were set based on parameters such as the shape of the river cross-section, the feathering factor at the bottom of the river, and the slope of the river. The outlet node of the drainage to the river was set as the end node of each pipe. The water flow exchange between pipe and river followed the one-dimensional Saint-Venant equation.

Five main rivers run through the study area: Fuyang River, Qin River, Zhu River, Zhizhang River, and Shuyuan River. Qin River flows from west to east; Fuyang River flows from south to north; Zhizhang River bypasses the urban area from southwest to northeast around the central area. From the southernmost part of the river system upstream of the city, there is a confluence of the three rivers at Zhangzhuang Bridge, where there is a sluice gate; when heavy rains come, the sluice gate is shut to prevent incoming floods, and the upstream water flows into Zhizhang River to the southeast to protect the urban area from upstream flood. Thus, the upstream water parameter was set to 0 during the severe rainstorms in the SWMM.

To perform hydraulic calculations of the pipe network, according to the principle of area equivalence, multiple pipes were merged into one pipe, and those with an inner diameter of less than 300 mm were omitted to simplify the original structure. The simplified drainage network of the study area is shown in Figure 3.

Figure 3. Simplified drainage network of Handan.

Drainage networks include the natural or manual removal pathways of ground and underground water from an area. The canal system was designed for 100-year frequency floods; there are six storage facilities with a total volume of 14 million m^3. Figure 4 shows the drainage network of the study area in 1987, 1997, 2007, and 2017, while Table 3 lists the sub-catchments, nodes and links, and total pipe length of different years. The drainage network was initially constructed using GIS data provided by the East Sewage Treatment Plant of Handan.

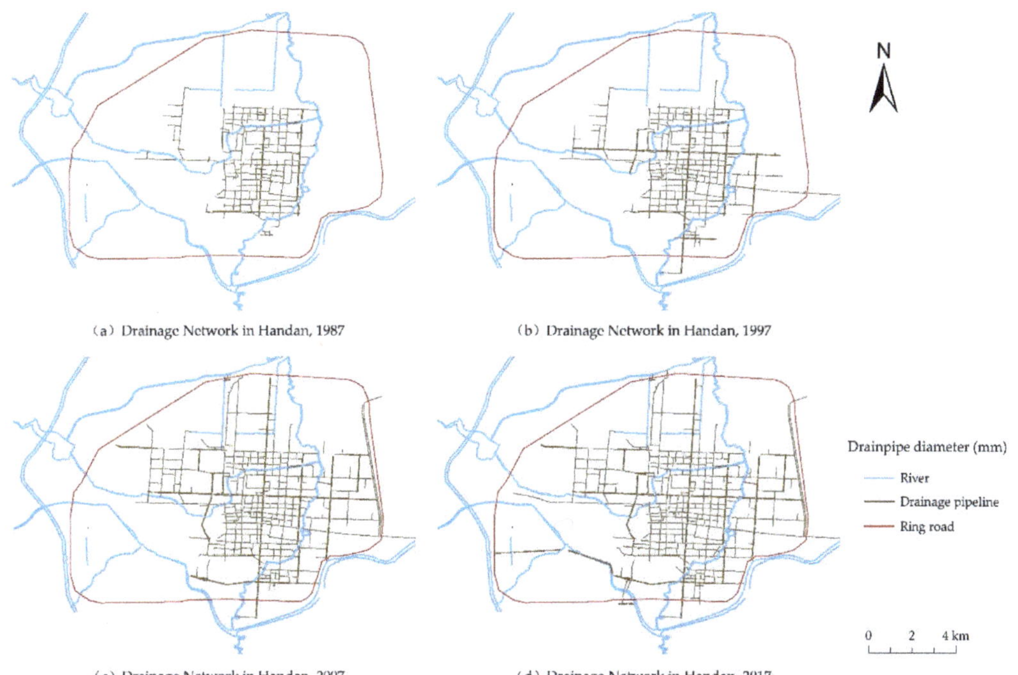

Figure 4. Simplified drainage network distribution of Handan in (**a**) 1987, (**b**) 1997, (**c**) 2007, and (**d**) 2017.

Table 3. Statistics of the drainage pipe networks.

Years	No. of Sub-Catchments	No. of Nodes	No. of Links	Total Pipe Lengths (km)
1987	2671	2781	2711	99.77
1997	3456	3748	2578	148.93
2007	5725	5987	5541	234.05
2017	6258	6984	6742	251.05

The construction of the drainage system has rapidly expanded during the last 30 years. As shown in Table 3, the total pipe lengths grew from 99.77 km in 1987 to 251.05 km in 2017. In this study, the development of the drainage system was applied in urban flood simulations, and the relationships between land cover changes and spatial variations in waterlogging were analyzed.

2.3. Method

2.3.1. Stormwater Management Model

The stormwater management model (SWMM), which has the advantages of being open-source and fast to compute, developed by the United States Environmental Protection Agency (US-EPA) in 1971, is a dynamic rainfall–runoff–subsurface runoff simulation model used for single-event to long-term (continuous) simulation of the surface/subsurface hydrology quantity and quality from primarily urban/suburban areas. For a large area such as Handan City, the SWMM can calculate the accumulated waterlogging amount of the node in each catchment. In order to visualize the results, the obtained data were equally distributed to each sub-catchment, and then the water surface was smoothed to determine the spatial distribution of waterlogging.

Surface runoff can be generated by rainfall, snowfall, or the melting of snow or glaciers. According to the land use and surface drainage trend, a watershed can be divided into several sub-catchment areas, and the runoff process is calculated according to the characteristics of each sub-catchment area. Runoff in the SWMM is simulated for single or continuous rainfall and snowmelt [45]. The dynamic wave method is used to consider flow characteristics during surface flooding and the pressure effects caused by the inverse slope in some parts of a catchment [46]:

$$R = (S_1 + S_2 + S_3) \int_0^T \left(i_{(t)} - e_{(t)} \right) d_t - S_1 \int_0^T f_{(t)} d_t - (D_1 + D_2) \tag{3}$$

where R is the runoff of a sub-catchment; S_1 is the area of the pervious surface; S_2 is the area of the impervious surface with depression storage; S_3 is the area of the impervious surface with no depression storage; $i_{(t)}$ is the function of rainfall intensity with time; $e_{(t)}$ is the function of evaporation intensity with time; $f_{(t)}$ is the function of infiltration intensity with time; and D_1 and D_2 are the depression storage of the pervious surface and impervious surface with depression storage, respectively.

The kinematic wave method cannot calculate backwater, backflow, or pressurized flow; thus, it is only applicable to tree drainage networks. The dynamic wave method can describe the storage, sink, inlet and outlet losses, backflow, pressurized flow, and backwater of the pipe; thus, it is suitable for complex annular drainage networks. The elevation difference between the outfall bottom of the drainage network and the river bottom in this study was generally between 0.5 m and 1.0 m in Handan City; backflow and full pipe flow phenomena can occur during heavy storms. Therefore, this study used the dynamic wave method to simulate the drainage confluence in the main urban area of Handan.

2.3.2. Determination of Model Parameters

The SWMM can simulate the runoff process generated from sub-catchments, flow depth in open channels and closed pipes, and the ponding quantity for inlet nodes. The model requires three types of parameters: runoff-producing parameters, overland flow concentration parameters, and pipeline concentration parameters.

Sub-catchments are hydrologic units of land, the topography and drainage systems of which dispose of direct surface runoff to a single discharge point. The properties associated with a sub-catchment in the SWMM are as follows: area, width, slope (%), imperviousness (%), soil properties, Manning's n for impervious and pervious surfaces for overland flow, depth of depression storage in pervious and impervious areas, and the proportion of impervious area with no depression storage (%).

The 3D Analyst tools in ArcGIS 10.2 (ArcGIS, ESRI, Redlands, CA, USA) were used to generate the slope of each point from the digital elevation model (DEM) data.

Width is defined as sub-catchment divided by the length of stream flow. The length of overland flow is difficult to determine accurately; therefore, sub-catchment width parameters are subject to calibration to obtain a close match between observation and calculation. The sub-catchments are classified into two types by observing their shapes: right-angled trapezoids and others. Most smaller and numerous sub-catchments, located in the central zone of the study area, are right-angled trapezoids, whereas the larger and fewer sub-catchments, located in the edge zone, exhibit various shapes. Hence, the length of the smaller sub-catchments can be calculated by computing geometry, and the length of larger sub-catchments can be measured in ArcGIS 10.2.

The sub-catchment impervious roughness is taken from the user's manual, and pervious roughness is calculated by area-weighted averaging for varying surfaces. The Manning's roughness values of each surface are listed in Table 4.

Table 4. Manning coefficient of different land use types.

Land Use Types	Manning's Coefficient
Impervious area	0.012
Grass land	0.06
Wood land	0.15
Farm land	0.03
Bare land	0.02

The rate of infiltration is a function of soil properties in the sub-catchment area. The maximum infiltration rate, minimum infiltration rate, and decay constant α of the Horton infiltration model were 76.2 (mm/h), 3.81 (mm/h), and 0.0006, respectively, according to research performed in Tianjin [47], a municipality near the study area.

Inverse elevation and maximum depth are two pipeline parameters in the SWMM. The inverse elevations of each node were provided by the local sewage treatment plant, and the maximum depth was calculated by subtracting the invert elevation from the ground elevation of DEM data.

2.3.3. Validation

The rapid nature of urban flooding from storms makes the accurate surveying of water depths difficult to observe, thus hindering the validation of urban flood models [48]. Nevertheless, social media is now an open platform for collecting emergency information [49,50]. Scraping data through a Web Spider bot is an important way to harvest verifiable data. A heavy rainstorm in Handan triggered floods on 19 July 2016, leaving many validation-supporting footprints on the Internet. Weibo, a widely used social media platform in China, was chosen as the scraping dataset. The crawling targets were "rain" and "waterlogging," and the time range was defined as 19 July 2016, 0:00 to 20 July 2016, 12:00. The 646 related posts contained valid information, such as photographs, text, upload locations, and time stamps. In order to protect the privacy information of uploaders, the exchangeable image files were deleted while posting the message, resulting in the absence of the original shooting location of the photo, which meant that the uploading location could have been different from the waterlogging location. In order to confirm the location, it was necessary to identify landmarks in the image. Geographic locations are essential to validate the model. The geographic position could be located based on unique landmarks in the photograph through a mobile mapping application, such as Google Maps or Gaode Maps.

The input data of the SWMM contained the rainfall process of 19 July 2016, the land use map in 2016, and the drainage pipeline in 2016. The Web Spider harvested several thousand related photographs, but only 11 included precise location information. The model simulation results showed that all 11 points had accumulated water (Figure 5). This indicates that the model inversion results are generally consistent with the actual situation.

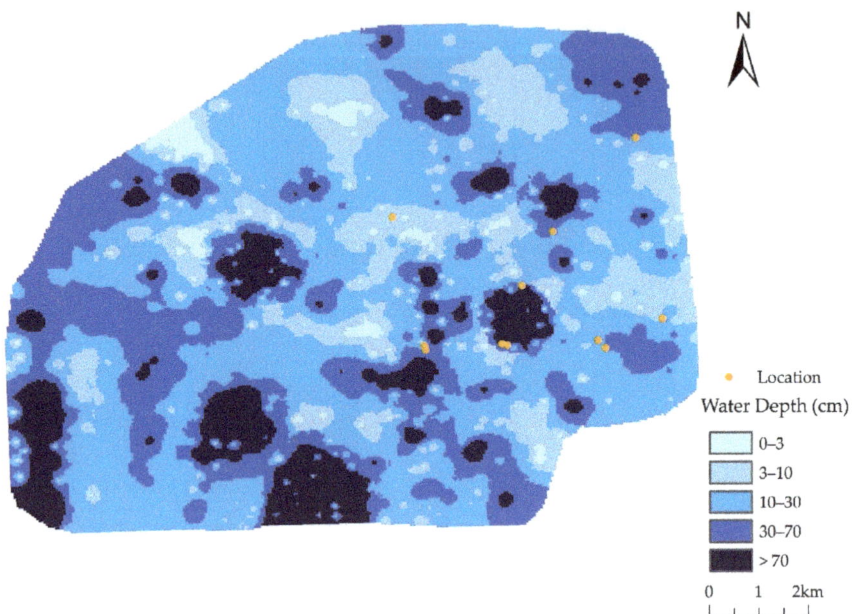

Figure 5. Simulated maximum water depth distribution and the validation locations in Handan flood, 19 July 2016.

2.3.4. Parameter Sensitivity Analysis

Parameter sensitivity analysis involves identifying the input parameters that exert considerable impact on the outcomes [51]. The Morris method, a widely used approach for parameter screening, is applied to determine the influence of land use changes and drainage pipeline changes on flood risk. The modified Morris method calculates the average value of the Morse coefficient to obtain the sensitivity factor S, as shown in Equation (4):

$$S = \sum_{i=0}^{n-1} \frac{(Y_{i-1} - Y_i)/Y_0}{(X_{i-1} - X_i)/X_0} / n \tag{4}$$

where S is the sensitivity index; n is the number of times the SWMM was run; Y_i is the outcome of the SWMM run for the i^t time; X_i is the input for the i run; Y_0 is the initial value of the outcome; and X_0 is the initial value of the input. According to previous research [52], the sensitivity indices are ranked into four classes: when $|S| \geq 1.00$, the parameter has very high sensitivity; when $0.20 \leq |S| < 1.00$, the parameter has high sensitivity; when $0.05 \leq |S| < 0.02$, the parameter has medium sensitivity; and when $0.00 \leq |S| < 0.05$, the parameter has low sensitivity [53].

2.3.5. Urban Waterlogging Risk Level

The impacts of water depth on different traffic subjects, such as pedestrians, trams, motor vehicles, electric vehicles, and automobiles, were used as the basis for classifying the risk. In total, five levels of risk are classified (Table 5). The thickness of an athletic shoe sole is generally about 3 cm; thus, it is believed that a water depth of less than 3 cm does not affect pedestrians in walking. The height from the ankle to the ground is generally 10 cm; therefore, it is believed that a water depth between 3 cm and 10 cm will affect pedestrian walking. The critical value of water wading depth for electric vehicles is 30 cm [53], and the heights of the air outlets and inlets of automobiles are around 30 cm and 70 cm, respectively.

When the inlet port is submerged, traffic is basically paralyzed. Thus, 30 cm and 70 cm were set to be the critical heights of different flooding risk levels.

Table 5. Urban flooding risk level classification.

Risk Level	Water Depth (cm)	Extent of the Threats to Different Means of Transportation
I	0–3	There is little impact on traffic.
II	3–10	The water depth is higher than the insteps, thus slowing down pedestrians.
III	10–30	When the rail surface water depth exceeds 30 cm, electric vehicles must be shut down.
IV	30–70	The water depth is higher than the outlet, thus impeding motor vehicles.
V	>70	The water depth is higher than the inlet port, basically paralyzing traffic.

3. Results and Discussion

3.1. Effect of Land Use Change

The distributions of flooding areas simulated using different land use in 1987, 1997, 2007, and 2017 are shown in Figure 6. The results were obtained directly from SWMM simulations, which used a two-year rainfall event and the drainage system in 2017.

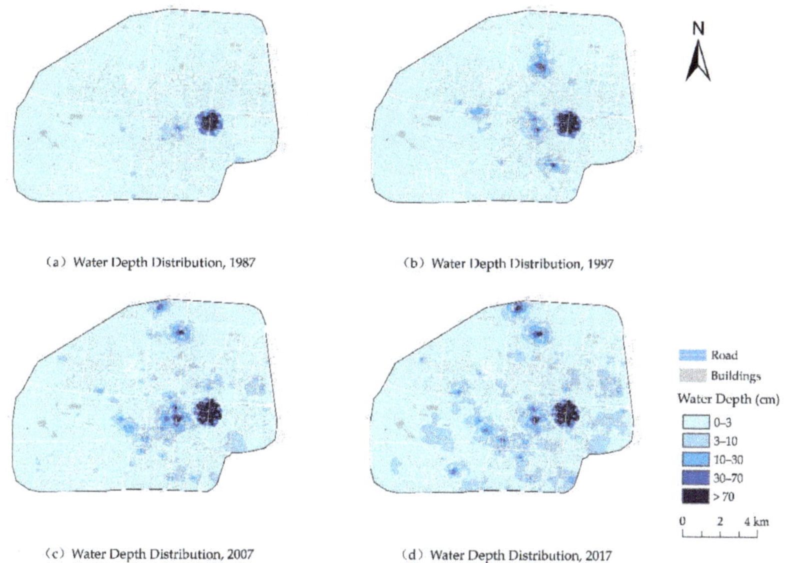

Figure 6. Distribution of flooding area and water depths showing different land uses of Handan in (**a**) 1987, (**b**) 1997, (**c**) 2007, and (**d**) 2017.

The distributions of flooding water depths are also shown in Figure 6. The high-risk areas corresponding to level V were mainly located in the center of Handan City in 1987, then grew to the north and south in 1997, and spread to the whole study area in the subsequent 20 years. Increasing areas of waterlogging shared the same growing region of the impervious surface (Figure 2).

As shown in Table 6, the areas with flood depths of more than 3 cm were positively correlated with the percentage of the impervious surface. The total flood volume exhibited the same trend as the water depth. The flooding volume is an indicator of the destruction of municipal infrastructure [54], and the locations of flood nodes represent high-risk areas. With the proportion of impervious surface increasing by 136.11% every 10 years, the total flood volume increased by 387.87%, from 7.89×10^4 m^3 to 3.85×10^5 m^3. The land use change, represented by the expansion of the impervious surface, exerts a considerable impact on the spatial distribution of waterlogging and the volume of flooding.

Table 6. Flooding volume under different land uses of Handan in 1987, 1997, 2007, and 2017.

Year	Proportion of Area at Different Water Depths (%)					Impervious Surface (%)	Total Flood Volume (m³)
	0–3 cm	3–10 cm	10–30 cm	30–70 cm	>70 cm		
1987	97.07	1.39	0.49	0.50	0.55	32.43	7.89×10^4
1997	91.88	4.89	1.61	0.70	0.91	44.73	1.52×10^5
2007	85.40	9.97	2.67	0.81	1.14	58.41	2.82×10^5
2017	75.24	19.76	2.99	0.93	1.08	76.57	3.85×10^5

3.2. Effect of Drainage System Change

Figure 7 shows the spatial variations in waterlogging using the drainage system. As detailed in Section 3.1, the results were obtained from SWMM simulations with biennial rainfall events and land use situations in 2017. The deep waterlogging areas occupied most of Handan City in 1987 but shrunk to the central zone in 2017, corresponding to the gradual reduction in areas with high-risk levels. The results reveal that a more developed drainage system may improve flood conditions in most areas, especially those located far from rivers. As more and more new pipelines transport water to rivers, the distribution of floods in the study area has become more concentrated near the outlets.

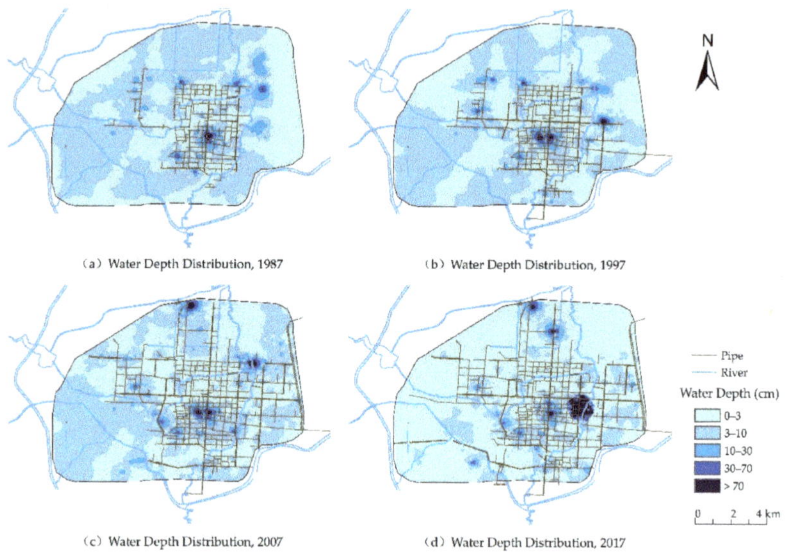

Figure 7. Distribution of flooding areas and water depths using different drainage system of Handan in (**a**) 1987, (**b**) 1997, (**c**) 2007, and (**d**) 2017.

The changes in different water depth areas in Table 7 show that, with the improvement in drainage conditions, the shallowest and deepest areas increased rapidly, from 44.24% to 75.24% and from 0.12% to 1.08%, respectively. Additionally, the areas with the deepest water depths are distributed at riverside areas or pipe outlets. As shown in Table 7, the total flood volume is negatively correlated with the total pipe length. With the total pipe length increasing by 151.63% to 251.05 km, the total flood volume decreased by 27.56% from 5.31×10^5 m³. The results reveal that a more effective drainage system may reduce the total flood volume, although it exacerbates the flooding risk at riverside areas and the pipe outlet.

Table 7. Flooding volume using different drainage systems in Handan in 1987, 1997, 2007, and 2017.

Year	Proportion of Area at Different Water Depths (%)					Total Pipe Lengths (km)	Total Flood Volume (m^3)
	0–3 cm	3–10 cm	10–30 cm	30–70 cm	>70 cm		
1987	44.24	51.17	3.97	0.51	0.12	99.77	5.31×10^5
1997	46.12	49.83	3.15	0.69	0.21	148.93	4.9×10^5
2007	45.57	48.90	4.30	0.86	0.36	234.05	4.54×10^5
2017	75.24	19.76	2.99	0.93	1.08	251.05	3.85×10^5

3.3. Combined Effect of Dual Changes in Land Use and Drainage System

Figure 8 shows the changes in flood areas of different land use structures and drainage systems over the past three decades. The SWMM uses biennial rainfall events to simulate the amount of flooding at each node. When combining the effects of land use change with drainage network change, the study results show that the flood conditions have deteriorated, especially in the central zones, and most of the flooding nodes are located in areas with impervious surfaces. This result is partially similar to a previous study [55], which demonstrated that the flood volume is more sensitive to urban development and drainage construction in central urban zones. The outlet in the SWMM is a node through which water can flow away from a pipeline to a river. Due to poor outlet conditions, flooding near the Fuyang River is growing worse in the eastern region. The topography of Handan is relatively gentle and could not meet the requirements for a drainage slope; therefore, the outlet of the drainage pipelines may be submerged by river floods during heavy rainfall. Under such conditions, the drainage system cannot discharge into the river; thus, the water is accumulated around the riverside where the outlet cannot transfer water to the river.

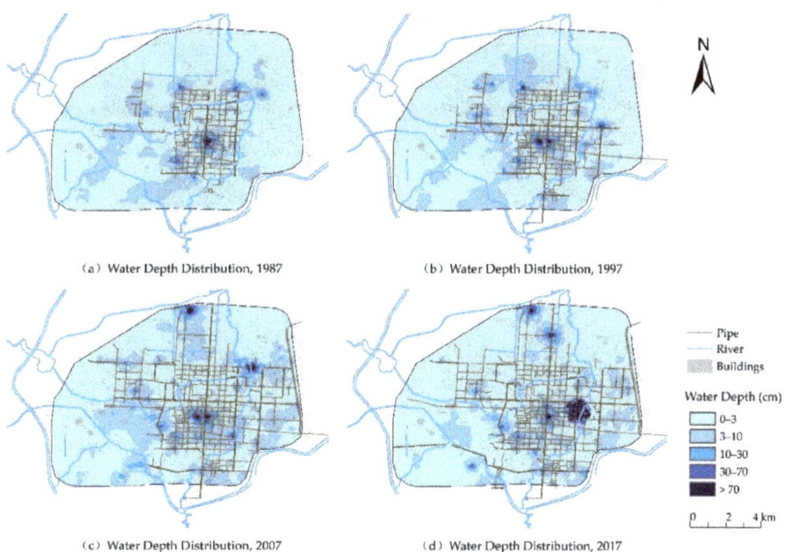

Figure 8. Distribution of flooding areas and water depths using both the land use and drainage system of Handan in (**a**) 1987, (**b**) 1997, (**c**) 2007, and (**d**) 2017.

As shown in Table 8, the flooding volume increases slightly year by year, from 3.31×10^5 m^3 to 3.85×10^5 m^3. The proportion of area with a water depth above 70 cm increased from 0.11% to 1.08%, and that between 30 cm and 70 cm increased from 0.41% to 0.93%. The changes in the total flooding volume over the years indicate that the construction of drainage systems is not consistent with the land use changes. Meanwhile, the spatial distributions of waterlogging indicate that the improved drainage system may trigger more severe flooding over impervious surfaces and the riverside due to rapidly draining pipes and poor outlet conditions.

Table 8. Flooding volume using both the land use and drainage system in Handan in 1987, 1997, 2007, and 2017.

Year	Proportion of Area at Different Water Depths (%)					Total Flood Volume (m^3)
	0–3 cm	3–10 cm	10–30 cm	30–70 cm	>70 cm	
1987	79.37	18.22	1.88	0.41	0.11	3.31×10^5
1997	73.03	23.90	2.44	0.51	0.12	3.53×10^5
2007	65.96	29.29	3.68	0.75	0.31	3.84×10^5
2017	75.24	19.76	2.99	0.93	1.08	3.85×10^5

3.4. Urban Flooding Volume Distribution Pattern Analysis

The percentage of impervious surface and length of drainage pipe represents the land use and the drainage system, with sensitivity indices of 1.09 and −1.08, respectively. The sensitivity of land use change parameters is slightly higher than the drainage pipe change parameters, but both are highly sensitive. This result is consistent with the results of Li et al. [56], who observed that the N-Imperv (percent impervious %) and Manning's roughness coefficient of pervious and impervious areas had a greater effect on the total runoff volume. The development of drainage systems contributes similarly to the land use pattern, which means that more effort is needed for the overall drainage system development.

In order to describe the flooding spatial distribution patterns shown in Figures 6–8, the correlations of imperviousness, total pipe lengths, and flooding volume are plotted in Figure 9. It can be seen in Figure 9a that as the imperviousness of the city increases from 32.43% to 76.57%, the flood volume in the city increases from 7.8×10^4 m^3 to 3.85×10^5 m^3, and there is a positive correlation between impervious surface and total flood volume; the linear correlation coefficients reached 0.9892. Changes in land use type and imperviousness have direct impacts on surface runoff [57].

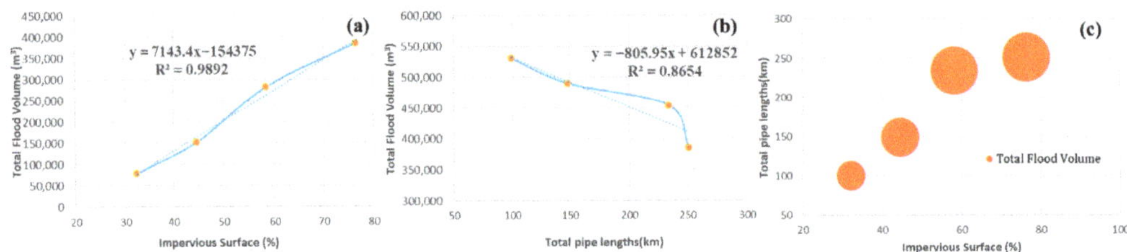

Figure 9. Correlation of ground–underground factors for exploring urban flood changes. (**a**) The correlation between impervious surface and total flood volume; (**b**) the correlation between total pipe lengths and total flood volume; (**c**) the relationship between the joint variation in impervious surface, total pipe lengths, and total flood volume.

In Figure 9b, it can be seen that as the total pipe length increases from 99.1 km to 251.05 km, the total flood volume decreases from 5.31×10^5 m^3 to 3.85×10^5 m^3, which shows a negative correlation; the linear correlation coefficients reached 0.8654. This result is different from previous studies [55]; however, we believe that this finding is reasonable because, with the construction of urban pipe networks, stagnant urban water will drain more smoothly, resulting in a decrease in the flood volume. Another possible reason is that the effects of land use changes and urban pipe networks on urban stormwater are discussed separately in this paper using the control variables approach.

In Figure 9c, the effects of urban land use change and urban pipe network construction on urban stormwater are not linear, and the combined effect of both increases the amount of urban waterlogging year by year, although the magnitude of change is significantly reduced compared with the single-factor case, especially between 2007 and 2017. The increase is very slight, which indicates that the aggressive expansion period of urban construction has ended, and urban construction from the perspective of water security has become more scientific and orderly.

For a better understanding of the information in Figure 8, all statistics of each sub-catchment for each year and the flooding volume were plotted. Figure 10 shows that the percentage of impervious area of each sub-catchment, the area of each sub-catchment, and the drainage network condition exert a considerable impact on the flood volume. With the land use development, the impervious area of each sub-catchment appears to be rising, and with the drainage pipeline network construction, the area of each sub-catchment reveals a generally decreasing trend. Waterlogged sub-catchments of more than $1.5 \times 10^4 \, \text{m}^3$ appeared in 1987 and 1997 due to the larger area of each catchment. The flood amount for each waterlogged sub-catchment has fallen during the last 30 years due to the improved drainage network conditions. Nonetheless, this does not signify safer water security conditions because the total amount increases each year.

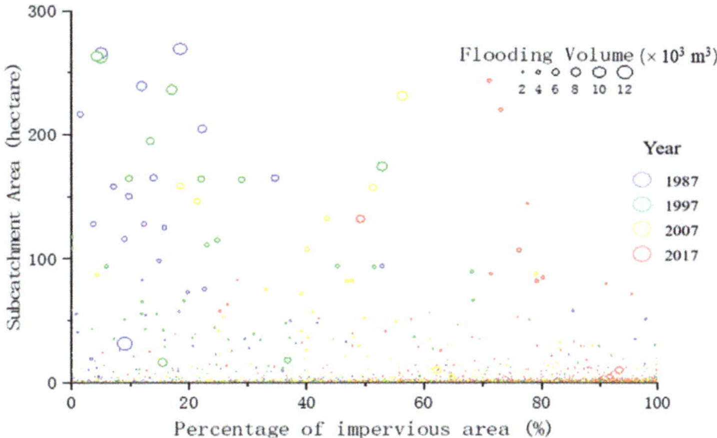

Figure 10. Scatterplot of all simulations: changes in flooding volume vs. changes in the percentage of effective impervious area and area for each sub-catchment; shapes according to the change in flooding volume. Colors denote different years.

As Figures 9 and 10 show, the flood volume increased greatly along with urbanization from 1987 to 2007. However, the flood volume increased with the same urbanization speed from 2007 to 2017. This indicates that the strategy of sustainable development in Handan has worked well in the field of urban planning, urban construction, and urban flood security.

Several limitations are acknowledged in this study. Although the land use development and drainage system construction data were obtained from multiple sources (e.g., remote sensing data and local drainage conditions) to ensure the reliability of the results, the simulation still has limitations due to a lack of input data (e.g., spatial distribution and hydrological conditions of the soil, high-resolution DEM data, hydrological measurements of ground and underground drainage systems, and operation conditions of pumps). The availability of soil spatial distribution is an important factor in choosing a suitable infiltration model, which is crucial for calculating flood volumes. With higher resolution, the DEM data and land use data provided detailed information for two-dimensional hydraulic models for a more realistic simulation of flood distributions. Additionally, the social media information was a faulty substitute for historical water depth observations and hydrological records to validate the simulation results. The lost information on accurate water depth, shooting time, and location could undermine the reliability of validation. Thus, the lack of measuring stations constrains the building and verifying model of urban flood simulation in rapidly developing cities in recent years in China.

4. Conclusions

In response to the question of how the dual changes in land use and drainage systems affect the distribution of flooding, this study interpreted remote sensing images of Handan City from the past three decades. Then, the dual changes in hydrological conditions caused by ground and underground development were investigated, a stormwater simulation model was developed, and the results were verified using social media data in order to identify the combined effects of spatial variations in waterlogging and volume.

In this case study of Handan City, against the backdrop of rapid economic growth, there has been a boom in both ground and underground construction. In addition, land use change, represented by the expansion of sealed surfaces, has exerted a positive impact on the distribution and amount of flooding. While the impermeable area has increased from 29.74 km^2 to 70.22 km^2, the proportion of area susceptible to submergence under water depths over 70 cm has increased from 0.55% to 1.08%, and the total volume of flooding has increased from $7.89 \times 10^4 \text{ m}^3$ to $3.85 \times 10^5 \text{ m}^3$. Furthermore, the construction of the drainage network has increased from 99.77 km to 251.05 km, and the flooding volume has reduced from $5.31 \times 10^5 \text{ m}^3$ to $3.85 \times 10^5 \text{ m}^3$; however, the proportion of area susceptible to submergence under water depths over 70 cm has increased from 0.12% to 1.08%. The dual changes in land use and drainage pipeline network have increased the total flooding amount from $3.31 \times 10^5 \text{ m}^3$ to $3.85 \times 10^5 \text{ m}^3$. Finally, the proportion of areas susceptible to submergence under water depths of more than 70 cm has increased from 0.11% to 1.08%, while those susceptible to submergence between 30 cm and 70 cm of water grew from 0.41% to 0.93%. These joint effects show that the drainage system can mitigate the effect of land use change, although at the same time also aggravate the flooding conditions in impervious and riverside areas, especially in central zones.

The main findings of the research are as follows: (1) both surface and underground construction are highly sensitive to the distribution of flooding; (2) the land use change is a positive factor influencing the total flood volume, whereas the drainage system is negative, and the waterlogging risk increases in impervious areas and riverside areas of central zones; (3) as discussed previously, the simulation results still come with uncertainties due to condition simplification and data limitations in this study. The results of this study suggest that sustainable planning for drainage systems in rapidly urbanizing cities requires greater consideration of the long-term developments of city-wide land use, which necessitates cross-sectoral and cross-region cooperation.

Author Contributions: Conceptualization, X.W.; methodology, X.W.; software, C.X.; validation, J.Y.; formal analysis, B.L.; investigation, B.L.; writing—original draft preparation, X.W.; writing—review and editing, X.W. and B.L.; visualization, B.L. and S.L.; supervision, S.L.; funding acquisition, B.L. All authors have read and agreed to the published version of the manuscript.

Funding: This research was funded by the National Key Research and Development Program of China (grant number 2018YFC1508806) and the Major Science and Technology Program for Water Pollution Control and Treatment (grant number 2014ZX07203-008).

Acknowledgments: Dong, S.X. of the East Sewage Treatment Plant of Handan, Hebei, China, provided the data on the drainage system and aided with the field research. The authors would like to extend their deepest gratitude to her.

Conflicts of Interest: The authors declare no conflict of interest.

References

1. National Economy and Society Developed Statistical Bulletin 2021. Available online: http://www.stats.gov.cn/tjsj/zxfb/202202/t20220227_1827960.html (accessed on 28 February 2022).

2. McDonald, R.I.; Green, P.; Balk, D.; Fekete, B.M.; Revenga, C.; Todd, M.; Montgomery, M. Urban growth, climate change, and freshwater availability. *Proc. Natl. Acad. Sci. USA* **2011**, *108*, 6312–6317. [CrossRef] [PubMed]

3. The national office of Ministry of Emergency Management pubished Zhengzhou, Henan "7-20" Extraordinarily Heavy Rainfall Disaster Investigation Report of 2022. Available online: https://www.mem.gov.cn/xw/bndt/202201/t20220121_407106.shtml (accessed on 27 January 2022).

4. Chen, W.; Huang, G.; Zhang, H. Urban stormwater inundation simulation based on SWMM and diffusive overland-flow model. *Water Sci. Technol.* **2017**, *76*, 3392–3403. [CrossRef] [PubMed]

5. Shi, X.; Li, Y.; Yan, D. Advances in the Impacts of Watershed Land Use/Cover Change on Hydrological Process. *Res. Soil Water Censervation* **2013**, *20*, 301–308.

6. Schirmer, M.; Leschik, S.; Musolff, A. Current research in urban hydrogeology: A review. *Adv. Water Resour.* **2013**, *51*, 280–291. [CrossRef]

7. Lee, J.; Pak, G.; Yoo, C.; Kim, S.; Yoon, J. Effects of land use change and water reuse options on urban water cycle. *J. Environ. Sci.* **2010**, *22*, 923–928. [CrossRef]

8. Ertan, S.; Çelik, R.N. The Assessment of Urbanization Effect and Sustainable Drainage Solutions on Flood Hazard by GIS. *Sustainability* **2021**, *13*, 2293. [CrossRef]

9. Zang, W.; Liu, S.; Huang, S.; Li, J.; Fu, Y.; Sun, Y.; Zheng, J. Impact of Urbanization on Hydrological Processes under Different Precipitation Scenarios. *Nat. Hazards* **2019**, *99*, 1233–1257. [CrossRef]

10. Deng, X.; Shi, Q.; Zhang, Q.; Shi, C.; Yin, F. Impacts of land use and land cover changes on surface energy and water balance in the Heihe River Basin of China, 2000–2010. *Phys. Chem. Earth Parts A B C* **2015**, *79*, 2–10. [CrossRef]

11. Fletcher, T.D.; Andrieu, H.; Hamel, P. Understanding, management and modeling of urban hydrology and its consequences for receiving waters: A state of the art. *Adv. Water Resour.* **2013**, *51*, 261–279. [CrossRef]

12. Del Giudice, G.; Padulano, R. Sensitivity Analysis and Calibration of a Rainfall-runoff Model with the Combined Use of EPA-SWMM and Genetic Algorithm. *Acta Geophys.* **2016**, *64*, 1755–1778. [CrossRef]

13. Lenhart, T.; Eckhardt, K.; Fohrer, N. Comparison of Two Different Approaches of Sensitivity Analysis. *Phys. Chem. Earth Parts A/B/C* **2002**, *27*, 645–654. [CrossRef]

14. Zhou, Q.; Leng, G.; Su, J.; Ren, Y. Comparison of Urbanization and Climate Change Impacts on Urban Flood Volumes: Importance of Urban Planning and Drainage Adaptation. *Sci. Total Environ.* **2019**, *658*, 24–33. [CrossRef] [PubMed]

15. Miller, J.D.; Kim, H.; Kjeldsen, T.R.; Packman, J.; Grebby, S.; Dearden, R. Assessing the Impact of Urbanization on Storm Runoff in a Pen-urban Catchment Using Historical Change in Impervious Cover. *J. Hydrol.* **2014**, *515*, 59–70. [CrossRef]

16. Quan, R.S.; Liu, M.; Lu, M.; Zhang, L.J.; Wang, J.J.; Xu, S.Y. Waterlogging risk assessment based on land use/cover change: A case study in Pudong New Area, Shanghai. *Environ. Earth Sci.* **2010**, *61*, 1113–1121. [CrossRef]

17. Mahmoud, S.H.; Gan, T.Y. Urbanization and Climate Change Implications in Flood Risk Management: Developing an Efficient Decision Support System for Flood Susceptibility Mapping. *Sci. Total Environ.* **2018**, *636*, 152–167. [CrossRef]

18. Kong, F.; Ban, Y.; Yin, H.; James, P.; Dronova, I. Modeling stormwater management at the city district level in response to changes in land use and low impact development. *Environ. Model. Softw.* **2017**, *95*, 132–142. [CrossRef]

19. Jiang, Y.; Zevenbergen, C.; Ma, Y. Urban pluvial flooding and stormwater management: A contemporary review of China's challenges and "sponge cities" strategy. *Environ. Sci. Policy* **2018**, *80*, 132–143. [CrossRef]

20. Yazdanfar, Z.; Sharma, A. Urban Drainage System Planning and Design-challenges with Climate Change and Urbanization: A Review. *Water Sci. Technol.* **2015**, *72*, 165–179. [CrossRef]

21. Skougaard, K.P.; Ravn, N.H.; Arnbjerg-Nielsen, K.; Madsen, H.; Drews, M. Comparison of the Impacts of Urban Development and Climate Change on Exposing European Cities to Pluvial Flooding. *Hydrol. Earth Syst. Sci.* **2017**, *21*, 4131–4147. [CrossRef]

22. Zhu, Z.H.; Chen, Z.H.; Chen, X.H.; He, P.Y. Approach for Evaluating Inundation Risks in Urban Drainage Systems. *Sci. Total Environ.* **2016**, *553*, 1–12. [CrossRef]

23. Makropoulos, C.K.; Butler, D. Distributed water infrastructure for sustainable communities. *Water Resour. Manag.* **2010**, *24*, 2795–2816. [CrossRef]

24. Astaraie-Imani, M.; Kapelan, Z.; Fu, G.; Butler, D. Assessing the combined effects of urbanisation and climate change on the river water quality in an integrated urban wastewater system in the UK. *J. Environ. Manag.* **2012**, *112*, 1–9. [CrossRef]

25. Zou, Q.; Zhou, J.Z.; Zhou, C.; Guo, J.; Deng, W.P.; Yang, M.Q.; Liao, L. Fuzzy Risk Analysis of Flood Disasters Based on Interior-Outer-Set Model. *Expert Syst. Appl.* **2012**, *39*, 6213–6220. [CrossRef]

26. Peng, H.Q.; Liu, Y.; Wang, H.W.; Ma, L.M. Assessment of the service performance of drainage system and transformation of pipeline network based on urban combined sewer system model. *Environ. Sci. Pollut. Res.* **2015**, *22*, 15712–15721. [CrossRef] [PubMed]

27. Zhang, L.M. Hydraulic Calculation of Drainage Network and Simulation Method of Stormwater logging Study. Master Degree, South China University of Technology, Guangzhou, China, 2015.

28. Arora, A.S.; Reddy, A.S. Conceptualizing a decentralized stormwater treatment system for an urbanized city with improper stormwater drainage facilities. *Int. J. Environ. Sci. Technol.* **2015**, *12*, 2891–2900. [CrossRef]

29. Qin, D.H. *China National Assessment Report on Risk Management and Adaptation of Climate Extremes and Disasters*, 1st ed.; Science Press: Beijing, China, 2015; pp. 55–58.

30. Li, X.X.; Sang, Y.F.; Xie, P.; Liu, C.M. Stochastic Characteristics of Annual Extreme Rainfall with Different Durations and Their Spatial Difference in China. *J. Geo-Inf. Sci.* **2018**, *20*, 1094–1101.

31. Lee, J.; Chung, G.; Park, H.; Park, I. Evaluation of the Structure of Urban Stormwater Pipe Network Using Drainage Density. *Water* **2018**, *10*, 1444. [CrossRef]

32. Shuster, W.D.; Dadio, S.; Drohan, P.; Losco, R.; Shaffer, J. Residential demolition and its impact on vacant lot hydrology: Implications for the management of stormwater and sewer system overflows. *Landsc. Urban Plan.* **2014**, *125*, 48–56. [CrossRef]

33. Kleidorfer, M.; Mikovits, C.; Jasper-Tönnies, A.; Huttenlau, M.; Einfalt, T.; Raucha. Impact of a changing environment on drainage system performance. *Procedia Eng.* **2014**, *70*, 943–950. [CrossRef]

34. Semadeni-Davies, A.; Hernebring, C.; Svensson, G.; Gustafsson, L. The impacts of climate change and urbanisation on drainage in Helsingborg, Sweden: Combined sewer system. *J. Hydrol.* **2008**, *350*, 100–113. [CrossRef]

35. Wang, Y.L.; Yang, X.L. Land use/cover change effects on floods with different return periods: A case study of Beijing, China. *Front. Environ. Sci. Eng.* **2013**, *7*, 769–776. [CrossRef]

36. Yang, X.L.; Chen, H.L.; Wang, Y.L. Evaluation of the effect of land use/cover change on flood characteristics using an integrated approach coupling land and flood analysis. *Hydrol. Res.* **2016**, *47*, 1161–1171. [CrossRef]

37. Handan City Statistical Bulletin of National Economic and Social Development in 2021. Available online: https://www.hd.gov.cn/hdzfxxgk/gszbm/auto23694/202203/t20220324_1549667.html (accessed on 21 March 2022).

38. Guo, F.; Guo, J.; Liu, Q.; Huo, Y. Characteristics of Water Quality Change in Different Underlaying Surface Rainfall Runoff of Handan City. *Water Sci. Eng. Technol.* **2012**, *1*, 85–87.

39. Zhou, S.; Zhai, C. Rainfall and Runoff Relationship Analysis in the New Residential Area. *South-to-North Water Transf. Water Sci. Technol.* **2013**, *11*, 22–24.

40. Wang, X.; Yang, X.L.; Xu, C.W.; Liu, Z.Y. Land use-based analysis of waterlogging traffic risk. *J. Nat. Disasters* **2018**, *05*, 197–204.

41. Yan, Z.; Sha, J.; Liu, B.; Tian, W.; Lu, J. An Ameliorative Whale Optimization Algorithm for Multi-Objective Optimal Allocation of Water Resources in Handan, China. *Water* **2018**, *10*, 87. [CrossRef]

42. Wang, L.; Yang, X.L. Estimation of Environmental Water Requirements via an Ecological Approach: A Case Study of Yongnian Wetland, Haihe Basin, China. *Sustain. Dev. Water Resour. Hydraul. Eng. China* **2019**, *5*, 377–386.

43. Fu, H.; Yang, X.L. Effects of the South-North Water Diversion Project on the Water Dispatching Pattern and Ecological Environment in the Water Receiving Area: A Case Study of the Fuyang River Basin in Handan, China. *Water* **2019**, *11*, 845. [CrossRef]

44. Zhang, D.; Zhao, D.; Chen, J. Application of Chicago hyetograph method in the drainage system simulation. *Water Wastewater Eng.* **2008**, *34*, 354–357.

45. Xiong, L.; Huang, F. Characteristics of rainfall and runoff in urban drainage based on the SWMM model. *Chin. J. Appl. Ecol.* **2016**, *27*, 3659–3666.

46. Zhang, P.L. Research on Simulation Model of Storm Runoff in Tianjin Urban District. Master Degree, Tianjin University, Tianjin, China, 2007.

47. Chen, S.S. Study on the Simulation and Utilization of storm Water in Urban Area. Master Degree, College of Hydrology and Water Resource, Nanjing, China, 2007.

48. Smith, L.; Liang, Q.; James, P.; Lin, W. Assessing the utility of social media as a data source for flood risk management using a real-time modelling framework. *J. Flood Risk Manag.* **2017**, *10*, 370–380. [CrossRef]

49. Wang, Y.; Wang, T.; Ye, X.Y.; Zhu, J.Q.; Lee, J. Using Social Media for Emergency Response and Urban Sustainability: A Case Study of the 2012 Beijing Rainstorm. *Sustainability* **2016**, *8*, 25. [CrossRef]

50. Wang, B.; Loo, B.P.Y.; Zhen, F.; Xi, G.L. Urban resilience from the lens of social media data: Responses to urban flooding in Nanjing, China. *Cities* **2020**, *106*, 102884. [CrossRef]

51. Roy, A.G.; Roy, R.; Bergeron, N. Hydraulic geometry and changes in flow velocity at a river confluence with coarse bed material. *Earth Surf. Process. Landf.* **1988**, *13*, 583–598. [CrossRef]

52. Gaudent, J.M.; Roy, A.G. Effect of bed morphology on flow mixing length at river confluences. *Nature* **1995**, *373*, 138–139. [CrossRef]

53. Li, J.H.; Zhang, L.T.; Fu, Z.J.; Ji, M. The Experience of New Energy Vehicles Wading Test According to Shanghai Local Regulations. *Automob. Appl. Technol.* **2021**, *20*, 138–146.

54. Mikovits, C.; Rauch, W.; Kleidorfer, M. Importance of scenario analysis in urban development for urban water infrastructure planning and management. *Comput. Environ. Urban Syst.* **2018**, *68*, 9–16. [CrossRef]

55. Zhou, Q.; Luo, J.; Su, J.; Ren, Y. Impacts of changing drainage indicators on urban flood volumes in historical urbanization in the case of Northern China. *Urban Water J.* **2021**, *18*, 487–498. [CrossRef]

56. Li, C.; Wang, W.; Xiong, J.; Chen, P. Sensitivity Analysis for Urban Drainage Modeling Using Mutual Information. *Entropy* **2014**, *16*, 5738–5752. [CrossRef]

57. Su, W.; Duan, H. Catchment-based Imperviousness Metrics Impacts on Floods in Niushou River Basin, Nanjing City, East China. *Chin. Geogr. Sci.* **2017**, *27*, 229–238. [CrossRef]

Article

Predicting Factors Affecting Preparedness of Volcanic Eruption for a Sustainable Community: A Case Study in the Philippines

Josephine D. German [1,2], Anak Agung Ngurah Perwira Redi [3], Ardvin Kester S. Ong [1,*], Yogi Tri Prasetyo [1,4] and Vince Louis M. Sumera [5]

1   School of Industrial Engineering and Engineering Management, Mapúa University, 658 Muralla St., Intramuros, Manila 1002, Philippines
2   School of Graduate Studies, Mapúa University, 658 Muralla St., Intramuros, Manila 1002, Philippines
3   Industrial Engineering Department, Sampoerna University, Jakarta 12780, Indonesia
4   Department of Industrial Engineering and Management, Yuan Ze University, 135 Yuan-Tung Road, Chung-Li 32003, Taiwan
5   Department of Civil Engineering and Geological Engineering, Mapúa University, 658 Muralla St., Intramuros, Manila 1002, Philippines
*   Correspondence: aksong@mapua.edu.ph; Tel.: +63-(2)8247-5000 (ext. 6202)

Citation: German, J.D.; Redi, A.A.N.P.; Ong, A.K.S.; Prasetyo, Y.T.; Sumera, V.L.M. Predicting Factors Affecting Preparedness of Volcanic Eruption for a Sustainable Community: A Case Study in the Philippines. *Sustainability* 2022, 14, 11329. https://doi.org/10.3390/su141811329

Academic Editors: Chong Xu and Jian Chen

Received: 31 July 2022
Accepted: 4 September 2022
Published: 9 September 2022

Abstract: Volcanic eruption activity across the world has been increasing. The recent eruption of Taal volcano and Mt. Bulusan in the Philippines affected several people due to the lack of resources, awareness, and preparedness activities. Volcanic eruption disrupts the sustainability of a community. This study assessed people's preparedness for volcanic eruption using a machine learning ensemble. With the high accuracy of prediction from the ensemble of random forest classifier (93%) and ANN (98.86%), it was deduced that media, as a latent variable, presented as the most significant factor affecting preparedness for volcanic eruption. This was evident as the community was urged to find related information about volcanic eruption warnings from media sources. Perceived severity and vulnerability led to very high preparedness, followed by the intention to evacuate. In addition, proximity, subjective norm, and hazard knowledge for volcanic eruption significantly affected people's preparedness. Control over individual behavior and positive attitude led to a significant effect on preparedness. It could be posited that the government's effective mitigation and action plan would be adhered to by the people when disasters, such as volcanic eruptions, persist. With the threat of climate change, there is a need to reevaluate behavior and mitigation plans. The findings provide evidence of the community's resilience and adoption of mitigation and preparedness for a sustainable community. The methodology provided evidence for application in assessing human behavior and prediction of factors affecting preparedness for natural disasters. Finally, the results and findings of this study could be applied and extended to other related natural disasters worldwide.

Keywords: volcanic eruption; artificial neural network; random forest classifier; natural disaster; machine learning algorithm

1. Introduction

Volcanic eruptions have been widely monitored and assessed for mitigation and preparation worldwide [1,2]. There are 1508 active volcanoes across 86 countries, and a population of 29 million is estimated to live within a 10-km range of volvanoes, while 800 million live within 100 km [3]. With communities living in areas of active volcanoes, the limited assessment of behavior for preparation has been widely underexplored. Niroa and Nakamura [4] explained how frameworks had been developed to assess the behavioral aspects of volcanic risks, coping mechanisms, and perception of the impact of eruption aftermath. However, with beliefs, cultural differences, and ways of governance, an assessment for community preparedness should still be considered.

Brown et al. [3] have identified several risks with volcanic eruptions, which have caused 278,368 fatalities from 1500AD to 2017. Several studies have started defining and assessing the characteristics of the population living in active volcanoes, their socio-economic status, and their perception of risk and knowledge [5]. With studies such as Reyes-Hardy et al. [6] and Barone et al. [7], it could be posited that information regarding volcanic eruption has been developed and is still developing mitigation plans. On the other hand, volcanic risk assessment for preparedness has been a trend in developing countries [2]. It was stated that community vulnerability has scarce data, and people's behavioral intentions should be explored more [7].

In the Republic of Congo, Michellier et al. [8] assessed the community's vulnerability to risks of a volcanic eruption. Their study focused on the population vulnerability assessment using the Social Vulnerability Index and community exposure to lava flow. Their results showed a mitigation plan using the Operational Vulnerability Index to communicate with the government for action. In Chile, Reyes-Hardy et al. [6] explored the vulnerability, volcanic hazard, and overall risk assessment using GIS-based volcanic hazard. It was seen that their model was able to assess areas of high vulnerability. However, their study focused on the physical, territorial, and social contexts. In Indonesia, Thouret et al. [5] analyzed hazard knowledge, socio-demographic characteristics, and community adaptation to volcanic threats. Their study considered hierarchical clustering (HC) to assess livelihood, demographics, and sustainability among resilient communities living in the active volcano area. However, it was evident that HC was limited to smaller datasets, user set cluster numbers, and, once done, HC cannot be returned to the original state. In Vanuatu, Niroa and Nakamura [4] assessed the indigenous disaster risk reduction framework for volcanic hazards. Their results showed that culture and belief played the most significant role in why the population would plan to mitigate the risk of a volcanic eruption. In addition, the study of Dogar and Sato [9] highlighted the consideration of the regional climate response of El Niño-Southern Oscillation (ENSO), which foregoes with volcanic eruption. They presented how ENSO phenomena occur due to volcanic explosion and that understanding of this should be considered. Due to high climate variability in the present time, all volcanic eruptions and related phenomena should be part of the discovery for human preparedness. In the study, climate variability and how it can greatly affect the phenomena of natural disasters was uncovered. Similarly, regional climate has been discussed as a sensitive aspect affecting climate trends [10]. These coincidences should be considered upon evaluation of volcanic eruptions, which lead to behavioral changes once all aspects have been considered [9,10]. Similar patterns of regional climate change were seen in the study by Dogar et al. [11]. Their study expounded on the hemispheric climate and temperature anomalies since the eruption of Mt. Pinatubo in 1991. Despite these findings, challenges in assessing behaviors for volcanic eruption preparedness were evident.

In the Philippines, Kurata et al. [2] assessed the preparation beliefs of people towards the Taal volcanic partial eruption. Their study also presented that the Philippines has been ranked third highest in the disaster risk index where active volcanoes are evidently present, as seen in Figure 1. Figure 1 represents the active and potentially active volcanoes in the Philippines [2]. With their study, several relationships were seen to be insignificant. In addition, the results showed that the causal relationship was seen to be a challenge. This is because structural equation modeling (SEM) was utilized. Fan et al. [12] expounded on the limitations of utilizing SEM as the multivariate tool to assess behaviors. It was explained that the farther the independent variable, the lower might significance be present. Moreover, the presence of mediating effects also hinders the value of the beta coefficient, representing the significance level of the independent latent variable [13]. Thus, the current studies have utilized a machine learning algorithm hybrid with SEM. Duarte and Pinho [14] have presented how the framework integration suffices within the limitations of the sole SEM methodology.

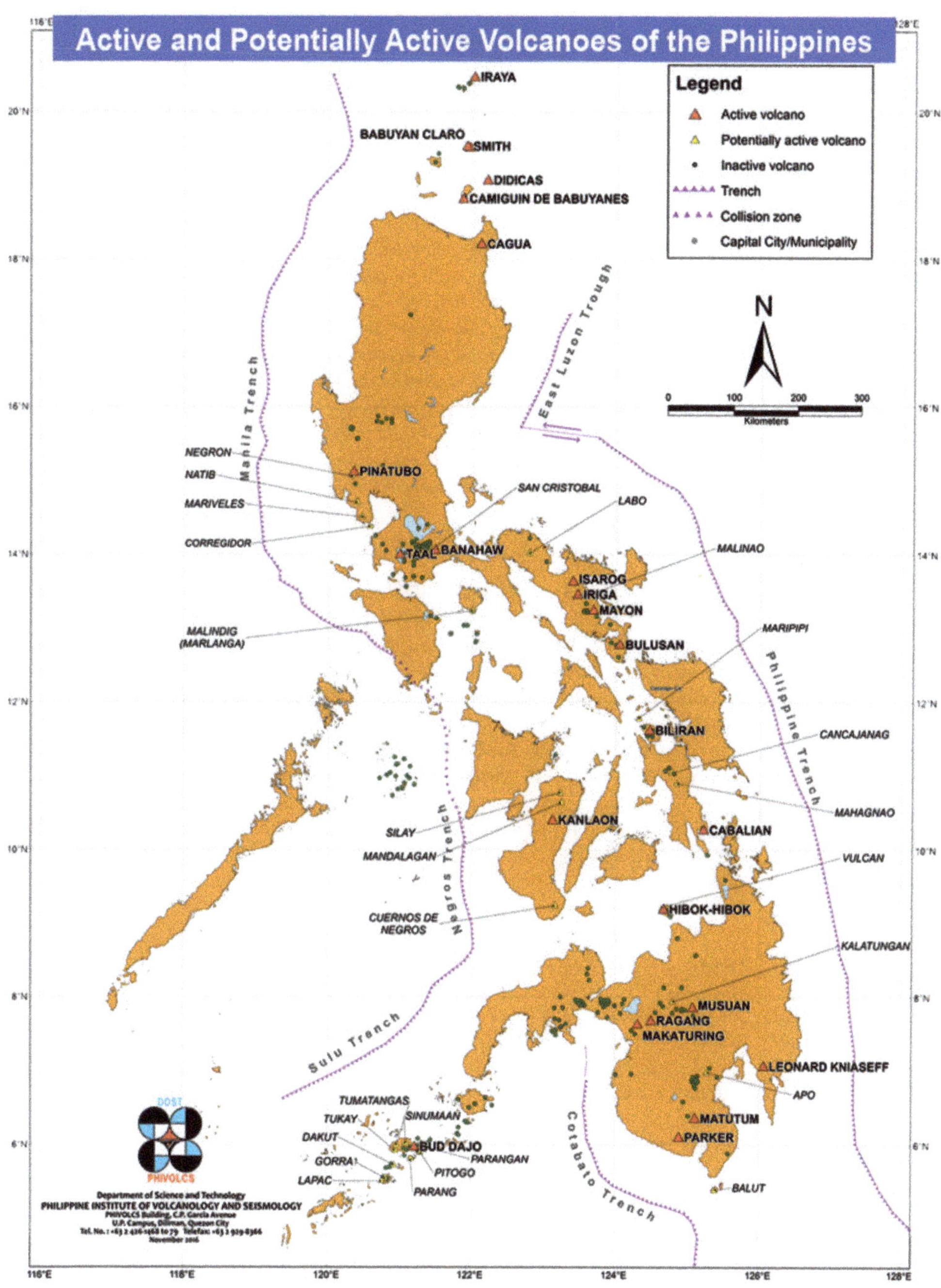

Figure 1. Philippine Geographic Location of Active and Potentially Active Volcanoes.

As of March 2022, the Taal volcano is still active with small earthquakes and seismic movement [15]. A short burst of 1500 m of ash means residents near the area are advised to leave the premises. Another active volcano in the Philippines, Mount Bulusan in

Sorsogon, spewed volcanic ash which covered the whole town in June 2022 [16,17]. The need for evacuation among people was immediately coordinated as the community was left surprised. The World Data [18] presented that there has been a total of 44 eruptions in the past 400 years, leaving 7400 casualties. People have been seen to be unprepared for volcanic eruptions that may happen in their proximity. The integrated Theory of Planned Behavior (TPB) and Protection Motivation Theory (PMT) may be applied to assess the intention and preparedness of people for a possible volcanic eruption.

Several studies have considered utilizing extended PMT [19] or the integration of TPB with PMT [20–22] to assess calamities, health-related mitigation, and natural disaster preparedness. It was seen that the integrated framework could holistically assess human behavior for intention, mitigation, and preparedness for natural disasters. Gumasing et al. [19] extended the PMT to measure the effects of response efficacy of people towards typhoons. However, limitations in the measurement of the behavioral aspect was seen. Ong et al. [20] assessed the intention to prepare for mitigation of "The Big One" earthquake that is expected to happen in the Philippines, utilizing SEM. Kurata et al. [20] measured the effectiveness of response for typhoons utilizing the integrated framework with SEM. In addition, Prasetyo et al. [22] utilized the integrated framework to measure the effectiveness of community quarantine during the COVID-19 pandemic utilizing SEM. It was seen among all studies that the integrated framework would holistically measure peoples' intentions and mitigation behavior towards health-related disasters and natural disasters. In addition, the extension by adding latent variables was effective with the framework.

The study by Kurata et al. [2] assessed people's preparedness for the Taal volcano. This study aimed to assess the limitations and decipher the most significant contributing factor to the preparedness for a volcanic eruption in the Philippines. Several factors under PMT, such as hazard knowledge, perceived risk proximity, perceived severity, and perceived vulnerability, were considered. In addition, factors of TPB, such as perceived behavioral control, subjective norm, and attitude, were also considered to assess intention to evacuate and preparedness, with media as an extension of the integrated framework. Machine learning algorithm (MLA) ensembles, such as artificial neural network (ANN) and random forest classifier (RFC), were used to analyze the latent variables simultaneously. Similar to the study by Ong et al. [23,24], utilizing a machine learning ensemble was claimed to be more effective in analyzing factors affecting human behavior relating to the use of technology. However, no studies have utilized MLA ensemble to assess preparedness for natural disasters such as volcanic eruptions.

The present research is considered the first to analyze and predict factors affecting volcanic eruption preparedness using MLA. This study's results would benefit researchers considering a new method of assessing human behavior worldwide. In addition, governments could utilize this study's findings to create mitigation plans for an emergency response to volcanic eruption events worldwide. The content of the manuscript is as follows: (1) introduction, (2) conceptual framework, (3) methodology and data analysis, (4) results, (5) discussion and recommendation, and (6) conclusion.

2. Conceptual Framework

Presented in Figure 2 is the conceptual framework utilized in this study through the integration of PMT and TPB. The TPB and PMT factors are separated, as shown by broken lines in the figure. In this framework, TPB was considered as a whole since several studies [20–22] have shown how PMT affects TPB through Perceived Severity and Perceived Vulnerability. This study utilized an MLA ensemble, which could more effectively measure the non-linear relationship present in the framework [23,24]. A total of nine hypotheses were considered to evaluate factors affecting the preparedness for volcanic eruption of people within proximity of active volcanoes. The further building of hypotheses is discussed in this section.

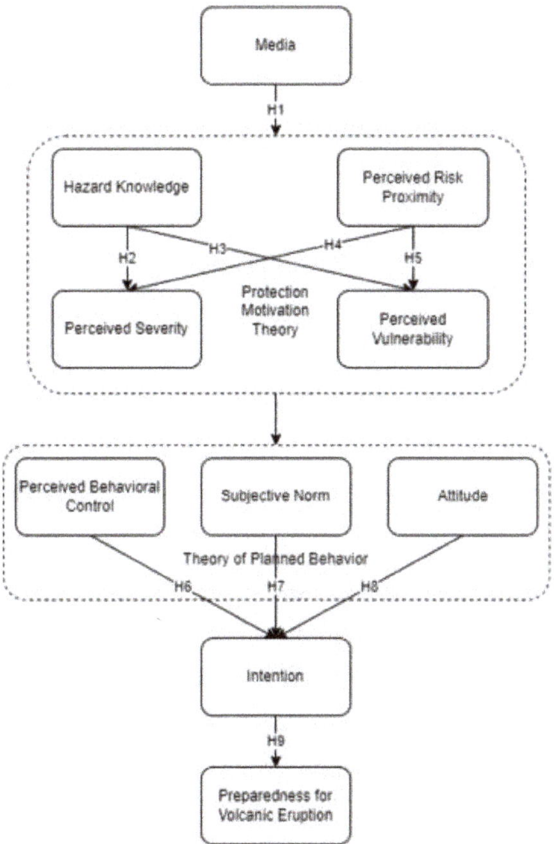

Figure 2. Conceptual Framework.

Media as the primary information source was considered an extended latent variable in this study. Media was seen to affect factors under PMT, as indicated by Ong et al. [20]. Ong et al. [24] stated that the community's primary sources of news and information would come from different media sources, such as television, newspapers and articles, and social media. The study of Kurata et al. [2] also presented how media is the main contributing factor that affects information concerning a natural disaster. Weichselgartner and Pigeon [25] also expounded on the importance of media as a source of knowledge and information regarding disasters to promote mitigation and action plans. Phengsuwan et al. [26] presented how media is a significant factor in enhancing management of disaster risks. It was seen how it delivers perception and knowledge regarding monitoring, risks, and susceptibility of natural disasters, which would lead to the intention and preparation for what may happen during, and in the aftermath of, a natural disaster. Therefore, it was hypothesized that:

H1. *Media is a significant factor affecting Preparedness for Volcanic Eruption.*

Hazard Knowledge is the information obtained, together with people's prior experiences, regarding natural disasters. Dube and Munsaka [27] presented that knowledge from prior experience would lead to the perception of severity and vulnerability among the community when a natural disaster happens. Through patterns and recognition, it was seen that the community would be ready for natural disasters such as floods. With volcanic eruptions, it was seen that Hazard Knowledge played a significant role in the Perceived Severity and Perceived Vulnerability, which would lead to the enhanced preparedness of people [2]. If the community has a heightened perception of the risks that may affect their health, they would be more inclined to prepare for it [28]. Thus, the following were hypothesized:

H2. *Hazard Knowledge is a significant factor affecting Perceived Severity for Preparedness for Volcanic Eruption.*

H3. *Hazard Knowledge is a significant factor affecting Perceived Vulnerability for Preparedness for Volcanic Eruption.*

The proximity of the source for natural disasters plays a significant role, especially for volcanic eruptions [3]. Arias et al. [29] explained how the population living near natural disasters has a higher perception of risk (i.e., perceived severity and perceived vulnerability). Yagoub and Al Yammahi [30] justified how the spatial distribution of hazard proximity dramatically affects the severity and vulnerability of the community. Knowledge and experience affect the perceived severity and perceived vulnerability due to how close people are to the source of disasters [29]. However, it was seen that some people would perceive the risk to be lower living near the source, due to their heightened preparedness from experience. In addition, Rana et al. [31] presented how proximity to the source of disaster increases the perceived risks, such as severity and vulnerability, which, in turn, leads to an increase in preparation. Therefore, the following were hypothesized:

H4: *Perceived Risk Proximity is a significant factor affecting Perceived Severity for Preparedness for Volcanic Eruption.*

H5: *Perceived Risk Proximity is a significant factor affecting Perceived Vulnerability for Preparedness for Volcanic Eruption.*

Factors under TPB, such as Perceived Behavioral Control, Subjective Norm, and Attitude, are the main latent variables that could be used to assess behavior altogether [32]. Ong et al. [20] have presented how AT, followed by SN, and PBC consequently affected the intention to prepare for natural disasters, preceded by factors under PMT. Their study expounded on how the indirect effect of people's understanding and knowledge led to an increase in perceived risks, affecting the intention to prepare for a natural disaster. The study by Kurata et al. [21] showed how factors of TPB were all highly significant to the perceived effectiveness of preparation for a natural disaster that may happen. In addition, Vinnell et al. [33] also considered TPB latent variables to assess preparation for natural hazards. Their study showed that distinct features of these latent variables holistically measure the community's preparedness, highlighting how TPB alone lacks dimensions for a total measurement of behavior. As supported, this study preceded TPB with PMT to holistically measure health-related behaviors for natural disasters. Thus, it was hypothesized that:

H6. *Perceived Behavioral Control is a significant factor affecting Preparedness for Volcanic Eruption.*

H7. *Subjective Norm is a significant factor affecting Preparedness for Volcanic Eruption.*

H8. *Attitude is a significant factor affecting Preparedness for Volcanic Eruption.*

Intention to prepare has been seen to significantly and highly affect preparedness for natural disasters [20]. Najafi et al. [34] expounded on how TPB factors greatly affected intention, which led to people's preparedness for natural disasters. The motivation of people through their intentions would significantly affect their preparedness. Bronfman et al. [35] discussed how levels of preparedness increased people's intention, especially if the community is exposed to significant hazards from natural disasters. Bourque et al. [36] also explained how the behavioral aspect of people affected their intention and significantly increased measures of preparedness. Heller et al. [37], Kurata et al. [21], and Bourque et al. [36] explained that people's preparedness is affected by their experience, knowledge, perception of risk, and behavioral aspect. When people know the adverse effects of hazards on health, both intention and preparation were considered directly proportional [38]. Therefore, it was hypothesized that:

H9. *Intention is a significant factor affecting Preparedness for Volcanic Eruption.*

3. Methodology

3.1. Demographics

A total of 653 valid responses were collected for this study resulting in 39,833 datasets. The data was collected from January to March 2022 through social media platforms using Google Forms with the survey questionnaire adopted from the study of Kurata et al. [2]. Through convenience sampling, no nonresponse and missing data were seen from the collected responses. Performing the Common Method Bias, 25.21% was attained, indicating no CMB in the responses [39]. The data presented 49.46% male and 50.54% female within an age range of 25–34 years old (54.52%), followed by 18–24 years old (28.02%), 35–45 years old (10.87%), and the rest were older than 45. From the responses, the majority were employed (60.16%), students (34.24%), and unemployed (5.60%) with monthly incomes within 15,000–30,000 PhP or 30,001–45,000 PhP. Most of the respondents were single and living in rural areas (50.66%) rather than urban areas (49.34%). In addition, some respondents were insured (55.67%), while 44.33% answered that they are not. A total of 79% answered that they had their own house and lot, rather than living in a condominium or renting.

3.2. Random Forest Classifier

Data pre-processing was conducted to analyze the dataset for input among the machine learning ensemble. Missing data and outliers were checked upon pre-processing and correlation analysis was conducted to determine insignificant indicators. From the correlation analysis, a threshold following the study of Ong et al. [23,24] was set with a p-value of 0.05 and coefficient of 0.20. Following which, data was aggregated using mean values representing the different factors. These data were set as the input for the machine learning ensemble run through Python 5.1.

A random forest classifier (RFC) is a classification model that considers a simple algorithm with higher prediction accuracy. RFC has been widely utilized in decision-making, natural disaster, and human behavior studies. Kim et al. [40] considered RFC to classify seismic facies. It was proven from their study that RFC produces better accuracy compared to the basic decision tree. Flood damage analysis was run using RFC by Snehil and Goel [41] and showed how an increase in prediction based on accuracy has been achieved with the simple RFC algorithm. Chen et al. [42] considered the MLA ensemble of RFC and artificial neural network (ANN) in predicting disaster risk from a flood in China. It was proven that an MLA ensemble could predict human behavior and risk assessment regarding natural disasters, specifically floods.

In addition, Yang and Zhou [43] analyzed carbon emission among residents with different tree classification techniques and presented how RFC would be primarily applicable with its capabilities to generate the best decision tree among other classifiers. Its ability to classify the optimum output from its criterion, splitter, and depth provides its advantage over the other tree classifiers. Similarly, Ong et al. [24] compared RFC and the basic DT in terms of calculation and output. It was seen that RFC dominated with a high accuracy rate. A similar study in [23] provided the pseudocode for the RFC and represented how advantageous RFC is even at the optimization stage. All possible parameters, such as the criterion, splitter, tree depth, training, and testing ratios, could be processed and analyzed in RFC, which presents a tool that can produce the optimum result.

Therefore, this study considered RFC and optimized the parameters to produce the best tree to serve as a classification model. Several parameters for the criterion and splitter, such as gini or entropy and random or best, were considered, similar to the study of Ong et al. [23,24]. In addition, the tree depth was considered among four to seven and run through different training and testing ratios. A total of 6400 runs for 100 combinations each were employed.

3.3. Artificial Neural Network

ANN has a more complex calculation and algorithm compared to the RFC. ANN has been utilized among other studies considering natural disasters and human behavior. Moustra et al. [44] utilized this algorithm for earthquake prediction. Their study focused on historical datasets to analyze the geographic locations of earthquakes. Yariyan et al. [45] considered it for risk assessment mapping for the vulnerability of earthquakes. Oktarina et al. [46] studied earthquake casualties and damages in Indonesia with ANN. Their study presented higher pattern recognition with high accuracy upon utilizing ANN. Similarly, Ong et al. [24] considered RFC and ANN to predict factors affecting the acceptance of Bataan reopening, a decommissioned nuclear power plant in the Philippines. Due to their limitations, these studies presented how ANN could be considered instead of multivariate and traditional statistical tools.

ANN, according to Jamshidi et al. [47], Ong et al. [23,24], and Yuduang et al. [38], could be a classification tool best suited to analyze factors affecting human behavior. Most studies consider ANN before proceeding to other types of neural networks (e.g., Deep Learning). It was stated that if ANN's accuracy and complexity power produces a low output, then deep learning may be considered. Nonetheless, ANN is considered sufficient when high predictive power is obtained. General neural networks are considered cutting-edge algorithms despite level. ANN was compared to other classification techniques, such as K-Nearest Neighbors and Naïve Bayes, which also have high predictive powers. However, Ong et al. [24] reiterated the applicability of ANN towards human behavior for relatively simple frameworks, similar to this study. Moreover, powerful algorithms should be applied if the study considers complex frameworks for analysis.

The input data for RFC, optimization for the parameters, several activation functions for the hidden layer, such as sigmoid, tanh, and relu, and the number of nodes for the input layer were considered for the generation of the ANN model utilizing Python 5.1. In addition, the activation functions for the output layer considered sigmoid and swish. Lastly, adam, SGD, and RMSProp were considered for the optimizers. In total, 16,200 runs were employed for ten combinations each at 150 epochs [48]. From the results, the best classification model considered parameters such as tanh for the hidden layer and sigmoid for the output and ran with adam with 50 nodes at the hidden layer. Further optimization for different training and testing ratios were conducted.

4. Results

4.1. Descriptive Statistics of the Items

The items utilized in this study were adopted from the study by Kurata et al. [2], as seen in the Appendix A of the manuscript. Presented in Table 1 are the descriptive statistical results of the items from the collected responses. It could be seen that average standard deviation and mean values were within the range of normality, ±1.96 using the Harman's Single Factor Analysis [20]. In addition, Cronbach's alpha results presented acceptable values, greater than 0.70, which indicated that the collected data could be used to assess factors affecting volcanic eruption preparedness among Filipinos.

Table 1. Indicators statistical analysis.

Variable	Item	Mean	StD	Cronbach's Alpha
Media	MP1	4.303	0.833	0.803
	MP2	4.383	0.799	
	MP3	4.663	0.614	
	MP4	4.381	0.761	
Hazard Knowledge	HK1	3.512	1.153	0.737
	HK2	4.809	0.485	
	HK3	4.778	0.527	
	HK4	4.643	0.698	
	HK5	4.682	0.650	
Perceived Risk Proximity	GP1	2.380	1.260	0.773
	GP2	2.753	1.290	
	GP3	3.599	1.142	
Perceived Severity	PS1	4.052	0.840	0.829
	PS2	4.443	0.725	
	PS3	3.490	0.993	
	PS4	3.452	1.256	
	PS5	3.299	1.256	
Perceived Vulnerability	PV1	3.054	1.200	0.851
	PV2	2.925	1.214	
	PV3	2.510	1.262	
	PV4	4.084	1.100	
Perceived Behavioral Control	PBC1	3.512	1.005	0.758
	PBC2	3.564	0.944	
	PBC3	3.675	1.001	
	PBC4	3.561	0.974	
Social Norm	SN1	3.216	1.010	0.837
	SN2	3.006	1.104	
	SN3	2.852	1.091	
	SN4	3.153	1.066	
	SN5	3.544	1.030	
Attitude Toward the Behavior	ATB1	3.392	1.107	0.796
	ATB2	4.138	0.965	
	ATB3	4.092	0.993	
	ATB4	4.283	0.820	
	ATB5	3.276	1.047	
Intention	IF1	4.496	0.702	0.844
	IF2	4.193	0.925	
	IF3	4.133	0.933	
	IF4	4.158	0.961	
Perceived Preparedness	PP1	3.674	0.975	0.822
	PP2	4.126	0.804	
	PP3	4.331	0.746	
	PP4	4.406	0.783	
	PP5	4.554	0.687	

4.2. Random Forest Classifier

Figure 3 represents the best tree with RFC among all results tested. From the figure, it could be seen that the parent node indicated media (X1) as the factor dictating preparedness for a volcanic eruption, with a value less than, or equal to, 0.386. Satisfying this would consider perceived severity (X0) with values less than, or equal to, 2.13. Satisfying this would consider X0, X1, and perceived vulnerability (X2), leading to high preparedness for a volcanic eruption. However, if X0 was not satisfied, it would consider X1 and X2, leading to high preparedness for a volcanic eruption.

On the other hand, if the parent node would not be satisfied, it would consider X0 with values less than, or equal to, 0.123. Satisfying this condition would consider X1 and X2 leading to high preparedness for a volcanic eruption. If this was not satisfied, it would consider X1 and intention (X3) with values less than, or equal to, −0.074 leading to very high preparedness. If the child node was not satisfied, it would consider X2, X3, and X2 leading to high preparedness, indicating that X0, X1, and X3 were the highly significant factors affecting preparedness for a volcanic eruption. X2, on the other hand, was considered a significant factor that would only highly affect preparedness.

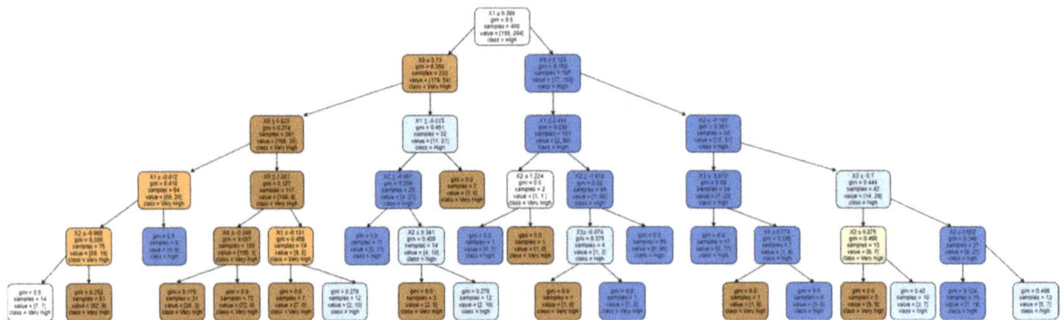

Figure 3. Classification Model from Random Forest Classifier. Legends: X0—Perceived Severity (PS); X1—Media (M); X2—Perceived Vulnerability (PV); X3—Intention (IN).

Table 2 represents the summary of accuracy results for the optimization. At depth 5, the best tree with RFC was produced. A 93% accuracy with a 0.00 standard deviation was seen as the highest among the results. Utilizing ANOVA, no significant difference was seen; thus, the highest accuracy with the lowest standard deviation was considered for the model. Gini and Best as the criteria and splitter at 80:20 training and testing ratio produced the best tree.

Table 2. Decision Tree Mean Accuracy (Depth = 5).

Category	60:40	70:30	80:20	90:10
		Random		
Gini	88.00	88.60	90.00	88.60
Std. Dev	0.000	3.848	0.000	0.548
Entropy	90.20	91.00	86.60	81.60
Std. Dev	1.096	0.000	2.510	1.140
		Best		
Gini	92.00	90.20	93.00	92.80
Std. Dev	0.000	0.837	0.000	2.280
Entropy	89.60	88.20	92.00	89.80
Std. Dev	0.894	3.114	0.000	1.483

Figure 4 represents the scatter plot for the RFC accuracy results among Gini Index Criterion. It could be seen that the scatter plot peaked at the highest average accuracy of 93%.

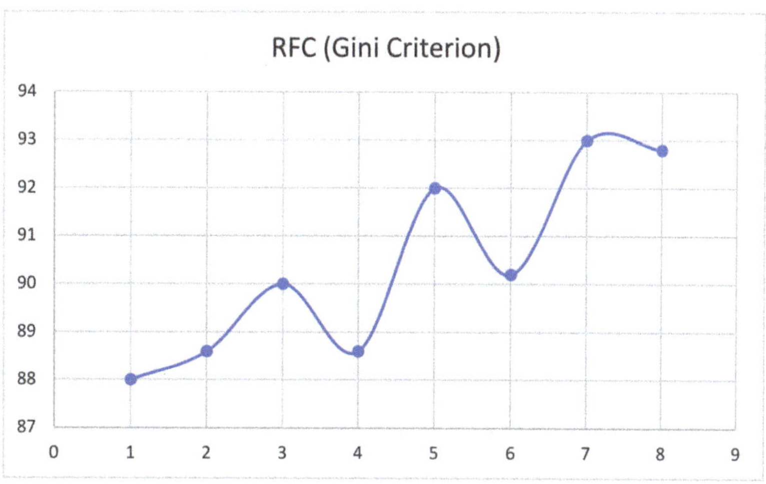

Figure 4. RFC Scatter Plot.

4.3. Artificial Neural Network

Table 3 presents the summary of ANN after the final optimization run. At 200 epochs, the final results from the 80:20 training and testing ratio are presented. According to the study of Ong et al. [23,24], the average testing sequence dictates the significance ranking among the latent variables considered. It could be deduced that media (M) presented as the contributing factor affecting preparedness for a volcanic eruption, followed by perceived severity (PS), perceived vulnerability (PV), and intention (IN). Other significant factors identified were perceived risk proximity (PR), subjective norm (SN), hazard knowledge (HK), perceived behavioral control (PBC), and attitude (AT). The threshold was 60%, and anything lower was not considered significant [34].

Table 3. Summary of ANN.

Latent	Average Training	Standard Deviation	Average Testing	Standard Deviation
Media	96.11	3.137	97.89	0.837
Perceived Severity	91.30	3.224	97.33	1.782
Perceived Vulnerability	84.30	2.515	97.17	1.286
Intention	81.93	0.377	96.95	1.839
Perceived Risk Proximity	88.93	1.400	95.65	2.133
Subjective Norm	86.79	0.979	95.34	2.173
Hazard Knowledge	86.49	3.130	87.10	6.200
Perceived Behavioral Control	67.86	6.015	83.44	5.752
Attitude	66.04	2.342	79.92	4.580

Figure 5 represents the scatter plot for the ANN average testing accuracy. It could be seen that the sequence of the results presented Media as the highest latent variable, followed by Perceived Severity, Perceived Vulnerability, Intentions, Perceived Risk Proximity, Subjective Norm, Hazard Knowledge, Perceived Behavioral Control, and Attitude as the lowest.

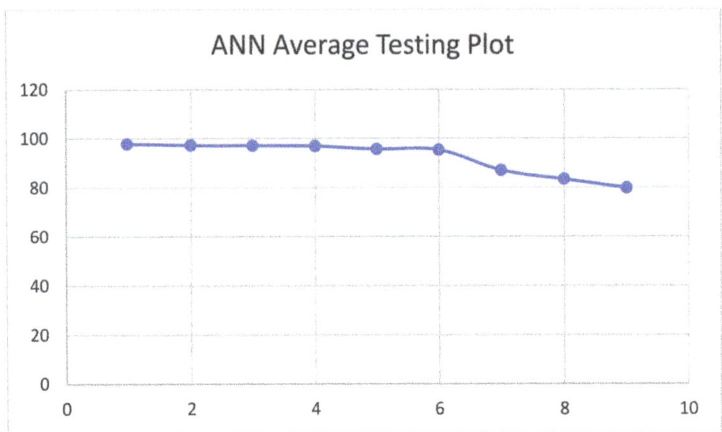

Figure 5. ANN Average Testing Accuracy Scatter Plot.

The score of importance was also computed to verify the findings of ANN. Similar results of significant factor ranking were seen, as presented in Table 4. In addition, the training and validation loss rate was also assessed, as presented in Figure 6. Following the suggestion of Lara et al. [49], the figure indicated no overfitting when the training and validation loss rates were relatively close with an area coinciding. If the rates were far above (below) each other, it indicated overfitting (underfitting).

Table 4. Score of Importance.

Latent	Importance	Score (%)
M	0.202	100
PS	0.199	98.6
PV	0.199	98.5
IN	0.196	97.2
PR	0.196	96.8
SN	0.190	94.3
HK	0.181	89.8
PBC	0.168	83.2
AT	0.161	79.9

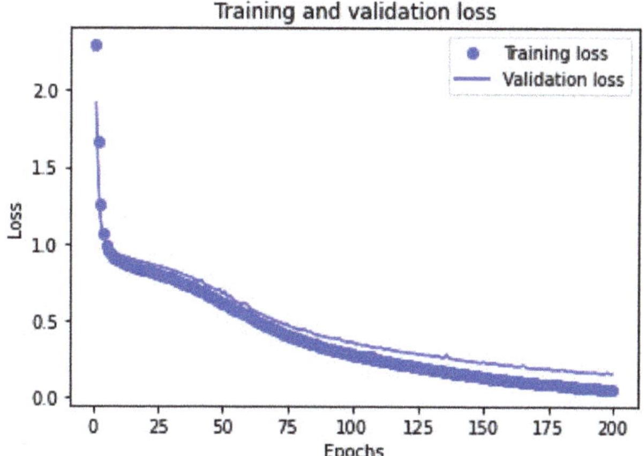

Figure 6. Training and Validation Loss Rate.

With 50 nodes in the hidden layer, using Tanh and Sigmoid as activation functions for the hidden and output layers, and considering adam as the optimizer, produced an accuracy of 98.86%. Presented in Figure 7 is the optimum ANN model produced from the parameters.

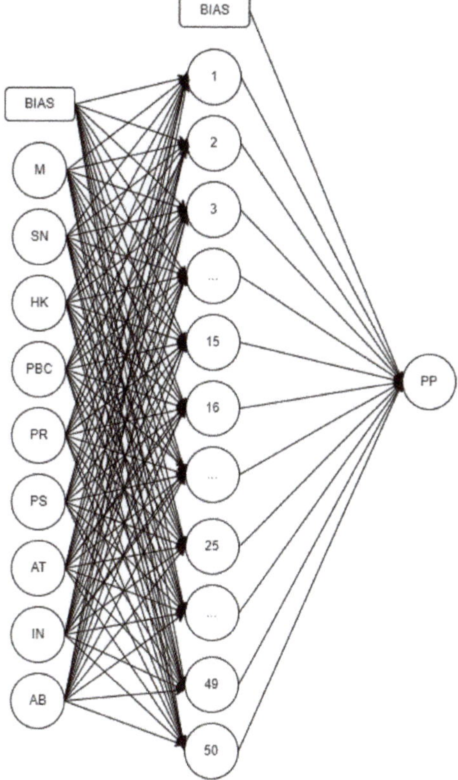

Figure 7. Optimum ANN Model.

4.4. Evaluation of Accuracy

To further evaluate the accuracy of the method used, ANN and RFC, different analyses, such as the Taylor Diagram, Violin Plot, and Box Plot, were conducted utilizing Python 5.1 using the seaborn package. Gholami et al. [50] explained how the Taylor Diagram assesses the performance of the model based on accuracy, showing the standard deviation and correlation. From their study, a threshold of Root Mean Square Error less than 20% was considered acceptable, while correlation greater than 90% was deemed significant. In this study, Figure 8 represents the Taylor Diagram showing the accuracy rate of RFC and ANN accuracies for different latent variables tested for their effect on preparedness of people. It could be deduced that RFC accuracy was within the threshold of highly significant results, together with latent variables M, PS, PV, IN, PR, SN, HK, and PBC. On the other hand, AT showed parameters outside the threshold, which indicated its low significance with regard to preparedness. In line with the results of individual analysis, AT was not part of the highly significant factors in RFC and least so in ANN. Thus, consistency for the accuracy rate was seen.

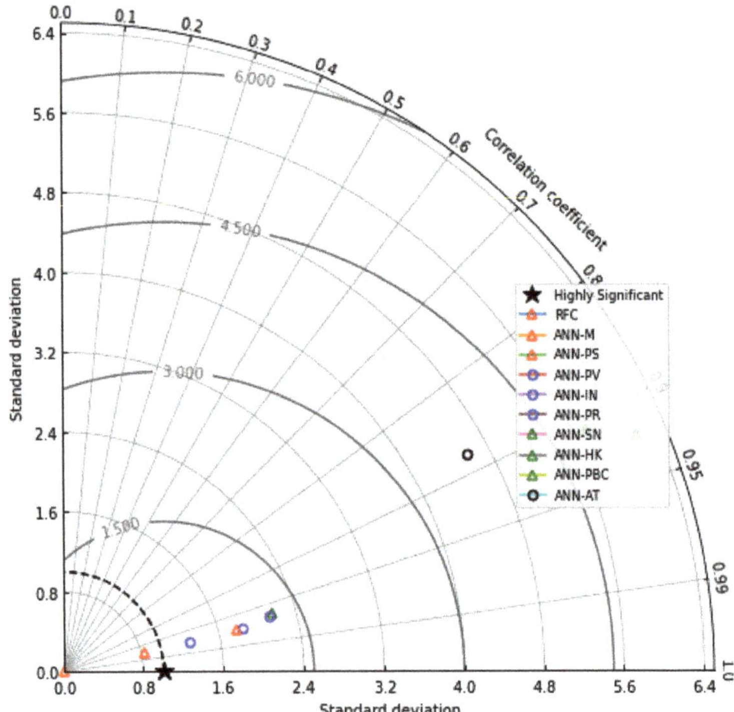

Figure 8. Taylor Diagram.

In addition, the Violin Plot to analyze the percentile ranking of the accuracies was conducted, together with the Box Plot, as seen in Figures 9 and 10. Figure 9 shows that that the accuracy mean for the RFC algorithm was at 90%. From the results, the highest accuracy obtained was 93%, which showed that it was within the higher level of interquartile range. Similarly, the ANN analysis showed a 94% accuracy mean. Until 83%, the accuracies would be considered in the interquartile range and anything below would be set for the lower adjacent values. In line with the results of the study, M, PS, PV, IN, PR, SN, HK, and PBC were among the interquartile range, while AT showed accuracy below. This showed how likely the AT latent variable was less significant compared to the others, which was similar to the findings of ANN and the Taylor Diagram. Similarly, the Box Plot in Figure 10 presented consistent results. Thus, it could be deduced that there was a uniform distribution seen from the Violin Plot results [51].

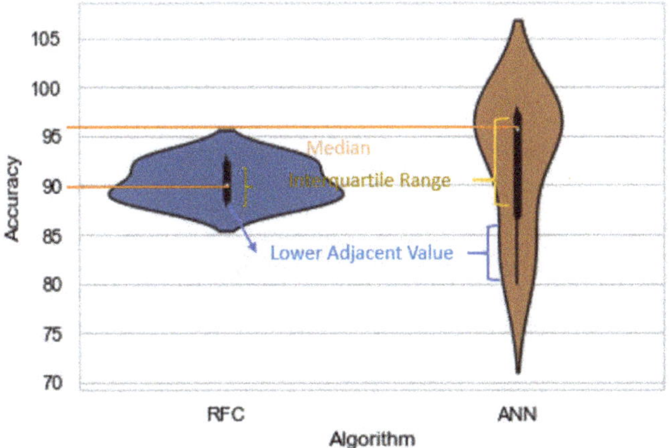

Figure 9. Violin Plot.

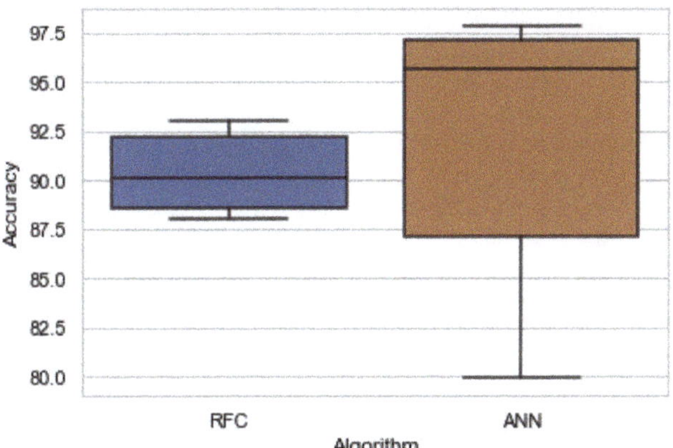

Figure 10. Box Plot.

5. Discussion

The evident increase in natural disasters, such as volcanic eruptions, has been widely prominent [2]. With people experiencing casualties brought by volcanic eruptions, this study considered an MLA ensemble with ANN and RFC to predict factors affecting preparedness for a volcanic eruption. The RFC produced a classification model with 93% accuracy resulting in media (M) being the most significant factor affecting preparedness for a volcanic eruption. Following which was perceived severity (PS), perceived vulnerability (PV), and intention (IN), which were all highly significant factors. ANN presented the same ranking of results with 98.86% accuracy, indicating that other factors, such as perceived risk proximity (PR), subjective norm (SN), hazard knowledge (HK), perceived behavioral control (PBC), and attitude (AT), were also considered significant.

Media (M) was seen to be a significant factor (100%), with its indicators presenting that people use media to quickly understand and obtain knowledge and correct information regarding threats of volcanic eruption=. It could also be deduced that people know which media platforms promptly report on threats brought by natural disasters, such as volcanic eruptions. Ong et al. [20] indicated that media is one of the most significant factors that indirectly affect the intention to prepare among communities when a threat of a natural disaster is present. Kurata et al. [21] support the findings by indicating how mass media and the development of technology have helped spread information about natural calamities relatively faster. The presence and availability of media have therefore been widely utilized to gain information for the community regarding disasters or calamities and to create positive social information dissemination [52]. Media is the highest contributing factor, as seen from the resultsa, since volcanic eruptions are among the most unpredictable natural disaster which could be active at any time. Use of media, allows people to monitor how active a volcano is, leading to more time for mitigation and preparation.

Relative to the current event of the Taal volcano being active and Mt. Bulusan in the Philippines, the media, through television, radio, and even social media, were able to indicate the level of activity, even giving precautions about activity before it began. Thus, the community relies heavily on media for them to have prepared for a volcanic eruption. However, studies like that of Ramakrishnan et al. [53] highlighted the applicability of Media in general. Of course, those without access to it are unable to utilize it, especially those living in underserved communities. It was also highlighted that the individual intends to know the significance through media access, similar to Armstrong et al.'s findings [54]. If people have difficulty in access, then Media would be considered an insignificant latent variable [53,54].

Second, PS was seen to have a highly significant effect (98.6%) on preparedness, much higher than the RFC result. People are highly aware of the effects of volcanic eruption on livelihood, fatalities, severity of aftermath, and the fact that volcanic eruptions can cause economic crisis. In line with this, SN was seen to have a highly significant effect on preparedness (94.3%). Andreastuti et al. [55] explained how perceptions would still be the critical factor affecting actions and attitude towards preparation and mitigation. When people know how an eruption will affect people who are important to them, their health, and their lives, people are more inclined to know how to deal with the cause [28]. Cahigas et al. [56] also indicated that crisis management would be one of the highest factors affecting an individual's behavior when a threat to health is present. Barclay et al. [57] highlighted that the community would focus on wellbeing and livelihood when natural disaster threats are present and would be highly incentivized by their perception of severity during and after a natural disaster like a volcanic eruption. Thus, this supports how PS presents a high indicator of preparedness for a volcanic eruption. However, if people perceive the natural disaster to have relatively low severity, then the preparedness and intention are low [57].

In line with PS, PV significantly affected preparedness for volcanic eruption (98.5%). With the recent events of the Taal volcano eruption in the country, people experienced how volcanic eruptions would affect a wide range of areas. The Taal was within a 100-km radius of the country's capital, which felt the movement and ash fall despite the minimal eruption. People indicated they were highly vulnerable if the Taal volcano fully erupted. Mt. Bulusan, on the other hand, is within a 570-km radius of the capital, but the community was also seen to be vigilant. This indicates how the experience of volcanic eruption promotes PV. It was indicated by Weichselgartner and Pigeon [25] that people are more inclined to prepare when they have knowledge, understanding, and experience of natural disasters such as volcanic eruptions. On another note, Warsini et al. [58] presented that more people would have camaraderie and mitigation plans due to the knowledge acquired from experience of volcanic eruptions due to their PV.

Fourth, IN was seen to be a significant factor affecting preparedness (97.2%). People willingly followed evacuation plans and safety measures set by the government if volcanic eruptions were bound to happen. The primary factors, such as PS and PV, would support this finding. Since people have the knowledge and experience, they would have high PS and PV, leading them to follow safety protocols. Aside from their mitigation plans, the evacuation from homes would be challenging, especially with 79% answering that they own their homes. Hershkovich et al. [59] and Kurata et al. [2] highlighted that with high PS and PV, people's IN would be more aligned with the government plans. It was also indicated that the government and other authorities are highly influential, due to their practical actions with evacuation and mitigation plans. Thus, the community would have high IN for preparation during a volcanic eruption.

Fifth, PR (96.8%) and HK (89.8%) were significant factors. It could be deduced that people understand and know about volcanic eruptions. In addition, people are familiar with how close or far they are to the proximity of volcanoes, and their preparedness is highly influenced when they are much closer. Martinez-Villegas et al. [60] showed the direct relationship between evacuation response during a volcanic eruption. Gaillard [61] showed that people living near active volcanoes are more likely to respond to preparation and mitigation due to economic factors. Baxter et al. [62] explored mitigation and preparation plans for volcanic eruptions due to limited experience and infrequent events. This relationship showed that intention and preparedness are insignificant if risks and hazard knowledge are low [62]. However, the past years in the Philippines have proposed high eruption rates due to the constant threat posed by the Taal volcano and now Mt. Bulusan. Perry and Lindell [63] showed how the effect of natural disasters near a community would enhance their preparedness. Thus, it presents a higher threat among people in proximity, leading to higher preparation.

It was seen that PBC (83.2%) and AT (79.9%) were significant factors affecting preparation for a volcanic eruption. With regard to t behavioral aspect of an individual's control over eruption hazards, how they can be avoided, having preventive measures, and knowledge of appropriate action presented significant indicators. A positive AT regarding security, confidence, and concern regarding safety in terms of health and life was evident. It was explained by Ong et al. [20] that TPB factors such as PBC and AT have a highly significant direct effect on the intention to prepare for mitigation of a natural disaster. Similar results were seen from the study by Morganstein and Ursano [64] and Abella et al. [65], wherein they highlighted and explained that awareness of a natural disaster would lead to an increase in response due to gaining knowledge of disaster preparation. Kusumastuti et al. [66] also highlighted how an individual's attitude and experience affected control over behavior and would develop and enhance alertness and preparation. Concerning the previous discussion, knowledge, PS, and PV would lead to a positive AT and PBC among individuals if they had experience and understanding of the natural disaster, its effects, and consequential aftermath [20]. Therefore, the more likely an individual is to be affected by a natural disaster, the more they would be inclined to mitigation and preparation.

It could be deduced that the government's protocol, policies, plans, and preemptive measures are highly effective in preparing for a volcanic eruption. In addition, the threat posed by the natural disaster, perceived severity and vulnerability, and knowledge would lead to remarkably high preparedness, among other factors. If people have experienced the effects of volcanic eruptions, they would be more inclined to prepare and mitigate. The effects would lead to a more positive behavioral control and positive attitude. Thus, the government may highlight the adverse effects of volcanic eruptions to help people adapt and prepare for volcanic eruptions.

5.1. Theoretical and Practical Implications

The results presented the high accuracy of classification models created from machine learning algorithm ensembles. Despite the non-linear relationship presented in the framework, a highly significant relationship was evident among all factors. In addition, in contrast to the findings of Kurata et al. [2], it was seen that several relationships prompted insignificant results, which signified that the claim by Fan et al. [12] and Woody [13] should be highly considered upon the utilization of structural equation modeling (SEM). In addition, the farther the dependent variable (e.g., media), the lower the relationship value. It was presented in Kurata et al. [2,21] that media was the fourth highest significant factor;

however, this study proved it to be the most significant among others. This shows that the arrangement of latent variables affected their results with SEM. Thus, utilizing an MLA ensemble could provide better output than the multivariate analysis promoted by SEM. This study contributed to the MLA ensemble usage, specifically in the field of natural disaster and human behavior related studies. It was indicated by Hanel et al. [67] that there is a difference in overall behavior in every country. In addition, different intentions and practices would be evident with different types of natural disasters, due to perceived severity, perceived vulnerability, and knowledge.

The findings suggest that PS and PV could be critical factors for preparing for a volcanic eruption. It could be posited that media may be highly utilized to spread awareness and information regarding natural disasters. Therefore, the government may utilize these attributes to enhance their mitigation and preparation plans. In addition, the relative behavioral aspects may also be considered when people have knowledge and experience. Thus, the government may consider a theoretical approach to address the aftermath of a volcanic eruption. The findings could also be classified among the community for them to visualize and understand what volcanic eruptions may cause during and after the said event. Since the Philippines is currently monitoring volcanic activities through the Philippine Institute of Volcanology and Seismology (PHIVOLCS), the government may consider the reports and information obtained for mitigation plans. Preparing before the disaster would help save livelihoods, reduce economic crisis, and save lives.

5.2. Limitations

Despite the relative findings, several limitations are still considered. First, this study considered a self-administered online questionnaire. Other factors were, therefore, not considered since the study had a theoretical-framework-based approach. It is suggested to conduct interviews among citizens, especially those living in proximity to volcanoes to create qualitative results. This way, researchers could dive deeper into the findings and it would help the government create action and mitigation plans. Second, it is suggested to consider specific respondents, such as those with and without access to media, the less educated, and rural versus urban areas. The results may present comparative findings of differences in educational attainment, hazard knowledge, and access to information. Third, other MLA could be utilized, such as clustering techniques, since this study created a classification model. The clustering may help segregate the demographic factors of the respondents. Lastly, it is suggested to test higher calculation complexity algorithms to promote the utility of an MLA ensemble regarding natural disasters and human behavior. Other classification tools may be considered, such as clustering, other types of neural networks, such as K-nearest neighbors and Naïve Bayes, and the like to compare and contrast the results, create a benchmark, and provide algorithms applicable in natural disaster-related and human behavior studies.

6. Conclusions

The increasing number of volcanic eruptions in the Philippines has been evident, with the Taal volcano erupting in early 2020 and Mt. Bulusan in 2022. Moreover, various countries have reported increasing deaths and damage from natural disasters (e.g., volcanic eruptions) over the past few years. However, the need to assess preparedness for volcanic eruption has been underexplored. This study evaluated the factors that could influence preparedness for volcanic eruptions. Utilizing a machine learning ensemble, a high accuracy rate was seen with RFC (93%) and ANN (98.86%).

The results indicated that the media, being the main source of information, had the highest effect on preparedness, followed by PS, PV and IN as contributing factors to high preparedness for a volcanic eruption. The effect on lives, economy and livelihood led to increased community preparedness. The proximity was seen to be a significant factor, followed by SN, and HK, which indicated how knowledge and experience, the effect on the individual and the people important to them caused a significant relationship. Lastly,

PBC and AT showed how people with control over their behavior would have positive attitudes to preparation for volcanic eruption, following the government's mitigation and preparation plan.

It was seen that people would likely follow protocols and plans presented by the government when natural disasters such as volcanic eruptions occur. Thus, the government may capitalize on this and present a volcanic eruption's theoretical aftermath and effects to enhance people's activeness and support towards preparedness. With its highly accurate results, the methodology, framework, and results of this study may be extended to evaluate other natural disasters. In addition, this could also be applied to other related studies across the world. With the threat of climate change, the need to reevaluate behavior and mitigation plans is needed. Thus, this study contributes to the safety and livelihood of people during natural disasters, such as volcanic eruptions.

Author Contributions: Conceptualization, J.D.G., A.A.N.P.R., A.K.S.O. and Y.T.P.; methodology, J.D.G., A.A.N.P.R., V.L.M.S. and A.K.S.O.; software, A.K.S.O., Y.T.P. and V.L.M.S.; validation, J.D.G. and A.A.N.P.R.; formal analysis, A.K.S.O. and Y.T.P.; investigation, J.D.G. and A.A.N.P.R.; resources; A.K.S.O. and Y.T.P.; data curation, Y.T.P.; writing-original draft preparation, J.D.G.; A.A.N.P.R.; writing-review and editing, J.D.G., A.A.N.P.R. and A.K.S.O.; visualization, A.K.S.O. and Y.T.P.; supervision, A.K.S.O. and A.A.N.P.R.; project administration, Y.T.P.; and funding acquisition, Y.T.P. All authors have read and agreed to the published version of the manuscript.

Funding: This research was funded by Mapua University Directed Research for Innovation and Value Enhancement (DRIVE).

Institutional Review Board Statement: This study was approved by Mapua University Research Ethics Committees (FM-RC-22-17).

Informed Consent Statement: Informed consent was obtained from all subjects involved in this study (FM-RC-21-54).

Data Availability Statement: The data presented in this study are available on request from the corresponding author.

Acknowledgments: The authors would like to thank all the respondents who answered our online questionnaire. We would also like to thank our friends for their contributions in the distribution of the questionnaire.

Conflicts of Interest: The authors declare no conflict of interest.

Appendix A

Table A1. Questionnaire (with permission from Kurata et al. [2]. 2022, Elsevier).

Construct	Items	Measures
Media	MP1	I think there are a variety of sources for media information about volcanic activities.
	MP2	I believe that the information among media is easily shared.
	MP3	I believe that social media contributes to the quick spreading of information regarding volcanic eruptions.
	MP4	I know how to distinguish irrelevant information (fake news) from social media platforms.

Table A1. *Cont.*

Construct	Items	Measures
Hazard Knowledge	HK1	I am familiar with the disaster sirens and warning signals.
	HK2	I believe I should wear masks to protect me from inhaling the ashes during volcanic eruption.
	HK3	I know that I should limit myself from going outdoors to reduce my ash exposure.
	HK4	I know that I should wear protective clothing during volcano eruption if I go outdoors.
	HK5	I know that there is poor air quality before, during, and after a volcanic eruption.
Perceived Risk Proximity	GP1	I think my location is within the danger zone from a volcano.
	GP2	I am familiar with the nearest evacuation facility I can go to if a volcano erupts.
	GP3	I am aware of the risks I have from volcaniv eruption based on my location.
Perceived Severity	PS1	I believe that a Taal volcano eruption is severe.
	PS2	I believe that a Taal volcano eruption may lead to deaths among people.
	PS3	I believe that a Taal volcano eruption is much more severe than other volcanic eruptions.
	PS4	I find that a Taal volcano eruption may affect my livelihood.
	PS5	I think that it will cpst me much to rebuild my resources affected by a Taal volcano eruption.
Perceived Vulnerability	PV1	I think my community is vulnerable to experience the effects of Taal volcanic eruption.
	PV2	I am likely to experience the effects of a Taal volcano eruption based on my experience.
	PV3	I have an experience of being vulnerable to volcano eruptions.
	PV4	I know I am more vulnerable to severe effects of volcanic eruption if I have breathing problems.
Perceived Behavioral Control	PBC1	I am in control of the situation in protecting myself from eruption hazards.
	PBC2	I think it is easy to implement preventive measures in my vicinity.
	PBC3	I know that I can avoid experiencing the effects of a volcanic eruption.
	PBC4	I think I have sufficient knowledge in responding to effects of volcanic eruption.
Subjective Norms	SN1	Most people in my community follow the preventive measures given by the local government unit.
	SN2	People in my community receivedaid from the local government unit.
	SN3	People living in my community go to the assigned evacuation centers before the volcano erupts.
	SN4	Most people in my community observe safety measures when the volcano shows possibility of eruption.
	SN5	People in my community still work for their living even though the government raises the alert levels for volcanic eruption.

Table A1. *Cont.*

Construct	Items	Measures
Attitude toward the Behavior	ATB1	I am stressed in a volcanic eruption.
	ATB2	I am scared if my family will be affected by a volcanic eruption.
	ATB3	I feel insecure when my community does not prepare for volcanic eruption.
	ATB4	I am concerned when the government raises the alert level of the Taal volcano.
	ATB5	I am confident that I know how to respond when a volcano erupts.
Intention to Evacuate	IF1	I am willing to adhere to the authorities' instructions if they tell me to evacuate.
	IF2	I am willing to leave anything behind to put myself in the safest place in the fastest way possible.
	IF3	My family is willing to leave anything behind to be in the safest place in the fastest way possible.
	IF4	I am willing to stay in the evacuation area with other people until it is safe to go back in my home.
Preparedness	PP1	I believe preemptive measures by the authorities for disaster response are effective.
	PP2	I believe that the emergency warning awareness will keep me safe during volcanic eruption.
	PP3	I believe that being updated by mass media will keep me safe during volcanic eruption.
	PP4	I think being connected to any friend or family member who does not live near the volcano is necessary in any case of a life-threatening circumstance.
	PP5	I think that it is essential to evacuate early if I am living in a community with the greatest risk.

References

1. Coppola, D.; Laiolo, M.; Cigolini, C.; Massimetti, F.; Delle Donne, D.; Ripepe, M.; Arias, H.; Barsotti, S.; Parra, C.B.; Centeno, R.G.; et al. Thermal Remote Sensing for Global Volcano Monitoring: Experiences from the mirova system. *Front. Earth Sci.* **2020**, *7*, 362. [CrossRef]
2. Kurata, Y.B.; Prasetyo, Y.T.; Ong, A.K.; Nadlifatin, R.; Persada, S.F.; Chuenyindee, T.; Cahigas, M.M. Determining factors affecting preparedness beliefs among Filipinos on Taal Volcano eruption in Luzon, Philippines. *Int. J. Disaster Risk Reduct.* **2022**, *76*, 103035. [CrossRef]
3. Brown, S.K.; Jenkins, S.F.; Sparks, R.S.; Odbert, H.; Auker, M.R. Volcanic fatalities database: Analysis of volcanic threat with distance and victim classification. *J. Appl. Volcanol.* **2017**, *6*, 15. [CrossRef]
4. Niroa, J.J.; Nakamura, N. Volcanic disaster risk reduction in indigenous communities on Tanna Island, Vanuatu. *Int. J. Disaster Risk Reduct.* **2022**, *74*, 102937. [CrossRef]
5. Thouret, J.-C.; Wavelet, E.; Taillandier, M.; Tjahjono, B.; Jenkins, S.F.; Azzaoui, N.; Santoni, O. Defining population socio-economic characteristics, hazard knowledge and risk perception: The adaptive capacity to persistent volcanic threats from Semeru, Indonesia. *Int. J. Disaster Risk Reduct.* **2022**, *77*, 103064. [CrossRef]
6. Reyes-Hardy, M.-P.; Aguilera Barraza, F.; Sepúlveda Birke, J.P.; Esquivel Cáceres, A.; Inostroza Pizarro, M. GIS-based volcanic hazards, vulnerability and risks assessment of the Guallatiri Volcano, Arica y Parinacota Region, Chile. *J. S. Am. Earth Sci.* **2021**, *109*, 103262. [CrossRef]
7. Barone, G.; De Giudici, G.; Gimeno, D.; Lanzafame, G.; Podda, F.; Cannas, C.; Giuffrida, A.; Barchitta, M.; Agodi, A.; Mazzoleni, P. Surface reactivity of Etna Volcanic Ash and evaluation of Health Risks. *Sci. Total Environ.* **2021**, *761*, 143248. [CrossRef]
8. Michellier, C.; Kervyn, M.; Barette, F.; Muhindo Syavulisembo, A.; Kimanuka, C.; Kulimushi Mataboro, S.; Hage, F.; Wolff, E.; Kervyn, F. Evaluating population vulnerability to volcanic risk in a data scarcity context: The case of Goma City, Virunga Volcanic Province (DRCongo). *Int. J. Disaster Risk Reduct.* **2020**, *45*, 101460. [CrossRef]

9.   Dogar, M.M.; Sato, T. Regional climate response of middle eastern, African, and South Asian monsoon regions to explosive volcanism and Enso forcing. *J. Geophys. Res. Atmos.* **2019**, *124*, 7580–7598. [CrossRef]

10. Dogar, M.M.; Sato, T. Analysis of climate trends and leading modes of climate variability for Mena Region. *J. Geophys. Res. Atmos.* **2018**, *123*, 13074–13091. [CrossRef]

11. Dogar, M.M.; Stenchikov, G.; Osipov, S.; Wyman, B.; Zhao, M. Sensitivity of the regional climate in the Middle East and North Africa to volcanic perturbations. *J. Geophys. Res. Atmos.* **2017**, *122*, 7922–7948. [CrossRef]

12. Fan, Y.; Chen, J.; Shirkey, G.; John, R.; Wu, S.R.; Park, H.; Shao, C. Applications of structural equation modeling (SEM) in Ecological Studies: An updated review. *Ecol. Processes* **2016**, *5*, 19. [CrossRef]

13. Woody, E. An SEM perspective on evaluating mediation: What every clinical researcher needs to know. *J. Exp. Psychopathol.* **2011**, *2*, 210–251. [CrossRef]

14. Duarte, P.; Pinho, J.C. A mixed methods UTAUT2-based approach to assess mobile health adoption. *J. Bus. Res.* **2019**, *102*, 140–150. [CrossRef]

15. PHIVOLCS Taal Volcano Bulletin 9 April 2022 08:00 am. Available online: https://www.phivolcs.dost.gov.ph/index.php/volcano-hazard/volcano-bulletin2/taal-volcano/14430-taal-volcano-bulletin-9-april-2022-08-00-am#:~{}:text=There%20has%20been%20no%20recorded,dropped%20on%203%20April%202022 (accessed on 31 May 2022).

16. Mangosing, M.A.M.-M.F. Bulusan Eruption Rains Ash, Forces Evacuation. Available online: https://newsinfo.inquirer.net/1606790/bulusan-eruption-rains-ash-forces-evacuation (accessed on 31 May 2022).

17. Manila, U.S.E. Natural Disaster Alert–Mount Bulusan at Alert Level 1, June 6, 2022. Available online: https://ph.usembassy.gov/natural-disaster-alert-mount-bulusan-at-alert-level-1/ (accessed on 25 June 2022).

18. The World Data Active Volcanoes and Eruptions in the Philippines. Available online: https://www.worlddata.info/asia/philippines/volcanos.php (accessed on 25 February 2022).

19. Gumasing, M.J.; Prasetyo, Y.T.; Ong, A.K.; Nadlifatin, R. Determination of factors affecting the response efficacy of Filipinos under Typhoon Conson 2021 (jolina): An extended protection motivation theory approach. *Int. J. Disaster Risk Reduct.* **2022**, *70*, 102759. [CrossRef]

20. Ong, A.K.; Prasetyo, Y.T.; Lagura, F.C.; Ramos, R.N.; Sigua, K.M.; Villas, J.A.; Young, M.N.; Diaz, J.F.; Persada, S.F.; Redi, A.A. Factors affecting intention to prepare for mitigation of "The big one" earthquake in the Philippines: Integrating protection motivation theory and extended theory of planned behavior. *Int. J. Disaster Risk Reduct.* **2021**, *63*, 102467. [CrossRef]

21. Kurata, Y.B.; Prasetyo, Y.T.; Ong, A.K.; Nadlifatin, R.; Chuenyindee, T. Factors affecting perceived effectiveness of typhoon vamco (Ulysses) flood disaster response among Filipinos in Luzon, Philippines: An integration of protection motivation theory and extended theory of planned behavior. *Int. J. Disaster Risk Reduct.* **2022**, *67*, 102670. [CrossRef]

22. Prasetyo, Y.T.; Castillo, A.M.; Salonga, L.J.; Sia, J.A.; Seneta, J.A. Factors affecting perceived effectiveness of COVID-19 prevention measures among Filipinos during enhanced community quarantine in Luzon, Philippines: Integrating Protection Motivation Theory and extended theory of planned behavior. *Int. J. Infect. Dis.* **2020**, *99*, 312–323. [CrossRef]

23. Ong, A.K.; Chuenyindee, T.; Prasetyo, Y.T.; Nadlifatin, R.; Persada, S.F.; Gumasing, M.J.; German, J.D.; Robas, K.P.; Young, M.N.; Sittiwatethanasiri, T. Utilization of random forest and deep learning neural network for predicting factors affecting perceived usability of a COVID-19 contact tracing mobile application in Thailand "Thaichana. " *Int. J. Environ. Res. Public Health* **2022**, *19*, 6111. [CrossRef] [PubMed]

24. Ong, A.K.; Prasetyo, Y.T.; Velasco, K.E.; Abad, E.D.; Buencille, A.L.; Estorninos, E.M.; Cahigas, M.M.; Chuenyindee, T.; Persada, S.F.; Nadlifatin, R.; et al. Utilization of random forest classifier and artificial neural network for predicting the acceptance of reopening decommissioned nuclear power plant. *Ann. Nucl. Energy* **2022**, *175*, 109188. [CrossRef]

25. Weichselgartner, J.; Pigeon, P. The role of knowledge in disaster risk reduction. *Int. J. Disaster Risk Sci.* **2015**, *6*, 107–116. [CrossRef]

26. Phengsuwan, J.; Shah, T.; Thekkummal, N.B.; Wen, Z.; Sun, R.; Pullarkatt, D.; Thirugnanam, H.; Ramesh, M.V.; Morgan, G.; James, P.; et al. Use of social media data in Disaster Management: A survey. *Future Internet* **2021**, *13*, 46. [CrossRef]

27. Dube, E.; Munsaka, E. The contribution of indigenous knowledge to disaster risk reduction activities in Zimbabwe: A big call to practitioners. *Jàmbá J. Disaster Risk Stud.* **2018**, *10*, 493. [CrossRef] [PubMed]

28. Chuenyindee, T.; Ong, A.K.; Prasetyo, Y.T.; Persada, S.F.; Nadlifatin, R.; Sittiwatethanasiri, T. Factors affecting the perceived usability of the COVID-19 contact-tracing application "Thai chana" during the early COVID-19 omicron period. *Int. J. Environ. Res. Public Health* **2022**, *19*, 4383. [CrossRef]

29. Arias, J.P.; Bronfman, N.C.; Cisternas, P.C.; Repetto, P.B. Hazard proximity and risk perception of tsunamis in coastal cities: Are people able to identify their risk? *PLoS ONE* **2017**, *12*, e0186455. [CrossRef] [PubMed]

30. Yagoub, M.M.; Al Yammahi, A.A. Spatial distribution of natural hazards and their proximity to Heritage Sites: Case of the United Arab Emirates. *Int. J. Disaster Risk Reduct.* **2022**, *71*, 102827. [CrossRef]

31. Rana, I.A.; Jamshed, A.; Younas, Z.I.; Bhatti, S.S. Characterizing flood risk perception in urban communities of Pakistan. *Int. J. Disaster Risk Reduct.* **2020**, *46*, 101624. [CrossRef]

32. German, J.D.; Redi, A.A.; Prasetyo, Y.T.; Persada, S.F.; Ong, A.K.; Young, M.N.; Nadlifatin, R. Choosing a package carrier during COVID-19 pandemic: An integration of pro-environmental planned behavior (PEPB) theory and Service Quality (SERVQUAL). *J. Clean. Prod.* **2022**, *346*, 131123. [CrossRef] [PubMed]

33. Vinnell, L.J.; Milfont, T.L.; McClure, J. Why do people prepare for natural hazards? developing and testing a theory of planned behaviour approach. *Curr. Res. Ecol. Soc. Psychol.* **2021**, *2*, 100011. [CrossRef]

34. Najafi, M.; Ardalan, A.; Akbarisari, A.; Noorbala, A.A.; Jabbari, H. Demographic determinants of Disaster Preparedness Behaviors amongst Tehran inhabitants, Iran. *PLoS Curr.* **2015**, *7*, 1–15. [CrossRef]

35. Bronfman, N.C.; Cisternas, P.C.; Repetto, P.B.; Castañeda, J.V. Natural disaster preparedness in a multi-hazard environment: Characterizing the sociodemographic profile of those better (worse) prepared. *PLoS ONE* **2019**, *14*, e0214249. [CrossRef] [PubMed]

36. Bourque, L.B.; Regan, R.; Kelley, M.M.; Wood, M.M.; Kano, M.; Mileti, D.S. An examination of the effect of perceived risk on preparedness behavior. *Environ. Behav.* **2012**, *45*, 615–649. [CrossRef]

37. Heller, K.; Alexander, D.B.; Gatz, M.; Knight, B.G.; Rose, T. Social and personal factors as predictors of earthquake preparation: The role of support provision, network discussion, negative affect, age, and EDUCATION1. *J. Appl. Soc. Psychol.* **2005**, *35*, 399–422. [CrossRef]

38. Yuduang, N.; Ong, A.K.; Vista, N.B.; Prasetyo, Y.T.; Nadlifatin, R.; Persada, S.F.; Gumasing, M.J.; German, J.D.; Robas, K.P.; Chuenyindee, T.; et al. Utilizing structural equation modeling–artificial neural network hybrid approach in determining factors affecting perceived usability of mobile mental health application in the Philippines. *Int. J. Environ. Res. Public Health* **2022**, *19*, 6732. [CrossRef] [PubMed]

39. Podsakoff, P.M.; MacKenzie, S.B.; Lee, J.-Y.; Podsakoff, N.P. Common method biases in behavioral research: A critical review of the literature and recommended remedies. *J. Appl. Psychol.* **2003**, *88*, 879–903. [CrossRef]

40. Kim, Y.; Hardisty, R.; Torres, E.; Marfurt, K.J. Seismic-facies classification using random forest algorithm. In *SEG Technical Program Expanded Abstracts 2018*; SEG Library: Anaheim, CA, USA, 2018.

41. Snehil; Goel, R. Flood damage analysis using machine learning techniques. *Procedia Comput. Sci.* **2020**, *173*, 78–85. [CrossRef]

42. Chen, J.; Li, Q.; Wang, H.; Deng, M. A machine learning ensemble approach based on Random Forest and radial basis function neural network for risk evaluation of Regional Flood Disaster: A case study of the yangtze river delta, China. *Int. J. Environ. Res. Public Health* **2019**, *17*, 49. [CrossRef]

43. Yang, W.; Zhou, S. Using decision tree analysis to identify the determinants of residents' CO2 emissions from different types of trips: A case study of guangzhou, China. *J. Clean. Prod.* **2020**, *277*, 124071. [CrossRef]

44. Moustra, M.; Avraamides, M.; Christodoulou, C. Artificial Neural Networks for earthquake prediction using time series magnitude data or seismic electric signals. *Expert Syst. Appl.* **2011**, *38*, 15032–15039. [CrossRef]

45. Yariyan, P.; Zabihi, H.; Wolf, I.D.; Karami, M.; Amiriyan, S. Earthquake risk assessment using an integrated fuzzy analytic hierarchy process with artificial neural networks based on GIS: A case study of sanandaj in Iran. *Int. J. Disaster Risk Reduct.* **2020**, *50*, 101705. [CrossRef]

46. Oktarina, R.; Bahagia, S.N.; Diawati, L.; Pribadi, K.S. Artificial Neural Network for predicting earthquake casualties and damages in Indonesia. *IOP Conf. Ser. Earth Environ. Sci.* **2020**, *426*, 012156. [CrossRef]

47. Jamshidi, M.B.; Lalbakhsh, A.; Talla, J.; Peroutka, Z.; Roshani, S.; Matousek, V.; Roshani, S.; Mirmozafari, M.; Malek, Z.; La Spada, L.; et al. Deep learning techniques and COVID-19 drug discovery: Fundamentals, state-of-the-art and Future Directions. *Stud. Syst. Decis. Control* **2021**, *348*, 9–31. [CrossRef]

48. Satwik, P.M.; Sundram, M. An integrated approach for weather forecasting and disaster prediction using Deep Learning Architecture based on memory augmented neural network's (Mann's). *Mater. Today Proc.* 2021, *in press*. [CrossRef]

49. Lara, F.; Lara-Cueva, R.; Larco, J.C.; Carrera, E.V.; León, R. A deep learning approach for automatic recognition of seismo-volcanic events at the Cotopaxi Volcano. *J. Volcanol. Geotherm. Res.* **2021**, *409*, 107142. [CrossRef]

50. Gholami, H.; Mohamadifar, A.; Sorooshian, A.; Jansen, J.D. Machine-learning algorithms for predicting land susceptibility to dust emissions: The case of the Jazmurian Basin, Iran. *Atmos. Pollut. Res.* **2020**, *11*, 1303–1315. [CrossRef]

51. Thrun, M.C.; Gehlert, T.; Ultsch, A. Analyzing the fine structure of distributions. *PLoS ONE* **2020**, *15*, e0238835. [CrossRef] [PubMed]

52. Sennert, S.S.; Klemetti, E.W.; Bird, D.K. Role of social media and networking in volcanic crises and Communication. *Adv. Volcanol.* **2015**, *1*, 733–743. [CrossRef]

53. Ramakrishnan, T.; Ngamassi, L.; Rahman, S. Examining the factors that influence the use of social media for disaster management by underserved communities. *Int. J. Disaster Risk Sci.* **2022**, *13*, 52–65. [CrossRef]

54. Armstrong, C.L.; Cain, J.A.; Hou, J. Ready for disaster: Information seeking, media influence, and disaster preparation for severe weather outbreaks. *Atl. J. Commun.* **2020**, *29*, 121–135. [CrossRef]

55. Andreastuti, S.; Paripurno, E.T.; Gunawan, H.; Budianto, A.; Syahbana, D.; Pallister, J. Character of community response to volcanic crises at Sinabung and Kelud Volcanoes. *J. Volcanol. Geotherm. Res.* **2019**, *382*, 298–310. [CrossRef]

56. Cahigas, M.M.; Prasetyo, Y.T.; Persada, S.F.; Ong, A.K.; Nadlifatin, R. Understanding the perceived behavior of public utility bus passengers during the era of COVID-19 pandemic in the Philippines: Application of social exchange theory and theory of planned behavior. *Res. Transp. Bus. Manag.* **2022**, *2022*, 100840. [CrossRef]

57. Barclay, J.; Few, R.; Armijos, M.T.; Phillips, J.C.; Pyle, D.M.; Hicks, A.; Brown, S.K.; Robertson, R.E. Livelihoods, wellbeing and the risk to life during volcanic eruptions. *Front. Earth Sci.* **2019**, *7*, 2296–6463. [CrossRef]

58. Warsini, S.; Buettner, P.; Mills, J.; West, C.; Usher, K. The psychosocial impact of the environmental damage caused by the Mt Merapi eruption on survivors in Indonesia. *EcoHealth* **2014**, *11*, 491–501. [CrossRef]

59. Hershkovich, O.; Gilad, D.; Zimlichman, E.; Kreiss, Y. Effective medical leadership in times of emergency: A perspective. *Disaster Mil. Med.* **2016**, *2*, 4. [CrossRef] [PubMed]

60. Martinez-Villegas, M.M.; Solidum, R.U.; Saludadez, J.A.; Pidlaoan, A.C.; Lamela, R.C. Moving for safety: A qualitative analysis of affected communities' evacuation response during the 2014 Mayon Volcano Eruption. *J. Appl. Volcanol.* **2021**, *10*, 6. [CrossRef]
61. Gaillard, J.-C. Alternative paradigms of volcanic risk perception: The case of Mt. Pinatubo in the Philippines. *J. Volcanol. Geotherm. Res.* **2008**, *172*, 315–328. [CrossRef]
62. Baxter, P.J.; Aspinall, W.P.; Neri, A.; Zuccaro, G.; Spence, R.J.S.; Cioni, R.; Woo, G. Emergency planning and mitigation at vesuvius: A new evidence-based approach. *J. Volcanol. Geotherm. Res.* **2008**, *178*, 454–473. [CrossRef]
63. Perry, R.W.; Lindell, M.K. Volcanic risk perception and adjustment in a multi-hazard environment. *SSRN Electron. J.* **2008**, *172*, 170–178.
64. Morganstein, J.C.; Ursano, R.J. Ecological disasters and mental health: Causes, consequences, and interventions. *Front. Psychiatry* **2020**, *11*, 1. [CrossRef] [PubMed]
65. Abella, A.A.; Prasetyo, Y.T.; Young, M.N.; Nadlifatin, R.; Persada, S.F.; Perwira Redi, A.A.; Chuenyindee, T. The effect of positive reinforcement of behavioral-based safety on safety participation in Philippine coal-fired power plant workers: A partial least square structural equation modeling (PLS-SEM) approach. *Int. J. Occup. Saf. Ergon.* **2022**, *1*, 1–27. [CrossRef] [PubMed]
66. Kusumastuti, R.D.; Arviansyah, A.; Nurmala, N.; Wibowo, S.S. Knowledge management and natural disaster preparedness: A systematic literature review and a case study of East Lombok, Indonesia. *Int. J. Disaster Risk Reduct.* **2021**, *58*, 102223. [CrossRef]
67. Hanel, P.H.; Maio, G.R.; Soares, A.K.; Vione, K.C.; de Holanda Coelho, G.L.; Gouveia, V.V.; Patil, A.C.; Kamble, S.V.; Manstead, A.S. Cross-cultural differences and similarities in human value instantiation. *Front. Psychol.* **2018**, *9*, 849. [CrossRef] [PubMed]

sustainability

Evaluation of the Predictive Performance of Regional and Global Ground Motion Predictive Equations for Shallow Active Regions in Pakistan

Muhammad Waseem [1], Zia Ur Rehman [2], Fabio Sabetta [3], Irshad Ahmad [1], Mahmood Ahmad [4,*] and Mohanad Muayad Sabri Sabri [5,*]

1. Department of Civil Engineering, University of Engineering and Technology Peshawar, Peshawar 25000, Pakistan; m.waseem@uetpeshawar.edu.pk (M.W.); irspk@yahoo.com (I.A.)
2. National Centre of Excellence in Geology, University of Peshawar, Peshawar 25000, Pakistan; ziaurrehman1993@uop.edu.pk
3. Department of Architecture, Roma Tre University, 00161 Rome, Italy; fabio.sabetta@uniroma3.it
4. Department of Civil Engineering, University of Engineering and Technology Peshawar (Bannu Campus), Bannu 28100, Pakistan
5. Peter the Great St. Petersburg Polytechnic University, 195251 St. Petersburg, Russia
* Correspondence: ahmadm@uetpeshawar.edu.pk (M.A.); mohanad.m.sabri@gmail.com (M.M.S.S.)

Abstract: Ground motion prediction equations are a key element of seismic hazard assessments. Pakistan lacks a robust ground motion prediction equation specifically developed using a Pakistan seismic ground motion databank. In this study, performance assessment of the ground motion prediction equations for usage in seismic hazard and risk studies in Pakistan, a seismically highly active region, is performed. In this study, an evaluation of the global ground motion prediction equations developed for the shallow active regions is carried out based on a databank of strong ground motion that was compiled in this study. Thirteen ground motion prediction equations were considered applicable, and their goodness of fit was evaluated using the databank of 147 peak ground acceleration of 27 shallow earthquakes in Pakistan. Residual analysis and three goodness of fit procedures were implemented in the evaluation of the equations. The results of this study suggest that global ground motion prediction equations can be applicable in the shallow active regions of Pakistan. These equations were developed based on data from Europe and the Middle East. Next Generation Attenuation West-2 equations were also applicable, but they did not perform as well as the European and Middle Eastern databank-derived equations. A total of four global equations were applicable in Pakistan. The best performing equation in this study should be applied with the highest weight, and the others should be applied with small weights on the logic tree to perform better. These equations can be employed in seismic hazard and risk assessment studies for disaster risk mitigation measures.

Keywords: Pakistan; peak ground acceleration; strong motion databank

Citation: Waseem, M.; Rehman, Z.U.; Sabetta, F.; Ahmad, I.; Ahmad, M.; Sabri, M.M.S. Evaluation of the Predictive Performance of Regional and Global Ground Motion Predictive Equations for Shallow Active Regions in Pakistan. *Sustainability* **2022**, *14*, 8152. https://doi.org/10.3390/su14138152

Academic Editors: Jian Chen and Chong Xu

Received: 29 May 2022
Accepted: 1 July 2022
Published: 4 July 2022

Publisher's Note: MDPI stays neutral with regard to jurisdictional claims in published maps and institutional affiliations.

1. Introduction

Pakistan has been experiencing major regional earthquakes regularly throughout its history, and these will continue to occur due to its peculiar tectonic setting concerning three major tectonic plates. Therefore, the seismic hazard is high in this part of the world. The most recent large earthquake that struck Pakistan occurred on 26 October 2015 with a magnitude (Mw) of 7.5 (focal depth = 210 km). Its epicenter was in the Hindukush region located northwest of Pakistan, and due to its deep focus, the shaking was felt all over Pakistan and in the neighboring countries. The Awaran earthquake with an M_w of 7.7, having a focal depth of 15 km, occurred on 24 September 2013 in southern Pakistan and is another example of a recent large earthquake. These two earthquakes signify that large earthquakes occur in and in close proximity to Pakistan.

Earthquakes in Pakistan have resulted in damage to infrastructure, loss of human life and monetary losses. Seismic risk in Pakistan is high due to the vulnerability of structures and the high hazard prominence. Poor seismic performance by structures during earthquakes in Pakistan is due to the construction of non-engineered structures and poor seismic design and construction practices.

Working toward a reliable estimate of the seismic hazard in Pakistan is one seismic risk mitigation measure. The seismic hazard map for Pakistan has been revised by Sabetta et al. [1] in line with the Italian Building Code guidelines. This approach provides three parameters for the control of the spectral shape in a dense grid of points and, at the same time, gives design spectra accurately representing the expected seismic motion at the site under consideration.

The seismic actions induced in the structures are computed from the seismic hazard maps. Among the most important inputs to seismic hazard studies are ground motion prediction equations (GMPEs). The development of GMPEs using local strong motion data has received very little attention in Pakistan due to the absence of a sufficient amount of data. If strong motion records were available for the region in abundance, local GMPEs could be developed by regression analysis techniques (e.g., Joyner and Boore [2,3]).

There exists only one GMPE available for Pakistan created by Shah et al. [4], which hereinafter will be called SH12. It was developed for peak ground acceleration (PGA) prediction based on data from northern Pakistan. The functional form of SH12 is given in Equation (1):

$$\ln (\text{PGA}) = -6.0985 + 1.4004\, M - 1.5357 \ln R \tag{1}$$

where M is the moment magnitude and R is the epicentral distance. The standard deviation of the equation has not been mentioned. It has been estimated to be 1.60 (in logarithmic units) from the data given in SH12. The standard deviation was estimated using Microsoft Excel software. The standard deviation is very high and possible due to the reason that for data in SH12 did not grouped into data-based on site class and the second reason maybe due to the sparsity in the availability of strong motion data.

If the data are not sufficient, then GMPEs developed for other regions can be adopted. However, their use and performance should be assessed before using them. Performance evaluation can be carried out by generating trellis plots of the GMPEs (plots of the predicted intensity values of the GMPEs for different earthquake scenarios) or using the available local strong motions in a data-driven testing. In the data-driven testing, values from real earthquake recordings are evaluated using procedures from the literature (e.g., Scherbaum et al. [5,6]). Thus far, the seismic hazard analyses carried out for Pakistan have used GMPEs adopted from regions outside of Pakistan. The strong motion data compiled for the Middle East region by Danciu et al. [7] include Pakistan. In that dataset, only 11 records were included from Pakistan. Due to this small contribution of Pakistan data, it is appropriate to carry out an independent study using all possible data.

This study refers to evaluation of the horizontal component of ground motion prediction equations. The main purpose of this work is to evaluate the feasibility of the global GMPEs in seismic hazard assessment studies for Pakistan using the local strong motion records in a data-driven testing. The residual analysis and goodness of fit measures proposed by Scherbaum et al. [5,6] and Kale and Akkar [8] are used. Thirteen different GMPEs developed for regions outside of Pakistan have been selected in this analysis, and a strong motion dataset for Pakistan consisting of 147 PGAs from 27 earthquakes has been used.

Global ground motion prediction equation evaluations for usage in seismic hazard analyses has also been a subject of interest in the neighboring countries of Pakistan using similar methods (e.g., Zafrani and Farhadi [9] in Iran and P. Anbazhagan et al. [10] in India). Seismotectonic and seismic hazard studies for this region are active, and some examples of such studies are (1) a seismic hazard map for the Middle East region, including Pakistan, by Girdani et al. [11], (2) macrozonation of the ground displacements in Iran by Farhani et al. [12], and (3) a study on plate deformations of the Eurasian and Arabian plates by Allen et al. [13].

The selected GMPEs include four Next Generation Attenuation (NGA)-West-2 GMPEs and nine other potential candidates. The applicability of NGA-West-2 equations has been demonstrated by various studies for their worldwide applications.

2. Tectonic Settings

Pakistan is located in one of the most seismically active regions of the world due to presence of the Indian-Eurasian and the Arabian-Eurasian plate boundaries. The mountain ranges of the Himalayas, Hindukush, Karakorum, Makran, Kirther, and Suleiman are the products and evidence of the ongoing activities between these plates. Due to the tectonic setting and geology, earthquakes are very frequent in this part of the world (Kazmi and Jan [14]). The Himalayan ranges located in northeast were formed by a head-on collision of the India and Eurasia plates and are seismically very active, with several devastating regional earthquakes generated by this belt. The Kirther and Suleiman ranges and the prominent Chaman Fault systems located in the southwest were also been by the collision of the Indian and Eurasian plates. The Arabian and Eurasian plates' interaction is represented by the Makran subduction zone located in the southwest of Pakistan.

The Indus platform and foredeep in southeastern Pakistan includes the Indus Plain and the Thar and Cholistan deserts. The Sulaiman range and Kirthar foredeep lie on its eastern flank. Its north-south structures are in contact with the fault of the east-west trending Himalayan fold and thrust belt. In synthesis, Pakistan may be classified into four broad tectonic regions: (1) an active shallow tectonic region, (2) a stable continental region, (3) a subduction interface tectonic region and (4) a deep crustal or in-slab subduction tectonic region (Figure 1).

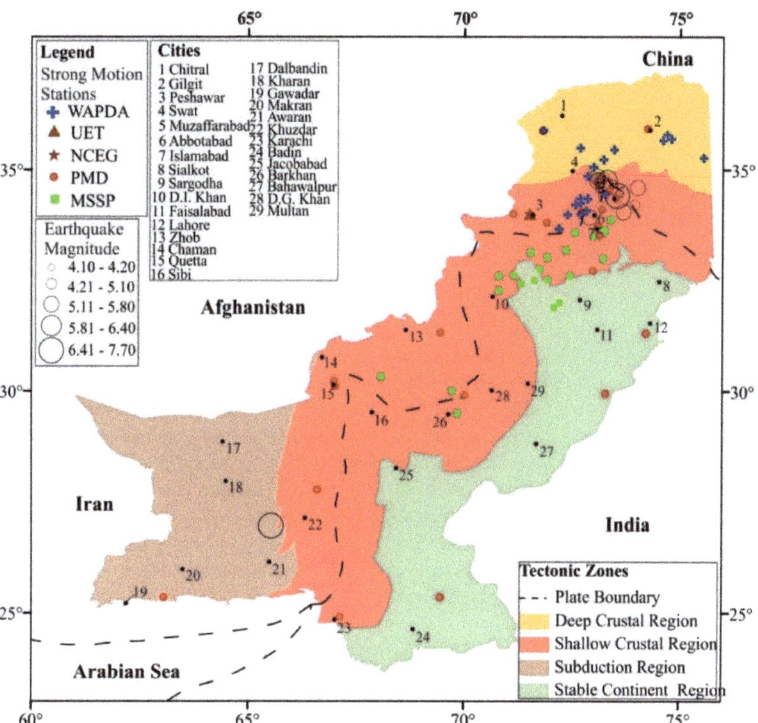

Figure 1. Tectonic regions in Pakistan along with the strong motion stations operated in the country and the epicenters of the recorded earthquakes.

3. Materials and Method

3.1. Data Collection

In Pakistan, strong motion arrays have been installed by various agencies: the Pakistan Meteorological Department (PMD), the Micro-Seismic Studies Project (MSSP) of the Pakistan Atomic Energy Commission (PAEC) and the Ministry for Water and Power Development Authority (WAPDA). Aside from these agencies, the National Centre of Excellence in Geology (NCEG) also operates a network of three stations in northern Pakistan. The University of Engineering and Technology (UET) operates one station. The information about the number of stations managed by each organization and their geographical coverage of these stations is shown in Figure 1.

WAPDA operates 19 stations at Tarbela Dam and 19 stations at the Bunje, Dasu and Basha observatories. The PMD array consists of 20 stations, while the MSSP operates 28 stations. WAPDA, PMD and the MSSP operate under different ministries, and there is no unified management system controlling the seismic arrays. As a result, there is a complete lack of coordination, and it is difficult to obtain data from any of the agencies. There is a strong need to bring all the agencies under a common platform to adequately manage and disseminate the data.

The strong motion data of Pakistan earthquakes have serious drawbacks due to the fact that they are mainly available from other studies, and few records come directly from the local arrays because the Pakistan agencies responsible for maintaining the recording stations do not share data. In this study, an effort was made to collect these data. The compiled data belong to small-to-moderate-magnitude earthquakes, except for 20 records of Kashmir (2005, Mw = 7.6, focal depth = 26 km) and Awaran (2013, Mw = 7.7, focal depth 15 km) earthquakes. The dataset has coverage from 2005 to 2018, and in Figure 1, only the epicenters of these earthquakes have been plotted. This dataset mostly consists of multiple recordings made at different locations for an earthquake evident form Appendix A. The dataset consists of 147 records with Mw between 4.1 and 7.7 and distances between 8 and 466 km. The datasets were retrieved only in terms of the PGA of the time histories, from the NCEG network (5 data), from Shah et al. [4] (128 data) and from the Douglas and Boore [15] paper (14 data). The focal mechanisms were extracted from the Global CMT catalog for 17 earthquakes, and for the rest, a reverse fault mechanism was assumed based on the prevalent faults where the epicenters were located.

Information about the earthquake magnitudes corresponding to the records were obtained from the United States Geological Survey (USGS) and are reported in Appendix A. When the magnitude was reported in different scales, the conversion to Mw was performed with the equations available in the literature concerning Pakistan.

The distributions of the earthquake magnitudes and PGA values are shown in Figure 2 as a function of the distance from the source. There were only five records with a PGA value greater than 100 cm/s^2, and the maximum value in the dataset was 221 cm/s^2. The reason for such low values is mainly that in many cases, they were recorded from far away distances, as shown in Figure 2. Even if distances larger than 100–200 km are generally out of the range of interest in GMPEs, strong motion recordings are rare for Pakistan, and we wanted to consider all the available data. The site conditions of the recording stations were defined using the shear wave velocity obtained through the approach of Allen and Wald [16]. For stations with unknown locations, a 310-m/s velocity was assumed. Site characterization information was only available for the NCEG observatory (i.e., VS 30 = 320 m/s).

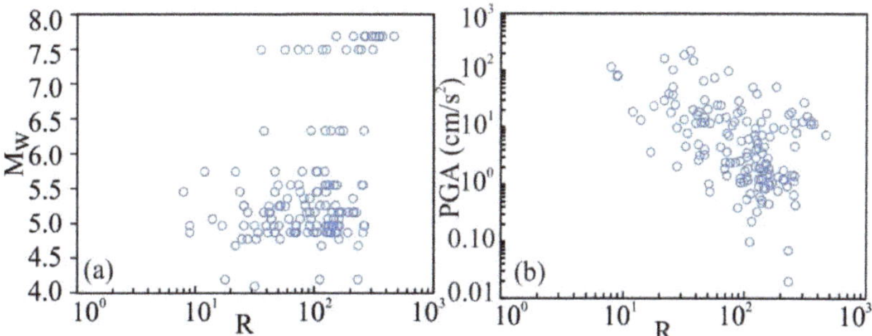

Figure 2. Distribution of the magnitude and peak ground acceleration recorded in Pakistan as a function of the epicentral distance from the source (**a**) showing magnitude distribution (**b**) showing PGA distribution.

3.2. Ground Motion Prediction Equation Evaluation Methods

A total of 13 GMPEs derived for the active shallow regions of the world were used in this study, which were checked for their suitability to predict PGA values for seismic action in Pakistan. The characteristics of these GMPEs are summarized in Table 1.

Table 1. Characteristics of considered GMPEs.

S. No	GMPEs	Distance Metrics	Target Regions	Magnitude Range	Distance Range (km)
1	Boore et al. [17] (BA14)	R_{JB}	Worldwide	3.0–7.9	0–400
2	Idriss et al. [18] (ID14)	R_{rup}	Worldwide	5.0–8.0	0–150
3	Campbell and Bozorgnia [19] (CB14)	R_{rup}	Worldwide	3.3–8.5	0–300
4	Chiou and Youngs [20] (CY14)	R_{rup}	Worldwide	3.5–8.5	0–300
5	Akkar et al. [21] (AK14)	R_{epi}, R_{JB}, R_{hyp}	Europe and Middle East	4.0–7.6	0–200
6	Akkar and Bommer [22] (AB10)	R_{JB}	Europe and Middle East	5.0–7.6	0–100
7	Bindi et al. [23] (BI14)	R_{epi}, R_{JB}, R_{hyp}	Europe and Middle East	4.0–7.6	<300
8	Zafarani et al. [24] (ZF18)	R_{epi}/R_{JB}	Iran	4.0–7.3	<200
9	Graizer and Kalkan [25] (GK15)	R_{rup}	Worldwide	5.0–8.0	0–250
10	Raghukanth and Kavitha [26] (RK14)	R_{hyp}	India	3.4–7.8	<300
11	Cauzzi et al. [27] (CZ15)	R_{rup}	Worldwide	4.6–7.9	<150
12	Shah et al. [4] (SH12)	R_{epi}	Northern Pakistan	4.1–7.6	9–265
13	Kanno et al. [28] (KAN06)	R_{rup}	Japan	5.2–8.2	0–300

3.3. Goodness of Fit Measure

The performance of all the selected GMPEs was evaluated using different measures of goodness of fit measures, which are described in following sections.

3.3.1. Residual Analysis

The residuals of each GMPE were analyzed using Equation (2):

$$\text{Res} = \ln(Y_{obs}/Y_{pre}) \tag{2}$$

where Res is the residual, Y_{pre} is the predicted PGA by the GMPE and Y_{obs} is the observed PGA value from the Pakistan database. The negative and positive values of the residuals represent over- and underprediction by the GMPE, respectively. Some examples of the residual plots of the GMPEs as a function of the magnitude and respective distance are shown in Figure 3. The epicentral distance was assumed to be the Joyner and Boore and hypocentral distance (R_{hyp}) equal to the distance metric R_{rup}. It can be observed that the ZF18, RK14 and KAN06 [24,26,28] residual plots are scattered with respect to the magnitude (Figure 3) and also with respect to the distance (Figure 4) while the CZ15, BA14 and AK14 [17,21,27] plots show balanced predictions with respect to the magnitude, but the residual distribution of AK14 [21] with respect to the distance is also not particularly balanced. The mean residual values and their standard deviation were calculated. These values indicate that AK14, CZ15, ID14, RK14 and SH12 [4,17,18,21,26,27] had the smallest mean residual values and the lowest standard deviations.

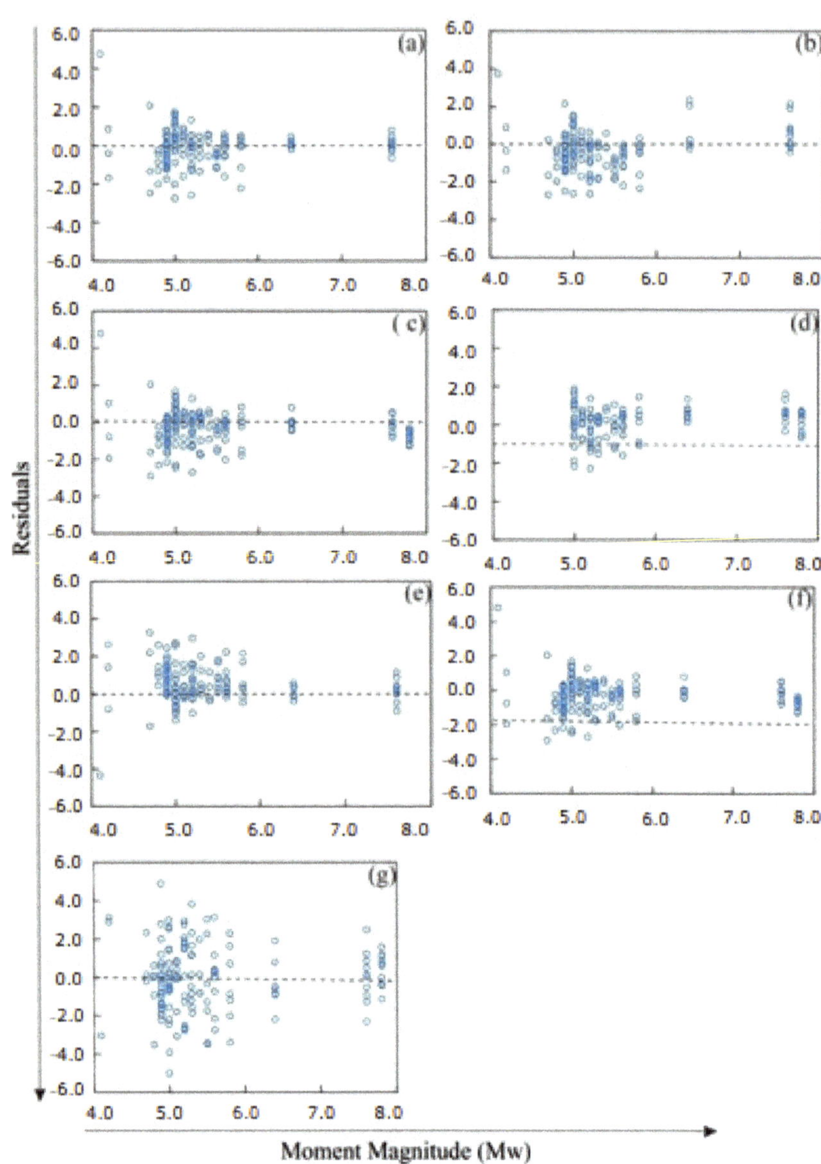

Figure 3. Residual plots of some selected GMPEs as a function of magnitude: (**a**) AK14, (**b**) BA14, (**c**) CZ15, (**d**) ID14, (**e**) RK14, (**f**) ZF18 and (**g**) KAN06.

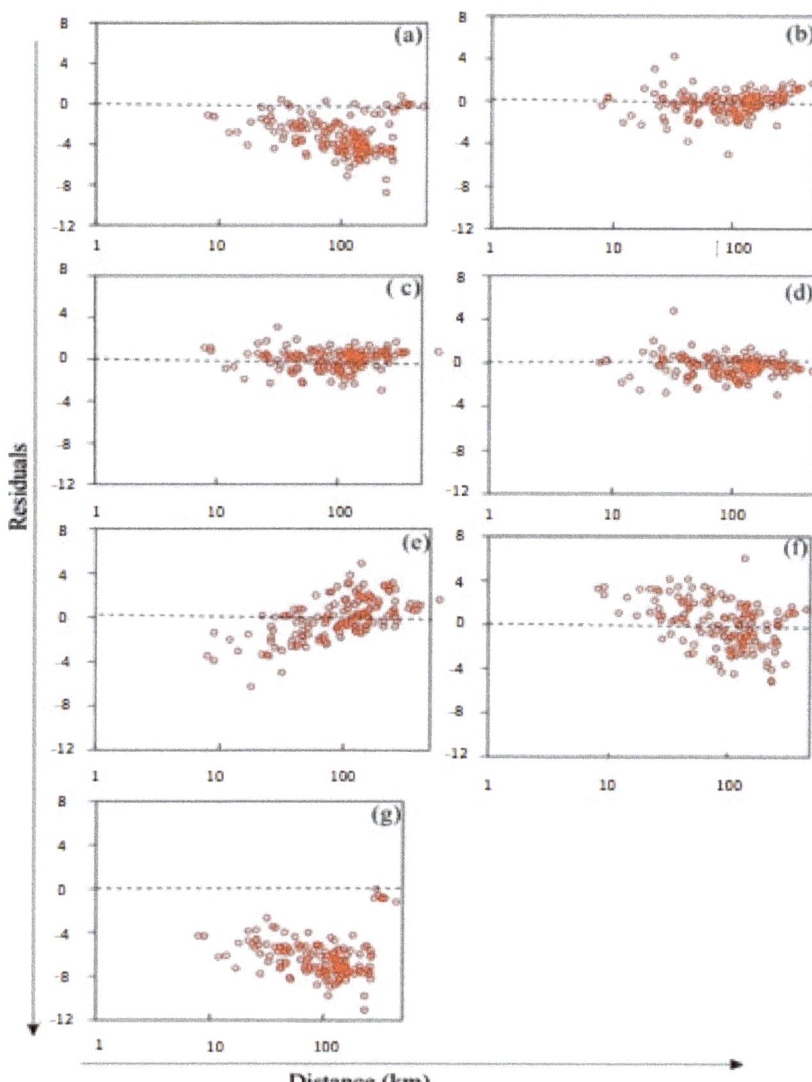

Figure 4. Residual plots of some selected GMPEs as a function of distance (lnR) in kilometers:
(**a**) AK14, (**b**) BA14, (**c**) CZ15, (**d**) ID14, (**e**) KAN06, (**f**) ZF18 and (**g**) RK14.

3.3.2. Likelihood (LH) Method

The method proposed by Scherbaum et al. [5] assesses the overall performance of
GMPEs in a complete sense (Scasserra et al. [29]). It was proposed for the selection and
ranking of GMPEs for a rational assignment of weights on a logic tree for a target region.
This method scales the performance of GMPEs as an exceedance probability known as the
likelihood (LH) value, ranging between 0 and 1.

An LH value of 0.5 is attained for situations where the predictive equation matches
the observed dataset perfectly in terms of the mean and standard deviation values. This
procedure is also known as the likelihood (*LH*) method, and it is based on the normalized
residual and the standard normal distribution. The final outcome of the *LH* method is
obtained as the median *LH* (MEDLH), median normalized residual (MEDNR), mean nor-

malized residual (MEANNR) and standard deviation of the normalized residual (STDNR) for the GMPEs.

The misfit between the predicted and observed values is determined by an error function (i.e., *LH* value) computed by Equations (3) and (4) as proposed by Scherbaum et al. [5]:

$$LH = \frac{2}{\sqrt{\pi}} \int_{|Z|/\sqrt{2}}^{\infty} e^{-t^2} dt = Erf\left(\frac{|Z|}{\sqrt{2}}, \infty\right) \tag{3}$$

$$Z = \frac{x_i - x_j}{\sigma} \tag{4}$$

The *LH* values are computed as an error function between the two limits given in the above equation. *Z* is the normalized residual, and *Erf* is the error function between the lower and upper limit. x_i is the observed value, x_j is the predicted one, and σ is the standard deviation. The results of the *LH* method, given in Table 2, indicate that AK14 [21] and SH12 [4] obtained the highest grade (i.e., category A), followed by CZ15 [27] in category B and ID14, GK15 and RK14 in category C. SH12 [4] also attained a high score because the data used in the test came from Pakistan. The remaining GMPEs yielded the lowest ranking of category D. All the equations were tested within the magnitude and distance ranges for which they were derived. The ranking criteria of categories A, B, C and D were outlined by Scherbaum et al. [5] and are not repeated here.

Table 2. Ranking of the GMPEs by the *LH* method.

S. No	GMPEs	MEDLH	MEDNNR	MEANNR	STDNR	Grade
1	AK14	0.76	0.01	0.06	0.61	A
2	AB10	0.06	0.27	0.12	2.95	D
3	BA14	0.34	0.4	0.39	1.55	D
4	BI14	0.13	1.45	1.56	1.29	D
5	CY14	0.05	0.05	0.09	2.77	D
6	CB14	0.41	0.11	0.30	1.50	D
7	CZ15	0.57	0.33	0.51	1.21	B
8	GK15	0.55	0.38	0.54	0.71	C
9	ID14	0.36	0.32	0.59	1.41	C
10	KAN06	0.16	0.02	0.14	2.10	D
11	RV14	0.56	0.38	0.56	1.08	C
12	SH12	0.74	0.04	0.10	0.64	A
13	ZF18	0.11	1.18	1.30	1.70	D

3.3.3. Log Likelihood (*LLH*) Method

The method proposed by Scherbaum et al. [6] also known as the *LLH* method, is an informational theoretical approach based on the log likelihood method to measure the misfit between two probability density functions representing the observed and estimated ground motion data. In the *LLH* method, these two functions are represented by $f(x)$ and $g(x)$, where $f(x)$ represents the log normal distribution of the observed data and $g(x)$ is the log normal distribution of the data predicted by the GMPE.

The log normal distribution is assumed for both the observed and predicted values. The difference between the predicted and observed data is represented by the *LLH* values given by Equation (5):

$$LLH(g, x) = -\frac{1}{N} \sum_{i=1}^{N} \log_{2(g(x_i))} \tag{5}$$

where *N* is the total number of data points considered. A small *LLH* value indicates a good relationship between the predictive model and the observed data.

The average *LLH* value was computed using the above equation for all candidate GMPEs. The observed and predicted values were evaluated in natural log units. The standard deviations of the GMPEs reported in common log units were converted to natural log units. The ranking of all the equations by the *LLH* method is shown in Table 3.

Table 3. Ranking of the GMPEs by the *LLH* method.

S. No	GMPEs	LHH	Ranking
1	CZ15	1.42	I
2	AK14	1.63	II
3	CB14	1.67	III
4	SH12	1.68	IV
5	ID14	1.78	V
6	BA14	1.88	VI
7	GK15	2.59	VII
8	RK14	4.05	VIII
9	CY14	4.39	IX
10	AB10	4.79	X
11	BI14	10.96	XI
12	KAN06	13.19	XII
13	ZF18	14.56	XIII

3.3.4. Euclidean Distance-Based Ranking (EDR) Method

This method, proposed by Kale and Akkar [8], is also known as the EDR method and is based on the Euclidean distance given by Equation (6):

$$DE = \sqrt{\sum_{i=1}^{N}(p_i - q_i)^2} \tag{6}$$

It is the square root of the sum of the squares of the differences between the N data pairs of the log-observed (p_i) and log-predictive (q_i) values. The behavior of the median value predicted by the GMPE and the observed data is represented by the following Kappa term in Equation (7):

$$k = \frac{DE_{original}}{DE_{corrected}} \tag{7}$$

The k value in Equation (7) is the ratio of the original and corrected Euclidean distances (DE). Ideally, the value of k is 1.0, and its higher values indicate a bias in the median value predicted by the GMPE. There are two DE values required to determined k: $DE_{original}$ and $DE_{corrected}$. The k value measures the bias between the observed and predicted data.

The $DE_{orignal}$ value was computed from Equation (6), while $DE_{corrected}$ was obtained from the observed data, and the corresponding value estimation was obtained by a linear fitting between the observed and estimated values.

The EDR method assumes a log normal distribution and accounts for the influence of sigma on the estimated ground motion and the bias between the observed data and the median in Equation (8).

Mathematically, we have

$$EDR^2 = k\frac{1}{N}\sum_{i=1}^{N} MDE_i^2 \tag{8}$$

where the Modified Euclidean Distance (MDE) is the probability-based average that takes sigma into consideration while testing the GMPEs. We also have

$$MDE = \sum_{j=1}^{n}|d_j|P_r(|D|) < |d_j| \tag{9}$$

where, $P_r(|D|) < |d_j|$ is the occurrence probability of "$|D|$", which is less than d_j. "D" is a random variable, and the difference between the observed value and the prediction by the GMPEs and dj is a discrete value of "D". A smaller *EDR* value implies better representation of the observed data by the respective relationship. The EDR method results are given in Table 4.

Table 4. EDR method results.

S. No	GMPEs	EDR	Ranking
1	AK14	0.75	II
2	AB10	0.96	IV
3	CZ15	0.57	I
4	KAN06	0.94	III
5	BI14	0.95	V
6	CB14	1.12	VI
7	GK15	1.12	VII
8	ZF18	1.23	VIII
9	BA14	1.31	IX
10	ID14	1.34	X
11	RK14	1.35	XI
12	SH12	1.77	XII
13	CY14	2.25	XIII

4. Results and Discussion

The goodness of fit tools discussed above were adopted for the selection of appropriate GMPEs using the available earthquake dataset in the shallow active regions of Pakistan. Residual analysis of the GMPEs showed that ZF18, BI14 and GK14 [23–25] mostly underpredicted the given dataset, while the other GMPEs showed a balanced prediction performance. The residuals shown in Figures 3 and 4 were not normalized with the standard deviation values of the respective GMPEs. Due to this, we did not want to consider the effect of the standard deviations on the performance of the GMPEs.

Considering the *LH* method (Table 2), SH12 and AK14 [4,21] reached the category A, CZ15 [27] fell into category B, and ID14, GK15 and RK14 [18,25,26] reached category C, while all the rest were placed in category D. The GMPEs from the neighboring countries (i.e., ZF18 and RK14 [24,26]) did not perform well and were ranked extremely low on the list. The GMPEs from California (NGA) were also ranked extremely low, except for ID14 [18].

The results of the ranking by the *LHH* method are reported in Table 3, and they suggest the GMPEs of CZ15, AK14, CB14, ID14 and SH12 [4,17,18,21,27] as the most feasible candidates, having the lowest *LLH* values. The results of the *LHH* method did not fully complement the *LH* method. As an example, the GMPE CZ15 [27] performed quite well based on the *LLH* method, while the *LH* method placed it in class B. However, it was also observed that for the top five GMPEs, the difference in *LLH* values was very small, ranging from 1.42 to 1.78. The EDR method indicated CZ15 [27] as the best GMPE in predicting the observed dataset, followed by AK14, KAN06, BI14 and AB10 [21,22,28], with a score ranging from 0.57 to 0.96.

In conclusion, the GMPEs recommended by all three evaluation procedures were AK14 [21] and CZ15 [27]. They were placed in categories A and B by the *LH* method, respectively, first under the *LLH* method and second when using the EDR methods.

The mean values of the residuals of the GMPEs were generally negative, indicating overprediction of the GMPEs in the given dataset, except for CY14, KAN06 and RK14 [20,26,28]. AK14 [21] had the lowest absolute mean of the residuals.

The final recommendation about the GMPEs was made based on the following:

1. GMPEs qualified in all *LH*, EDR and *LHH* methods (category I);
2. GMPEs recommended by at least two methods (category II).

In category I, the EDR and *LLH* qualifying score means that the corresponding GMPE is ranked among the top five. The *LH* qualifying score means that the GMPE has not attained category D. Category II means that the GMPE is ranked in the top five grades by the *LHH* and EDR methods.

The GMPEs of CZ15 [27] and AK14 [21] were recommended by all three procedures (category I), while the GMPEs of CB14, ID14 and SH12 [4,18,19] were recommended by the *LH* and *LLH* methods, respectively (category II). SH12 [4] was also suggested by two procedures, but it was not included in the final recommendation due to its inappropriate functional form. A summary of the ranking of the selected GMPEs is presented in Table 5.

Table 5. Scores of the GMPEs according to the different methods considered.

S. No	GMPEs	EDR	LHH	LH (Grade)	Remarks
1	AK14	II	II	A	
2	AB10	IV	X	D	
3	BA14	IX	VI	D	Category I
4	BI14	V	X	D	
5	CY14	XII	IX	D	
6	CB14	VI	III	D	Category II
7	CZ15	I	Is	B	Category I
8	GK15	VII	VII	C	
9	ID14	X	V	C	
10	KAN06	III	XII	D	
11	RK14	XI	VIII	D	Category II
12	SH12	XII	IV	A	
13	ZF18	VIII	XIII	D	

A comparison of the four selected GMPEs (i.e., CZ15, AK14, CB14 and ID14 [18,19,21,27]) with the recorded strong motion data (SM) is shown in Figure 5. A total of 97 PGA values were plotted against the GMPEs for the rock site conditions (i.e., Vs = 800 m/s) and reverse fault mechanism. The median predicted values were used in the plots, and the GMPEs were applied within the magnitude and distance ranges for which they were derived. The selected GMPEs exhibited good agreement with the local strong motion data.

It can be observed that four distinct tectonic regions exist in Pakistan: (1) active shallow regions, (2) shallow subduction, (3) stable continental and (4) deep sub-crustal regions. Therefore, at least four different types of GMPEs are required to implement seismic hazard assessments in Pakistan. Earthquake records are mostly available in the shallow active regions, and for other tectonic regions in Pakistan, GMPEs must be selected from the literature. Since the Hindukush region is a very active deep crustal source, there are several data records through the NCEG network, but they are mainly far field records. Data-driven testing can be carried out with these data records to suggest feasible candidate GMPEs for Hindukush.

The limitation of this work is the usage of PGA data only for evaluation of the global round motion prediction equations, as spectral acceleration were not available for the recorded data in Pakistan. This was due to the absence of data sharing by the department responsible for the recording and sharing of data in Pakistan. This compiled databank is mostly from the literature, and it was subjected to some evaluation procedures for better understanding of the performance of global ground motion prediction equations in the shallow active regions in Pakistan. The absence of near field strong motion data is another key limitation of the compiled databank. These limitations can be overcome as more and more data become available in the future, and with the current state, this was the best databank that could be compiled for this exercise.

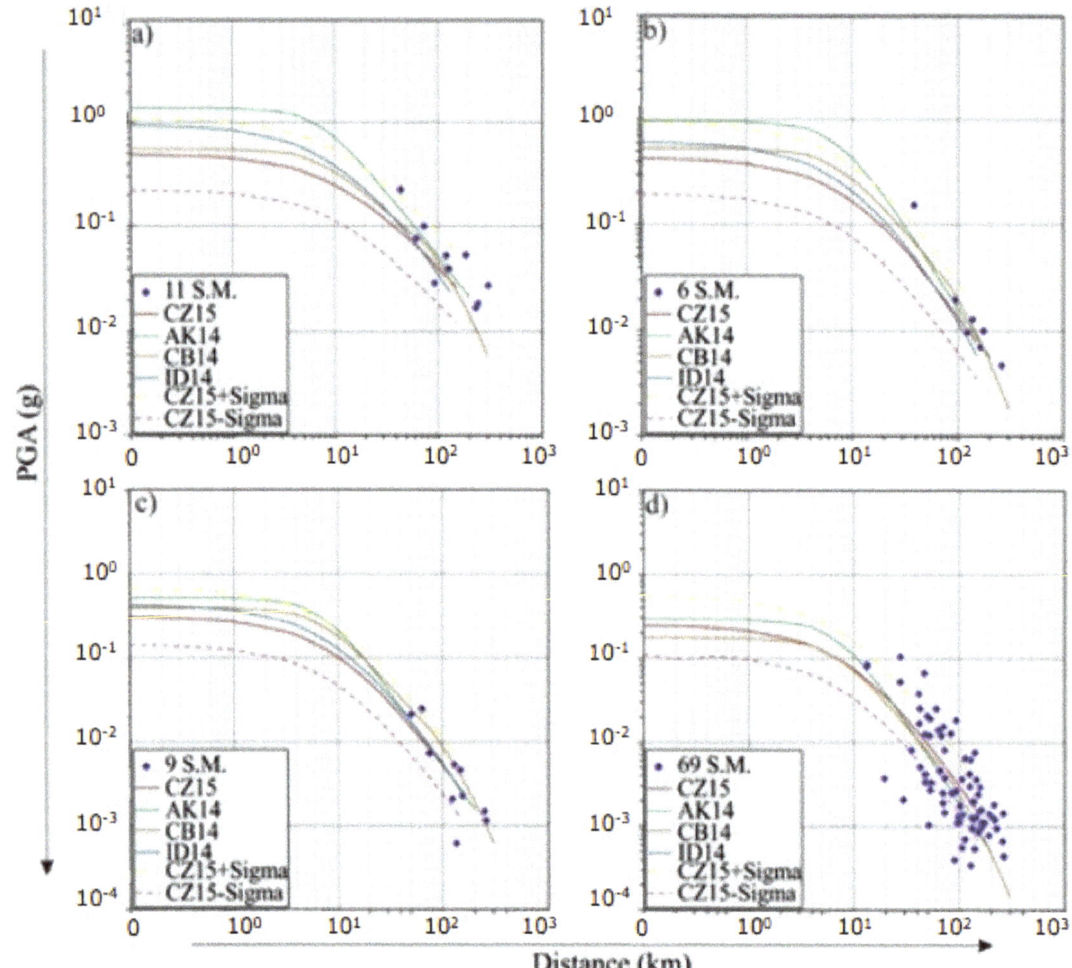

Figure 5. Comparison of the recommended GMPEs with the Pakistan PGA strong motion data (S.M.) in terms of distance and for different classes of magnitude: (**a**) Mw 7.6, (**b**) Mw 6.4, (**c**) Mw 5.6 and (**d**) Mw 4.9, 5.0, 5.2 and 5.3.

Seismic site classes based on direct measurement of the shear wave velocity from the recording stations were also not available, and an indirect procedure using the topographic slope proxy as measure of the shear wave velocity profile at the station location was created. This is one of the limitations of the current study. Moreover, direct measurement of the site classes of the recording stations will increase the accuracy of the strong motion data in Pakistan further.

Faults are also not characterized well in Pakistan, as their dip and strike directions and their values limit the uncertainties of data in Pakistan. The availability of this data will enhance the accuracy of the results in the use of the site to source distance metrics for studying GMPEs. In this study, the hypocentral and epicentral distances were considered equivalent metrics for the site-to-source distances used by the GMPEs under study.

The exact locations of some of the stations maintained by the MSSP were also missing this data, and its absence presents a limitation of this study.

Information on the combination rule for the two horizontal components was also not available for the strong motion data compiled in this study. These all are the limitations of this study.

5. Conclusions

A total of 147 PGA values recorded in the shallow active region of Pakistan were used to form a strong motion dataset for Pakistan. This dataset was used in this work for the selection and ranking of the global GMPEs available in the literature which were suitable to be adopted in Pakistan. The major contribution to the analyzed dataset was extracted from the literature because, due to a lack of coordination, strong motion data records are not shared by the agencies operating seismic arrays in Pakistan. This is also the reason why only PGA values are available, and they were used as a ground motion intensity measure. The methods of Scherbaum et al. [5,6] and Kale and Akkar [8] were used to perform data-driven testing on the available dataset. With the use of three different tests, we found that four global GMPEs could be employed in seismic hazard studies for the shallow active regions of Pakistan. Two of these GMPEs (i.e., CZ15 and AK14 [21,27]) had the best rankings and were classified into category I. In addition, a further two (ID14 and CB14 [18,19]) had acceptable rankings, even though they were classified into category II.

Since the Pakistan dataset is not very large, and the seismic site information was not directly available, our suggestion is to use all four of the selected GMPEs combined in a logic tree to appropriately consider their epistemic uncertainty. The GMPEs classified in category I could carry higher weights compared with the category II GMPEs. Category I GMPEs may be assigned weights of 70%, and those in category II may be assigned weights of 30% on the logic tree. The weights proposed are based on expert judgement.

Direct measurement of the seismic site class was not available, and a slope proxy was used to extract the values. Recording stations with unknown exact locations were assigned a shear wave velocity of 310 m/s. The strong motion data lacks near-source data. These are all the limitations of this study.

Author Contributions: Conceptualization, M.W., F.S..; data curation, M.W., Z.U.R. and I.A.; formal analysis, M.W., Z.U.R., I.A. and M.M.S.S.; investigation, F.S., Z.U.R., I.A. and M.A.; methodology, M.W., F.S., Z.U.R., M.A. and M.M.S.S.; project administration, M.A. and M.M.S.S.; resources, M.M.S.S.; software, M.W., Z.U.R. and I.A.; supervision, F.S.; validation, F.S.; visualization, F.S., I.A.; writing—original draft, M.W. and F.S., M.A. All authors have read and agreed to the published version of the manuscript.

Funding: This research is partially funded by the Ministry of Science and Higher Education of the Russian Federation as a part of World-Class Research Center Program: Advanced Digital Technologies (Contract No. 075-15-2022-311, dated 20 April 2022).

Institutional Review Board Statement: Not Applicable.

Informed Consent Statement: Not Applicable.

Data Availability Statement: The data that support the findings of this study are available within the article.

Conflicts of Interest: All the authors have seen the final version of the paper and have declared no potential conflicts of interest regarding the publication of this research work in its current form.

Appendix A

It must be remarked that that the site characterization of the recording stations was not available, and the corresponding definition was extracted from the topographic slope based on the approach of Allen and Wald (2009). The seismic site definition of records from Boore and Douglas (2018) was adopted as reported by their study. It was observed that the site conditions of the recording stations generally varied from soil (180 m/s < Vs < 360 m/s) to soft rock conditions (360 m/s < Vs < 760 m/s). The seismic site conditions are mentioned in the Appendix A to the paper.

Definitions of the site-to-source distances for the records in the databank were available as epicentral and hypocentral distances (i.e., R_{epi} and R_{hpo}). GMPEs with epicentral and hypocentral formulations were preferred over their formulation in terms of R_{JB} and R_{hpo} (i.e., AK14 and BI14). For the rest of the GMPEs, the epicentral distance was set equal to R_{JB}, and R_{hpo} was considered equivalent to R_{rup}.

The PGA values had a major contribution from SH12, though information was not available about the type horizontal component in SH12 concerning whether the reported PGA values were the largest of two components or the geometric mean of the two horizontal components. The NCEG dataset was the geometric mean of the horizontal components.

Table A1. Strong motion databank used in the evaluation of predictive equations.

S. No	Date	M_w	R_{epi} (km)	Station Code	Vs_{30} (m/s)	Observed PGA (g)	Focal Mechanism Solution
1	8 October 2005	7.6	36	ABT	360	0.226	
2	8 October 2005	7.6	57	MUR	760	0.076	
3	8 October 2005	7.6	90	NIL	760	0.029	
4	8 October 2005	7.6	118	FAG	560	0.052	
5	8 October 2005	7.6	185	PWR	310	0.053	
6	8 October 2005	7.6	246	THW	760	0.019	
7	8 October 2005	7.6	234	Thamewali	760	0.153	
8	8 October 2005	7.6	310	Chashma	760	0.020	
9	8 October 2005	7.6	74	Tarbela Dam	360	0.010	
10	8 October 2005	7.6	127	Barotha	310	0.013	
11	8 October 2005	6.4	38	BAF	310	0.007	
12	8 October 2005	6.4	95	MUR	760	0.010	
13	8 October 2005	6.4	124	NIL	760	0.005	
14	14 October 2005	6.4	138	FAG	560	0.067	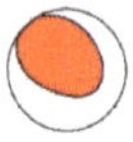
15	14 October 2005	6.4	164	CHT	900	0.015	
16	14 October 2005	6.4	176	PWR	310	0.004	
17	14 October 2005	6.4	262	THW	760	0.001	
18	14 October 2005	5.0	46	BAF	310	0.007	
19	14 October 2005	5.0	71	ABT	360	0.003	
20	14 October 2005	5.0	106	MUR	760	0.0014	
21	14 October 2005	5.0	132	NIL	760	0.000950	
22	14 October 2005	5.0	141	FAG	560	0.0021	
23	14 October 2005	5.0	158	CHT	900	0.0013	
24	14 October 2005	5.0	261	THW	760	0.0016	
25	17 October 2005	5.0	72	ABT	360	0.008	
26	17 October 2005	5.0	133	NIL	760	0.00198	
27	17 October 2005	5.0	144	FAG	560	0.00128	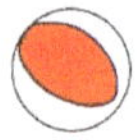
28	17 October 2005	5.0	163	CHT	560	0.00156	
29	17 October 2005	5.0	265	THW	760	0.00046	
30	19 October 2005	5.6	64	ABT	360	0.02466	
31	19 October 2005	5.6	124	NIL	760	0.00208	
32	19 October 2005	5.6	130	FAG	560	0.00528	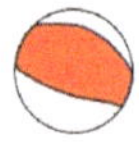
33	19 October 2005	5.6	147	CHT	900	0.0046	
34	19 October 2005	5.6	250	THW	760	0.00148	
35	19 October 2005	5.1	62	ABT	360	0.01444	
36	19 October 2005	5.1	95	MUR	760	0.0059	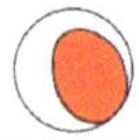
37	19 October 2005	5.1	131	FAG	560	0.00434	

Table A1. *Cont.*

S. No	Date	M_w	R_{epi} (km)	Station Code	Vs_{30} (m/s)	Observed PGA (g)	Focal Mechanism Solution
38	23 October 2005	5.6	50	BAF	310	0.021	
39	23 October 2005	5.6	75	ABT	360	0.00728	
40	23 October 2005	5.6	135	NIL	760	0.00062	
41	23 October 2005	5.6	155	CHT	900	0.00228	
42	23 October 2005	5.6	259	THW	760	0.0012	
43	23 October 2005	5.6	47	BAF	310	0.0124	
44	23 October 2005	5.6	72	ABT	360	0.0034	
45	23 October 2005	5.6	106	MUR	760	0.00135	
46	23 October 2005	5.6	138	FAG	560	0.00124	
47	23 October 2005	5.6	151	CHT	900	0.00086	
48	24 October 2005	4.9	47	BAF	310	0.00336	
49	24 October 2005	4.9	72	ABT	360	0.00146	
50	24 October 2005	4.9	106	MUR	760	0.00056	
51	24 October 2005	4.9	140	FAG	560	0.00054	
52	24 October 2005	4.9	41	BAF	310	0.016	
53	24 October 2005	4.9	67	ABT	360	0.0041	
54	24 October 2005	4.9	100	MUR	760	0.00112	
55	24 October 2005	4.9	128	NIL	760	0.00034	
56	24 October 2005	4.9	137	FAG	560	0.00196	
57	24 October 2005	4.9	156	CHT	900	0.00152	
58	26 October 2005	4.9	64	ABT	360	0.0064	
59	26 October 2005	4.9	69	BAF	310	0.0019	
60	26 October 2005	4.9	89	NIL	760	0.0004	
61	26 October 2005	4.9	132	FAG	560	0.00126	
62	26 October 2005	4.9	191	CHT	900	0.00078	
63	26 October 2005	4.9	258	THW	760	0.00066	
64	28 October 2005	5.3	26	BAF	310	0.0523	
65	28 October 2005	5.3	52	ABT	360	0.0119	
66	28 December 2005	5.3	85	MUR	760	0.0024	
67	28 December 2005	5.3	113	NIL	760	0.0007	
68	28 December 2005	5.3	125	FAG	560	0.0041	
69	28 December 2005	5.3	151	CHT	900	0.00238	
70	25 December 2005	5.8	12	BAL	310	0.19	
71	25 December 2005	5.8	22	BAF	310	0.164	
72	25 December 2005	5.8	47	ABT	360	0.03	
73	25 December 2005	5.8	80	MUR	760	0.01	
74	25 December 2005	5.8	105	ISL	360	0.013	
75	25 December 2005	5.8	108	NIL	760	0.00194	
76	25 December 2005	5.8	123	FAG	560	0.00744	
77	28 December 2005	5.5	8	BAL	310	0.1182	
78	28 December 2005	5.5	24	BAF	310	0.0399	
79	28 December 2005	5.5	47	ABT	360	0.0149	
80	28 December 2005	5.5	77	MUR	760	0.002	
81	28 December 2005	5.5	106	ISL	360	0.0031	
82	28 December 2005	5.5	108	NIL	760	0.00122	
83	28 December 2005	5.5	126	FAG	560	0.0029	
84	4 January 2006	5.0	9	BAF	310	0.085	
85	4 January 2006	5.0	17	ABT	360	0.0037	
86	4 January 2006	5.0	26	BAL	310	0.105	Not available
87	4 January 2006	5.0	51	MUR	760	0.00104	
88	4 January 2006	5.0	93	FAG	560	0.0011	
89	4 January 2006	5.0	126	CHT	900	0.0062	

Table A1. *Cont.*

S. No	Date	M_w	R_{epi} (km)	Station Code	Vs_{30} (m/s)	Observed PGA (g)	Focal Mechanism Solution
90	11 January 2006	5.2	28	MUR	760	0.002	
91	11 January 2006	5.2	42	ABT	360	0.005	
92	11 January 2006	5.2	102	FAG	560	0.00174	Not available
93	11 January 2006	5.2	164	CHT	900	0.00098	
94	11 January 2006	5.2	193	PWR	310	0.0014	
95	11 January 2006	5.2	227	THW	310	0.00094	
96	2 March 2006	4.8	25	BAL	310	0.0185	
97	2 March 2006	4.8	28	BAF	310	0.01	
98	2 March 2006	4.8	33	ABT	360	0.0047	Not available
99	2 March 2006	4.8	52	MUR	760	0.00076	
100	2 March 2006	4.8	161	CHT	900	0.00046	
101	19 March 2006	5.1	14	BAL	310	0.0135	
102	19 March 2006	5.1	44	ABT	360	0.00248	
103	19 March 2006	5.1	77	MUR	760	0.0126	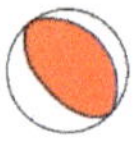
104	19 March 2006	5.1	146	CHT	900	0.00178	
105	19 March 2006	5.1	161	PWR	310	0.0018	
106	20 March 2006	5.4	60	BAL	310	0.025	
107	20 March 2006	5.4	83	ABT	360	0.01262	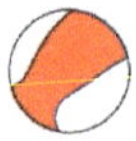
108	20 March 2006	5.4	93	MUR	760	0.00262	
109	20 March 2006	5.4	211	CHT	900	0.00126	
110	3 May 2006	4.9	9	BAF	310	0.08	
111	3 May 2006	4.9	34	ABT	360	0.008	Not available
112	3 May 2006	4.9	70	MUR	760	0.0024	
113	3 May 2006	4.9	140	CHT	900	0.00048	
114	12 August 2007	5.0	40	GHB	310	0.012	
115	12 August 2007	5.0	32	BAF	310	0.014	
116	12 August 2007	5.0	146	CHT	900	0.0009	
117	6 February 2008	4.1	32	GHB	310	0.0191	Not available
118	3 June 2009	4.2	18	BAF	310	0.023	
119	3 June 2009	4.2	112	FAG	560	0.0001	Not available
120	3 June 2009	4.2	233	THW	760	0.00009	
121	13 July 2009	4.7	22	BAF	310	0.03	
122	13 July 2009	4.7	116	FAG	560	0.00023	Not available
123	13 July 2009	4.7	234	THW	760	0.00002	
124	10 October 2010	5.2	38	FAG	560	0.0207	
125	10 October 2010	5.2	43	NIL	760	0.0033	
126	10 October 2010	5.2	88	CET	310	0.0183	
127	10 October 2010	5.2	93	DHN	310	0.0013	
128	10 October 2010	5.2	142	PAL	310	0.0036	Not available
129	10 October 2010	5.2	160	THW	760	0.0011	
130	10 October 2010	5.2	171	SAK	310	0.0011	
131	10 October 2010	5.2	213	CRB	310	0.0018	
132	10 October 2010	5.2	216	SAG	310	0.0012	

Table A1. *Cont.*

S. No	Date	M_w	R_{epi} (km)	Station Code	Vs_{30} (m/s)	Observed PGA (g)	Focal Mechanism Solution
133	24 September 2013	7.8	373	Bampoor	310	0.0121	
134	24 September 2013	7.8	352	Chabahar	310	0.0093	
135	24 September 2013	7.8	270	Gosht	310	0.0075	
136	24 September 2013	7.8	327	Qasr-e-Qand	310	0.0114	
137	24 September 2013	7.8	301	Negoor	310	0.0178	
138	24 September 2013	7.8	262	Rask	310	0.0117	
139	24 September 2013	7.8	212	Saravan	310	0.0131	
140	24 September 2013	7.8	466	Zar Abad	310	0.0081	
141	24 September 2013	7.8	351	Iran Shahr	310	0.0158	
142	24 September 2013	7.8	153	Sirkan	310	0.0127	
143	9 May 2018	5.29	108.8	NCEG	310	0.0121	
144	9 May 2018	5.29	42	NCEG	310	0.0093	
145	9 May 2018	5.29	27	NCEG	310	0.0075	Not available
146	9 May 2018	5.29	26	NCEG	310	0.0114	
147	9 May 2018	5.29	42	NCEG	310	0.0178	

References

1. Sabetta, F.; Waseem, M.; Khan, N.A.; Lodi, S.H.; Rafi, M.M.; Asif, M.A.; Briseghella, B.; Nuti, C. A proposal for a new formulation of the Pakistan seismic design action. In Proceedings of the 1st South Asia Conference on Earthquake Engineering (SACEE'19), Karachi, Pakistan, 21–22 February 2019. Paper No. 51.
2. Joyner, W.B.; Boore, D.M. Peak horizontal acceleration and velocity from strong motion records including records from the 1979 Imperial Valley California Earthquake. *Bull. Seismol. Soc. Am.* **1981**, *71*, 2011–2038. [CrossRef]
3. Joyner, W.B.; Boore, D.M. Methods of regression analysis of strong motion data. *Bull. Seismol. Soc. Am.* **1993**, *83*, 469–487. [CrossRef]
4. Shah, M.A.; Iqbal, T.; Qaiser, M.; Ahmed, N.; Tufail, M. Development of Attenuation for North Pakistan. In Proceedings of the 15th World Earthquake Conference on Earthquake Engineering (15WCEE), Lisbon, Portugal, 24–28 September 2012.
5. Scherbaum, F.; Cotton, F.; Smit, P. On the use of response spectral reference data for the selection and ranking of ground-motion models for seismic hazard analysis in regions of moderate seismicity: The case of rock motion. *Bull. Seismol. Soc. Am.* **2004**, *94*, 1–22. [CrossRef]
6. Scherbaum, F.; Delavaud, E.; Riggelsen, C. Model selection in seismic hazard analysis: An information-theoretic perspective. *Bull. Seismol. Soc. Am.* **2009**, *99*, 234–3247. [CrossRef]
7. Danciu, L.; Kale, Ö.; Akkar, S. The 2014 earthquake model of middle east: Ground motion model and uncertainties. *Bull. Earthq. Eng.* **2018**, *16*, 3497–3533. [CrossRef]
8. Kale, O.; Akkar, S. A novel procedure for selecting and ranking candidate ground motion prediction equations for seismic hazard analysis: Euclidean distance based ranking (EDR) method. *Bull. Seismol. Soc. Am.* **2013**, *103*, 1069–1084. [CrossRef]
9. Zafarani, H.; Farhadi, A. Testing ground motion prediction equations for small to moderate earthquakes magnitude data in Iran. *Bull. Seismol. Soc. Am.* **2007**, *107*, 912. [CrossRef]
10. Anbazhagan, P.; Sreenivas, M.; Ketan, B.; Moustafa, S.S.R.; Al-Arifi, N.S.N. Selection of Ground Motion Prediction Equations for Seismic Hazard Analysis of Peninsular India. *J. Earth Eng.* **2015**, *20*, 699–737. [CrossRef]
11. Giardini, D.; Danciu, L.; Erdik, M.; Şeşetyan, K.; Tümsa, M.B.D.; Akkar, S.; Gülen, L.; Zare, M. Seismic hazard map of the Middle East. *Bull. Earthq. Eng.* **2018**, *16*, 3567–3570. [CrossRef]
12. Farahani, S.; Behnam, B.; Tahershamsi, A. Macrozonation of seismic transient and permanent ground deformation of Iran. *Nat. Hazards Earth Syst. Sci.* **2020**, *20*, 2889–2903. [CrossRef]
13. Allen, M.; Jackson, J.; Walker, R. Late Cenozoic reorganization of the Arabia-Eurasia collision and the comparison of short-term and long-term deformation rates. *Tectonics* **2004**, *23*, 1–16. [CrossRef]
14. Kazmi, A.H.; Jan, M.Q. *Geology and Tectonics of Pakistan*; Graphic Publishers: Karachi, Pakistan, 1997.
15. Douglas, J.; Boore, D.M. Peak ground accelerations from large (M ≥ 7.2) shallow crustal earthquakes: A comparison with predictions from eight recent ground-motion models. *Bull. Earthq. Eng.* **2018**, *16*, 1–21. [CrossRef]
16. Allen, T.I.; Wald, D.J. *Topographic Slope as a Proxy for Global Seismic Site Conditions (VS 30) and Amplification around the Globe*; U.S. Geological Survey Open-File Report 2007-1357; USGS: Reston, VA, USA, 2007; 69p.
17. Boore, D.M.; Stewart, J.P.; Seyhan, E.; Atkinson, G.M. NGA West 2 equations for predicting PGA, PGV, and 5% damped PSA for shallow crustal earthquakes. *Earthq. Spectra* **2014**, *30*, 1057–1085. [CrossRef]
18. Idriss, I.M. An NGA-West 2 empirical model for estimating the horizontal spectral values 724 generated by shallow crustal earthquakes. *Earthq. Spectra* **2014**, *30*, 1155–1177. [CrossRef]

19. Campbell, K.W.; Bozorgnia, Y. NGA-West2 ground motion model for the average horizontal components of PGA, PGV, and 5%-damped linear acceleration response spectra. *Earthq. Spectra* **2014**, *30*, 1087–1115. [CrossRef]
20. Chiou, B.; Youngs, R.R. Update of the Chiou and Youngs NGA model for the average horizontal component of peak ground motion and response spectra. *Earthq. Spectra* **2014**, *30*, 1117–1153. [CrossRef]
21. Akkar, S.; Sandıkkaya, M.A.; Bommer, J.J. Empirical ground motion models for point-and extended-source crustal earthquake scenarios in Europe and the Middle East. *Bull. Earthq. Eng.* **2014**, *12*, 359–387. [CrossRef]
22. Akkar, S.; Bommer, J.J. Empirical equation for prediction of PGA, PGV and Spectral Acceleration in Europe, the Mediterranean region and the Middle East. *Seismol. Res. Lett.* **2010**, *81*, 173–215. [CrossRef]
23. Bindi, D.; Massa, M.; Luzi, L.; Ameri, G.; Pacor, F.; Puglia, R.; Augliera, P. Pan-European ground-motion prediction equations for the average horizontal component of PGA, PGV, and 5%-damped PSA at spectral periods up to 3.0 s using the RESORCE dataset. *Bull. Earthq. Eng.* **2014**, *12*, 391–430. [CrossRef]
24. Zafarani, H.; Luzi, L.; Lanzano, G.; Soghart, M.R. Empirical equations for the prediction of PGA and pseudo spectral accelerations using Iranian strong-motion data. *J. Seismol.* **2018**, *22*, 263–285. [CrossRef]
25. Graizer, V.; Kalkan, E.O. *Update of the Graizer-Kalkan Ground Motion Prediction Equations for Shallow Crustal Continental Earthquakes*; U.S. Geological Survey Open-File Report; USGS: Reston, VA, USA, 2015; 79p. [CrossRef]
26. Raghukanth, S.T.G.; Kavitha, B. Ground Motion Relations for Active Regions in India. *Pure App. Geop.* **2014**, *171*, 2241–2275. [CrossRef]
27. Cauzzi, C.; Faccioli, E.; Vanini, M.; Bianchini, A. Updated predictive equations for broadband (0.01–10 s) horizontal response spectra and peak ground motions, based on a global dataset of digital acceleration records. *Bull. Earthq. Eng.* **2015**, *13*, 1587–1612. [CrossRef]
28. Kanno, T. A New Attenuation Relation for Strong Ground Motion in Japan Based on Recorded Data. *Bull. Seism. Soc. Am.* **2006**, *96*, 879–897. [CrossRef]
29. Scasserra, G.; Stewart, J.P.; Bazzurro, P.; Lanzo, G.; Mollaioli, F. A comparison of NGA ground-motion prediction equations to Italian data. *Bull. Seismol. Soc. Am.* **2009**, *99*, 2961–2978. [CrossRef]

MDPI
St. Alban-Anlage 66
4052 Basel
Switzerland
www.mdpi.com

Sustainability Editorial Office
E-mail: sustainability@mdpi.com
www.mdpi.com/journal/sustainability